exploring
INNOVATION

exploring
INNOVATION

DAVID SMITH

The **McGraw·Hill** Companies

London	Boston	Burr Ridge, IL	Dubuque, IA	Madison, WI	New York
San Francisco	St Louis	Bangkok	Bogotá	Caracas	Kuala Lumpur
Lisbon	Madrid	Mexico City	Milan	Montreal	New Delhi
Santiago	Seoul	Singapore	Sydney	Taipei	Toronto

Exploring Innovation
David Smith
ISBN-13 978 007710861 8
ISBN-10 0-07-710861-2

Published by McGraw-Hill Education
Shoppenhangers Road
Maidenhead
Berkshire
SL6 2QL
Telephone: 44 (0) 1628 502 500
Fax: 44 (0) 1628 770 224
Website: www.mcgraw-hill.co.uk

British Library Cataloguing in Publication Data
A catalogue record for this book is available from the British Library

Library of Congress Cataloguing in Publication Data
The Library of Congress data for this book has been applied for from the Library of Congress

Acquisitions Editor: Kate Mason / Rachel Gear
Editorial Assistant: Natalie Jacobs
Marketing Manager: Alice Duijser
Production Editor: Eleanor Hayes / Beverley Shields

Text Design by SCW
Cover design by Ego Creative
Printed and bound in the UK by Bell and Bain

ISBN-13 978 007710861 8
ISBN-10 0-07-710861-2

brief table of contents

Detailed table of contents

DETAILED TABLE OF CONTENTS CONTD.

detaiLeD taBLe of contents CONTD.

preface

There are not many texts on innovation. Those that there are all too often include aspects of other things including strategic management, technology strategy, new product development or marketing. Similarly, most texts seem to be aimed at the postgraduate market, especially scientists, engineers and technologists with experience of technology-based firms. Hopefully this book is different. It is aimed at the undergraduate market, especially those taking degrees in Business Studies. Consequently, it assumes a familiarity with at least some of the key aspects of management such as marketing, human resource management and strategic management, but a lack of familiarity with most aspects of technology.

Confining the focus to innovation in this way provides an opportunity for a more exploratory approach aimed specifically at those who have only a limited acquaintance with the subject. Hence the inclusion of specific chapters on technological change, sources of innovation and theories of innovation. The use of a four-part structure is deliberate. The book begins with an explanation of the nature of innovation, which is categorised into different types and forms, and then contextualised in relation to the wider subject of technological change. This is followed by an in-depth examination of some of the activities associated with innovation. Having explored what innovation entails, the book moves on to consider how individuals and organisations manage innovation, particularly the people and financial aspects involved. Finally, innovation is placed in a wider context by looking at some of the macro-aspects of innovation, especially those aspects associated with policy at regional and national government levels.

The book is structured so that it can be used on a typical 12–15 week, single-semester course. The inclusion of a large case study at the end of every chapter together with related questions for discussion is designed to facilitate its use on conventional lecture and seminar courses, where the chapter covers a single lecture and the case study provides the basis of a seminar.

guided tour

OBJECTIVES
When you have completed this chapter you will be able to:
- Distinguish between science and technology
- Describe the nature of technology
- Analyse the link between technological change and the long wave cycle
- Describe the phases of the long wave cycle
- Differentiate and distinguish between technological paradigms and technological trajectories.

LEARNING OBJECTIVES
Each chapter opens with a set of learning objectives, summarising what you should learn from each chapter.

Date	Cycle/Wave	Technology
1780-1830	First	Cotton, Iron, Water Power
1830-1880	Second	Railways, Steam Power, Steamship
1880-1930	Third	Electricity, Chemicals, Steel
1930-1980	Fourth	Cars, Electronics, Oil, Aerospace
1980-	Fifth	Computers, Telecommunications and Biotechnology

Table 3.1 Long Wave Cycles

FIGURES AND TABLES
Each chapter provides a number of figures and tables to help you to visualise the various economic models, and to illustrate and summarise important concepts.

MINI-CASE: THE PERSONAL COMPUTER (PC)

The arrival of the personal computer represented a technological discontinuity. Previously computers had been large, specialised pieces of equipment operated by computer specialists. The PC changed all that. Suddenly there was a computer anyone could use. But to begin with there were large numbers of competing designs. They had very different configurations in terms of storage devices, video display units and input devices. For example, some PCs had twin disk drives, others single disk drives, some used 5½-in. disks, others 3½-in. disks, some had external disk drives others internal disk drives. Apple Computer went some way to creating a common product configuration with their Apple II machine, but it was the arrival of the IBM PC that created a configuration that was copied by others and effectively created the industry standard.

MINI CASE STUDIES
Throughout the book these real life examples help you to understand the concepts of innovation more easily and enable you to relate abstract ideas to actual products.

CASE STUDY: MCLAREN MP4

Despite being highly successful for most of the previous decade, by 1980 the McLaren Formula 1 team had slipped to the back of the grid, having been uncompetitive for the last couple of seasons. This lack of success led to the team being taken over by Ron Dennis, who had previously run the Project 4 team in Formula 2, and being re-formed as McLaren International. As an outsider who was not part of the established Formula 1 scene, Dennis looked not to existing teams for a designer but instead picked John Barnard who had previously worked in Indycar racing in the US.

At this time all Formula 1 cars were of monocoque construction with the body and the chassis forming an integral unit made from sheets of aluminium riveted together. This type of chassis construction had been pioneered by Colin Chapman of Lotus some

CASE STUDY
Each chapter contains a full-length case study with questions. These comprehensive studies demonstrate the material from each chapter and test your understanding of the theories and principles covered.

guided tour

QUESTIONS FOR DISCUSSION
These questions encourage you to review and apply the knowledge you have acquired from each chapter. They are pitched at different levels.

QUESTIONS FOR DISCUSSION

1 Distinguish between technological paradigms and technological trajectories.

2 Why do process innovations tend to occur during the later phases of the long wave cycle?

3 Where does the 'Dot.com' bubble fit within the long wave cycle? To what extent do you consider it to have been predictable?

4 What is technology? Use examples to illustrate your answer.

ASSIGNMENTS
A selection of more detailed questions is included, designed to cover the material in more depth.

ASSIGNMENTS

1 Take an account of innovation (this could be from a biography of an innovator, or a television programme or a film) and use any one theory of innovation to explain why and how innovation occurred.

2 Provide a detailed critique of any one of the theories of innovation.

3 What is the connection between notions of evolution, particularly technological evolution, and theories of innovation?

4 What is the value of theories of innovation for (a) would-be innovators and (b) policy-makers?

RESOURCES
A selection of resources is discussed at the end of each chapter, including web links, books and articles.

RESOURCES

Books

The resources available for technological change are limited. There is comparatively little on either the long wave cycle or Nicholai Kondratiev. Fortunately the field covered by technological trajectories and paradigms is rather better covered.

Many books on innovation mention the long wave cycle (Goffin and Mitchell, 2005), but the treatment is often little more than a brief introduction. Detailed coverage of the long wave cycle is provided by Freeman and Soete (1997), Tylecote (1992) and a comparatively recent offering from Freeman and Louçã, (2001). This last-named text is a really excellent source. Not only does it explain the theory of the long wave cycle, but it provides detailed accounts of each long wave cycle, with particular emphasis upon the current fifth Kondratiev.

technology to enhance Learning and teaching

VISIT WWW.MCGRAW-HILL.CO.UK/TEXTBOOKS/INNOVATION TODAY

Online Learning Centre (OLC)

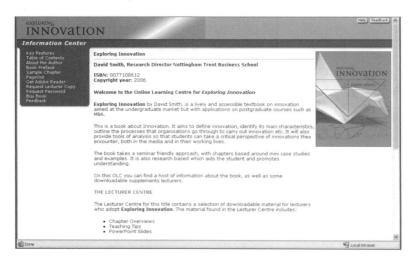

Lecturers adopting this textbook can access supporting resources to assist with their teaching.

Resources available for lecturers:

■ PowerPoint Presentations
■ Lecture notes

Please refer to website for up-to-date information.

For lecturers: Primis Content Centre

If you need to supplement your course with additional cases or content, create a personalised e-Book for your students. Visit www.primiscontentcenter.com or e-mail primis_euro@mcgraw-hill.com for more information.

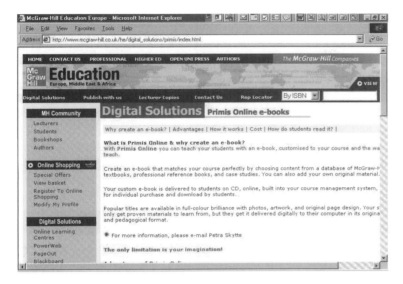

Study Skills

We publish guides to help you study, research, pass exams and write essays, all the way through your university studies.

Visit **www.openup.co.uk/ss/** to see the full selection and get £2 discount by entering promotional code **study** when buying online!

Computing Skills

If you'd like to brush up on your Computing skills, we have a range of titles covering MS Office applications such as Word, Excel, PowerPoint, Access and more.

Get a £2 discount off these titles by entering the promotional code **app** when ordering online at www.mcgraw-hill.co.uk/app

acknowLeDgements

Author's Acknowledgements:

My thanks to all my students over more than thirty years, but particularly those at Nottingham Trent University who, over some five years now have taken my second year undergraduate module, *Innovation and Strategy*. With a tenfold increase in numbers taking the module, hopefully something has gone right! I have not only had a very positive response to the ideas I have propounded, I have been introduced to some really interesting examples of innovation and been involved in some lively discussions that have helped to hone my own understanding of innovation.

Among my colleagues at Nottingham Trent, Paul Joyce first had the idea that I should do something on innovation, while other members of the Innovation and Strategy team, Weili Teng, Pete Horsley, Gamal Ibrahim and Thomas Fischer have quietly listened to my ideas on innovation and interpreted them in a manner that enabled us to deliver the module effectively. Thank you team! Among my former colleagues at University College Northampton, Richard Blundel, now of Oxford Brookes University, and Stephen Swailes, now of Hull University, and Foster Rogers, formerly of the University of Derby, re-kindled my interest in innovation and technology management. I owe a particular debt to Peter Robertson. He not only provided some lively and stimulating inputs to my classes, he is a real innovator. He's done it. And to the innovator in the OneClick Technologies case study I'm particularly grateful both for the data he provided that made the case a reality and for the time and trouble he took to help me get the detail right. Similarly my thanks to Phil Wadsley for helping to unravel the mysteries of sound recording in general and Dolby noise reduction technology in particular.

My thanks to librarians the world over, but especially Lisa Warburton and Lindsay Jones of Nottingham Trent University library, you are wonderful. You not only bought the books I liked, you found the books I wanted and most important of all you kept me ahead of my students.

Similarly my thanks to the security staff, porters and technicians at Nottingham Trent. Your efforts, your assistance and your humour have been much appreciated throughout.

My fellow walkers, Rorie Grieve, Bob Crompton and John Horton put up with me droning on about innovation across an assortment of mountains in Scotland on several occasions. Not only that, you handled the logistics of week long walks while I made no contribution whatsoever. You also showed an uncanny grasp of the finer points of incremental innovation!

Natalie Jacobs, editorial assistant at McGraw-Hill has helped me throughout the writing process. She kept me informed, kept me on track and explained all the intricacies of the publication process. Her enthusiasm and organisation has done wonders for my motivation and made the whole exercise far easier than I thought it was going to be.

I have really enjoyed writing this book. It has taken a year during which my family have had to endure even less engagement from me than usual. Tom, Jonny and Kate have not only endured this without a murmur, they have consistently taken an interest and enquired about progress. Di has had to endure evenings alone, solitary walks with the dog, and a reduction in my already limited interest in social events of any description. I can't promise I'll mend my sins of omission immediately, but I can promise that you make it all worthwhile.

Publisher's Acknowledgements:
Thanks to the following reviewers for their comments at various stages in the text's development:

Jaseem Ahmad, Middlesex

Neil Alderman, Newcastle

Michael Brennan, University of Ulster, Jordanstown

Gerry Edgar, Stirling

Dolores Anon Higon, Aston

Emile Horak, University of Pretoria

Ossie Jones, Manchester Metropolitan

Jimme Keizer, Eindhoven University of Technology

Ajit Nayak, UWE

Paul Oakley, Brighton

Elizabeth Sherry, Institute of Technology, Tallaght

Ian Walsh, Swansea

Gerald Watts, Gloucester

Dedication

John Keetley Smith

1917-2005

"My theory is that you do what your father wanted to do rather than what he did"
Alec Broers (2005)

part 1

WHAT IS
INNOVATION?

chapter 1

INtRODUCtION

OBJECTIVES

When you have completed this chapter you should be able to:

- Appreciate the importance of innovation for business and the national economy
- Distinguish between invention and innovation
- Describe the steps involved in undertaking innovation
- Distinguish between individual and corporate innovation
- Appreciate the factors that can cause innovations to fail.

MINI-CASE: SILICON VALLEY - THE HOME OF INNOVATION?

Silicon Valley in California is synonymous with innovation. It is the quintessential example of a place that is all about innovation. Nowhere else in the world is so readily identified with new products and new services. Certainly in the past 50 years no other place on earth has been the location of so many innovations. Among the better known are:

- Integrated circuit (Intel)
- Personal computer (Apple)
- 3D graphics (Silicon Graphics)
- Database software (Oracle)
- Web browser (Netscape)
- Online auction (eBay)

It is not merely the number of innovations, it is the fact that Silicon Valley has gone on producing wave after wave of innovations.

▶

part i: what is innovation?

There are other places in the world that have stronger records in scientific discoveries and scientific breakthroughs. Cambridge in the UK has a quite outstanding record in terms of scientific breakthroughs that include: discovery of the electron, splitting the atom, and the identification of the structure of DNA. Though these achievements are the stuff of Nobel prizes and as such are highly significant, they are scientific breakthroughs rather than innovations. Only Silicon Valley has an outstanding record in terms of innovation.

It was not always so. In the 1940s what is now Silicon Valley was a peaceful agricultural valley (Saxenian, 1983). Located at the southern tip of San Francisco Bay, Silicon Valley (or rather Santa Clara county) stretches from Palo Alto in the north to Gilray in the south. Covering 1500 square miles and currently home to 2.5 million people, today it is the densest concentration of high-technology business on earth.

Such is Silicon Valley's reputation for high technology and innovation that it has had many imitators:

- Silicon Glen (Scotland)
- Silicon Bog (Ireland)
- Silicon Fen (England)
- Silicon Beach (Vietnam)

While these places may share part of their name with their Californian namesake, it is probably about all that they do share. Many governments have attempted to re-create their own purpose-built versions of Silicon Valley in the form of vast technology parks. Notable examples include:

- Sophia Antipolis (France)
- Hsinchu (Taiwan)
- Multimedia Super Corridor (Malaysia)

Of these the most recent and the most spectacular is probably Malaysia's Multimedia Super Corridor (MSC). Occupying a strip of land 15km by 50km between the capital, Kuala Lumpur, and the country's new international airport, it aims to create the best environment in the world for customers across Asia and the world (Tidd and Brocklehurst, 1999).

Despite these imitators, Silicon Valley remains unique. While the imitators may have had some success in attracting high-technology firms and industries, it is innovation that makes Silicon Valley unique. As one recent study (Lee *et al.*, 2000: p. 3) noted: 'What sets Silicon Valley apart is not the technologies discovered there, but its record in developing, marketing and exploiting new technologies.'

Silicon Valley's unique feature is its capability for developing new technologies and exploiting them commercially – in short its capability for innovation. It is this capability that this book aims to explore.

Sources: Lee *et al.* (2000); Saxenian (1983); Tidd and Brocklehurst (1999)

WHAT IS INNOVATION?

In 2002 listeners to the *Today Programme* on BBC Radio 4 in a poll to mark 150 years of the UK Patent Office voted for their top ten inventions (in descending order):

1 Bicycle (Pierre Lallement, 1866)
2 Radio (Guglielmo Marconi, 1897)
3 Computer (Alan Turing, 1945)
4 Penicillin (Florey and Heatley, 1940)
5 Internal Combustion Engine (Nicolaus Otto, 1876)
6 World Wide Web (Tim Berners-Lee, 1989)
7 Light Bulb (Thomas Edison and Joseph Swan, 1829)
8 Cat's Eyes (Percy Shaw, 1936)
9 Telephone (Alexander G. Bell, 1876)
10 Television (John Logie Baird, 1923)

The bicycle, invented by Pierre Lallement in Paris in the 1860s, won by a landslide. Since the poll was organised to mark the opening of the Patent Office, it is not surprising that it was designed to elicit a top ten of inventions. The Patent Office is in the business of inventions. It issues patents to inventors whose ingenuity and perseverence has resulted in an 'inventive step' that gives rise to an invention.

Interestingly, all the inventions listed here are also innovations. Why? Because in each case the process did not end with invention. All the inventions became products that found markets and are still in use today.

This is the difference between invention and innovation. Not all inventions – even those that are registered for patents – get as far as being successful products in the marketplace.

Figure 1.1 Inventions and Innovations

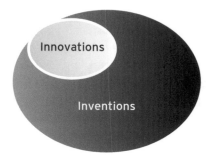

As Figure 1.1 shows, innovations represent a subset of a much bigger set of inventions. Why? Arduous though the task of invention is, innovation is equally so, and requires a different range of capabilities so that many prospective innovations do not make it to market successfully.

While this has helped differentiate innovation from invention, it has not made clear what innovation is and in particular what it involves. The following definitions of innovation can help to shed some light on this.

> ❝ An innovation is an idea, practice or object that is perceived as new by an individual or other unit of adoption. ❞ Rogers (1995: p. 11)

> ❝ Innovation is the successful exploitation of ideas. ❞ DTI (2004: p. 5)

> ❝ Innovations are new things applied in the business of producing, distributing and consuming products or services. ❞ Betje (1998: p. 1)

> ❝ the first commercial application or production of a new process or product. ❞
> Source: Freeman and Soete (1997: p. 1)

What we have got here is a number of perspectives, which is really quite useful because it serves to show that it is actually quite hard to be precise about innovation. Some of these perspectives are broad. For instance, they refer to ideas and they refer to things being new. This is fine, but these are both features of invention. We need a greater degree of precision because so far there is some overlap with invention. The references to business and commercial transactions help because innovation is about commercialisation of inventions – making them relevant to business. But it is the verbs in the later quotations that really come closest to the mark in defining innovation. Words like: *exploit* and *apply*. Innovation involves exploiting inventions so that they can be traded in a marketplace. Quite literally, innovation is about bringing inventions out of the workshop or the laboratory and getting them ready for the market. In this sense it is applying inventions to business, so that they can be bought and sold. Only when something new is available to consumers in a market or is being produced in a new way, has innovation occurred.

INVENTION – INNOVATION – DIFFUSION

From the definitions of innovation it becomes clear that innovation is closely linked to invention. In fact invention forms part of (or even a stage in) the process of innovation. However there is another important concept closely related to innovation, namely diffusion.

Invention involves new ideas, new discoveries and new breakthroughs. These are then developed via a process of often lengthy experimentation and testing to arrive at a workable invention. A key feature of inventions is their 'newness', which means that they incorporate some 'inventive step'. However, inventions are not normally ready for market at this stage. It is one thing to produce a single item, quite another to be able to produce it in large volumes at a cost and level of reliability that everyday use by consumers demands.

Innovation, therefore, includes not only invention but also activities such as design, manufacturing, marketing, distribution and product support. These activities form part of the exploitation/commercialisation which is such an essential part of innovation. Figure 1.2 makes clear that innovation involves both invention and commercialisation. The reason for the set of inventions shown in Figure 1.1 being much bigger than the subset of innovations is that commercialisation is often a lengthy process and many inventions, though they incorporate good ideas, never make it as far as innovation.

Figure 1.2 Invention, Innovation and Diffusion

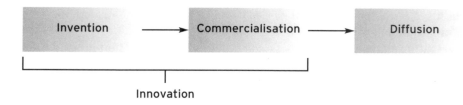

But, as Figure 1.2 shows, innovation is not the end of the story. There remains diffusion. As the term implies this is the stage where innovations became widely used and in time spread to other fields.

MINI-CASE: THE MINI

When it was introduced in 1959 the original Mini was a revolutionary car. It was the first small car able to transport four adults in comfort. One of its key features was that it had front-wheel drive. This innovation had been used before by manufacturers such as Citroën and Saab but only on cars that were produced in small numbers. It had not been used before on a small car with a conventional four-cylinder petrol engine. Front-wheel drive enabled the car's designer, Alec Issigonis, to provide an interior with room for four adults in a car that was little over 3 metres in length. It also gave the car much better handling than other cars of this period. Despite a shakey start Mini sales soon rose reaching almost 250 000 (Golding, 1984) a year by 1964. This success spurred the British Motor Corporation (BMC) to adopt front-wheel drive and fit it to bigger cars. These included the 1100 model and the Maxi. The process of diffusion spread as other major European car manufacturers then began to make use of the Mini's innovation of front-wheel drive. Models introduced in the early 1970s included the Renault 5 and the Volkswagen Golf. These models sold in very large numbers. Since then almost every European and Japanese manufacturer has developed a small car with a highly compact mechanical using front-wheel drive. So widely had front-wheel drive become diffused that in 2004, BMW, by then the makers of the latest version of the Mini, introduced a small/medium saloon, the BMW 1 Series, with an unusual feature for a car of this size – rear wheel drive.

Source: Golding (1984); Nahum (2004)

As the Mini mini-case study shows, front-wheel drive introduced on the original Mini in 1959 diffused to other models and classes of car in the BMC range in the 1960s and then to other manufacturers in the 1970s until after 20 years or so it had become the norm for small- and medium-sized cars throughout the world. A similar process of diffusion occurred with the personal computer. When Apple introduced the Apple II personal computer in the late 1970s, the computer market was dominated by large mainframe computers. During the 1980s personal computer sales took off as other manufacturers adapted and developed the personal computer idea. By the 1990s personal computers had diffused to the point where they had come to dominate the market, displacing large mainframe computers from most applications.

Nor is diffusion limited to products and services. Innovations often have the greatest impact on society when they are process innovations that begin to diffuse. F.W. Taylor pioneered 'time and motion' studies designed to identify the most efficient methods of working. Combined with what some have described as the 'American' system of manufacture based on standard interchangeable parts, this enabled the early car manufacturers to dispense with many of the 'craft' elements involved in making cars. However, it was Henry Ford's introduction of the moving assembly line that brought the cars to the workers instead of them going to it which was the decisive step in creating what we now know as 'mass production'. Introduced by Ford, so great was the improvement in productivity which it provided that it was adopted by virtually all the world's car manufacturers in the years after the First World War. Not only did Ford's process innovation lead to much lower prices for cars, it also radically altered working practices, consumption patterns and leisure activities.

WHO ARE THE INNOVATORS?

The list of top ten inventions throws up some well-known names. Even though their activities, for the most part, took place a hundred years ago, the names of Alexander Graham Bell, Thomas Edison, Guglielmo Marconi and John Logie Baird are relatively familiar. This reflects the relatively high public profile enjoyed by inventors, probably because inventing, that is to say creating something new, appears as challenging and exciting and almost personal. The commercial exploitation which lies at the heart of innovation is usually rather less exciting. Another factor is that, while invention is done by individuals, commercialisation is quite often the task of firms or corporations.

In fact, many of the inventors in the top ten list are also innovators. Examples are Alexander Graham Bell and Thomas Edison who not only invented but also innovated very successfully, exploiting and commercialising their inventions to great effect and creating large and successful businesses in the process. As such Bell and Edison are good examples of technical entrepreneurs/innovators.

Technical Entrepreneurs

Covered in detail in Chapter 9, these are individuals who invent a new product or new service and then proceed to innovation, thereby creating a new business in the process. Bell and Edison are classic examples of this type of innovator, and have their modern-day equivalents.

All of the individuals listed in the middle column of Table 1.1 are examples of successful innovators. They had an idea, developed it, produced it and then marketed it as a product that consumers could purchase. The individuals named in Table 1.1 created a business in order to innovate. From very small beginnings they watched it grow as the innovation achieved commercial success and the business then duly expanded. The case of Joseph Bamford is an outstanding example. Bamford developed the first hydraulic excavator in 1947 (Christensen, 1997). At the time large cable-operated machines dominated the excavator

Innovation	Technical Entrepreneur	Company
Light bulb	Thomas Edison	General Electric
Television	John Logie Baird	Baird Television Ltd
Instant Photography	Edwin Land	Polaroid Corporation
Telephone	Alexander Graham Bell	Bell Telephone Company
Backhoe hydraulic excavator (the JCB)	J.C. Bamford	J.C. Bamford Excavators Ltd
Cat's Eyes	Percy Shaw	Reflecting Roadstuds Ltd
Radio	Gugliemo Marconi	Wireless Telegraph & Signal Co.

Table 1.1 Technical Entrepreneurs

market. Bamford's innovation was to use hydraulic technology. His first product was a small 'backhoe' arrangement attached to the back of a conventional tractor. It had a capacity of a quarter of a cubic yard at a time when conventional cable-operated excavators had a capacity of 1 to 4 cubic yards. However Bamford persisted. Hydraulic technology improved steadily and his machines got bigger and more powerful until today when the company he founded is the fifth biggest construction equipment company in the world.

Women Innovators

All of the innovators referred to so far have been men. When innovators or inventors are mentioned they are nearly always men. In fact, records at the UK Patent Office show that there have always been women inventors and many of them have been highly successful as innovators. Five outstanding examples of women innovators and their innovations are:

- Stephanie Kwolek: Kevlar®
- Melitta Bentz: Coffee filters
- Maria Montessori: Teaching materials
- Bette Nesmith Graham: Correcting fluid
- Martha Lane Fox: Internet clearing house

Stephanie Kwolek worked at the US chemical firm Du Pont (Van Dulken, 2000). In the 1960s she discovered a new branch of synthetic materials called liquid crystallic polymers, one of which produced a new fibre which was stronger than steel. It was a form of carbon fibre and was eventually marketed as Kevlar®. Kevlar® has found a large number of diverse applications. Probably best known as the material used in body armour for soldiers, it is also used in fishing rods, brake pads, bridge cables and skis. Annual sales of Kevlar® today run into hundreds of millions of dollars.

Melitta Bentz from Leipzig developed a new method of making coffee using a paper filter (Jaffé, 2003). Her first filter was made of blotting paper placed in a brass pot in which she drilled holes. Her idea worked very effectively and she went on to found a company, whose brand name and products are synonymous with freshly brewed coffee.

Maria Montessori was an Italian doctor whose work in a children's psychiatric clinic led her to develop a range of educational methods designed to stimulate sensory development. She patented a number of items of teaching equipment including apparatus for teaching arithmetic and geometry. Her innovations in the teaching of young children particularly proved enormously influential in changing educational practice and in time were diffused throughout Europe (Jaffé, 2003).

Bette Nesmith Graham is a contemporary innovator. Working as a secretary in a bank in Texas in the days before word processors, she devised a means of correcting typing by painting out the errors. Unable to persuade corporate America to take up her idea, she set up on her own. Today the correcting fluid she devised is sold the world over.

Martha Lane Fox is another contemporary innovator. In 1998 she co-founded Lastminute.com, the Internet-based clearing house for everything from holidays to gifts. When Lastminute.com was floated on the stock market in March 2000, at the height of the Dot.com bubble, the company was valued at £750 million, making her 6.5 per cent stake worth some £50 million. Even though the company lost £17 million in its first six months as a PLC, it proved robust enough to survive the bursting of the Dot.com bubble at the end of 2000 (Parkhouse, 2001).

Corporate Innovators

MINI-CASE: VIAGRA®

Viagra, or it give it its chemical name, Sildenafil Citrate (Van Dulken, 2000: p. 236) was patented by the US pharmaceutical company Pfizer in 1993. The drug dilates blood vessels causing more blood to flow. Pfizer began experimenting with sildenafil citrate as a treatment for heart conditions such as angina and hypertension in 1990. It was only later that Pfizer began experimenting with the drug as a treatment for male erectile dysfunction (MED). Although patented in 1993 it was to be a further five years before the drug could be launched onto the market because of the large number of mandatory tests required for new drugs. When the drug finally did become available, Viagra® was for a time the fastest (legal) selling drug in history.

Source: Howells (2005)

Because of the attention given to inventions and inventors there is a tendency in the popular imagination to associate innovation with individuals. While in the past it was undoubtedly the case that innovation was an individual activity, in the twentieth century innovation increasingly became a corporate activity. At this time, as Mowery and Rosenberg (1998) have shown, the corporate R & D laboratory became an important source of innovation, reducing the importance of the individual innovator (i.e. technical entrepreneur) as a source of patents. (This view has been challenged in recent years (Rothwell, 1989) because, as the twenty-first century dawned, there were signs of a resurgence of small organisations as a source of innovation.)

Innovation	Corporation	Date
Viagra	Pfizer	1990s
Cloning Animals	Roslin Institute	1990s
First Direct	HSBC	1990s
Wide Chord Fan	Rolls-Royce	1980s
Widget	Arthur Guiness & Co	1980s
Post-it® Note	3M	1970s

Table 1.2 Corporate Innovations

Table 1.2 provides some recent examples. In the main these are innovations that resulted from a team effort and required a scale of resources that only a large corporation could provide.

The resources required for innovation today are extensive. First there is the cost of research and development. In 2004 a high-technology company like Rolls-Royce spent a sum amounting to 10 per cent of its turnover on research and development. But innovation requires much more than research and development. New drugs, for instance, require a massive expenditure on marketing. Then there is the cost of testing to ensure appropriate safety standards are met and the costs of distribution to get the product to consumers. Providing these sorts of resources has helped fuel the growth in corporate innovation over the past 50 or 60 years.

WHY DO INNOVATIONS FAIL?

It may seem paradoxical to be looking at the reasons why some innovations fail in a chapter devoted to explaining the nature of innovation. However, in order to understand what innovation is it can be highly instructive to look at innovation failures – in particular analysing the reasons for failure. These reasons can provide valuable insight into the nature of innovation.

MINI-CASE: CORFAM ARTIFICIAL LEATHER

In the 1960s the US chemical giant Du Pont undertook a huge research effort to develop a porous polyurethane material called Corfam that replicated almost all of the qualities of leather. The only qualities it did not share were generally undesirable ones like non-uniformity, stiffening when melted and high maintenance costs (i.e. a need for frequent polishing). Du Pont undertook extensive market analysis and the results were encouraging. Although the company was convinced that it had a product that was superior to leather, the manufacturing cost was high and Corfam had to be priced towards the top end of the market. Introduced in 1964, some 75 million pairs of shoes were sold over a 5-year period. A surprised Du Pont found itself losing some $70 million on Corfam over this period. Du Pont encountered higher than expected consumer resistance. One factor

▶

was that consumers maintained that the shoes were less comfortable. Another crucial factor was that at exactly this time European shoe styles using different types of leather became more fashionable. Between 1964 and 1969 shoe imports rose from 9 per cent to 26 per cent. Unfortunately Corfam was only available as one type of leather. This lack of flexibility forced Corfam out of the quality shoe market and resulted in losses. In 1971 Du Pont abandoned its innovation, selling the plant and the process for manufacturing Corfam to Poland.

Source: Hounshell and Smith (1988)

MINI-CASE: SINCLAIR C5

The Sinclair C5 was a battery-powered electric tricycle with a white plastic body, a top speed of 15mph and a range of 20 miles (Anderson and Kennedy, 1986). The C5 was the brainchild of Clive Sinclair who had previously enjoyed considerable success with electric calculators and a novel personal computer, the ZX81. The C5 was launched in January 1985 amid a barrage of publicity.

It was heralded as a revolution in personal transportation although it used a modified washing machine motor and a car battery. Unfortunately, it faced adverse publicity from the start, much of it initially from safety organisations. This was soon followed by adverse publicity from consumers disappointed to find that its maximum speed was 12mph and its range a mere 6.5 miles. With production capacity planned for 200 000 units per year, the C5 sold only 1000 units in the first month. It was soon clear that it was not going to be commercial success and Sinclair Vehicles went into receivership in October 1985.

Source: Anderson and Kennedy (1986)

What do these mini-case studies of Du Pont's Corfam and Sinclair's C5 tell us about why innovations fail? First, they indicate that its cause is rarely down to failure of the technology. Corfam performed satisfactorily and, while the Sinclair C5 did not perform as well as expected, this was not caused by the failure of new technology but rather by the use of old, and for the most part inadequate, technology. Second the Corfam case highlights the importance of production. Du Pont lost money on Corfam because from the outset it was a high-cost process.

Probably the most significant message of both cases highlights the importance of marketing and shows how the market, especially changes in the market, can cause an innovation to fail. In the case of the Sinclair C5 there appears to have been insufficient attention paid to marketing. The Sinclair C5 was a technology-led project with little attention paid to consumer requirements and market requirements in terms of safety. In the case of Corfam, there was extensive market analysis and the product was carefully marketed. What went wrong was that the market, certainly in terms of fashion trends, changed dramatically and Du Pont found itself with a product that could not meet the new market conditions. Finally, the case of the Sinclair C5 highlights the importance of public relations and distribution when launching an innovation. Sinclair's decision to launch an open-topped vehicle in the middle of winter was at best unwise, while the choice of mail order as the means of distribution was simply inadequate, when compared to the extensive dealer networks offered by competitors.

These innovation failures serve to show that innovation, unlike invention, involves a number of business functions, such as marketing, distribution and public relations, and relying purely on technology to deliver a successful innovation is normally inadequate. For success in innovation, it is necessary to get the technology and the business functions correct. Innovators, especially if they have a background in technology, have to appreciate the importance of the business aspects of innovation and ensure that, as well as the technology, they get the business right. A feature of successful innovators is that like Edison and Bell they have a strong intuitive feel for business, or at least have the sense to work with people who do.

CASE STUDY: THE WORKMATE® WORKBENCH

The last 30 years has seen a massive growth in the market for do-it-yourself (DIY) products. This growth has been fuelled by a rapid increase in the number of people who own their own homes and the current popularity of TV programmes like *Grand Designs* that are about house, homes and property. Stores selling DIY products are now to be found throughout the UK and firms like B & Q, Homebase and Focus offering everything for the DIY enthusiast are everywhere. Go round any one of them and you will come across a product that has become a household name: the Workmate®.

The Workmate® was the brainchild of Ron Hickman. Although he came up with the idea, he did not work for one of the manufacturers of power tools or DIY products. Rather, Hickman's background was in the more glamorous field of Formula 1 racing.

Hickman who was once quoted as saying 'you can't invent to order', was a South African who in the 1950s worked for Ford as a car stylist. In 1958 (*Automobile Magazine*, 1989), though he lacked formal engineering or design training, he moved to Lotus Cars (Crombac, 2001: p. 84), a small, London-based engineering company, led by racing car designer Colin Chapman. At this time Lotus was developing the first of its commercial road-going sports cars. For a time he was general manager of Lotus Developments (Tippler, 2002: p. 16) and was part of the team, along with Colin Chapman and Mike Costin (Robson, 1999: p. 41), that developed the innovative Lotus 25 racing car for Formula One. In 1962 Hickman was responsible for designing the highly successful Lotus Elan sports car.

Hickman left Lotus in the late 1960s to explore ideas of his own. A keen DIY enthusiast, he was infuriated when one day while sawing a piece of wood he found he had accidentally sawn the leg off a chair. He began sketching designs for a workbench that could hold pieces of wood while being sawn. Recognising that most DIY enthusiasts worked in their spare time, with limited resources and limited space, he was keen that the workbench would be foldable to aid portability and facilitate storage. He worked on several prototypes. His ideas evolved from a workbench with a single beam, similar to a

▶

conventional carpenter's bench (or 'horse' as it is sometimes known), to one employing two parallel beams. With this arrangement the beams combined the functions of work surface and a vice to hold the object being worked on. Hickman credited the development process through which the design evolved and the features of his workbench to his lack of formal design training and relative inexperience in the field, and facilitated 'lateral thinking' and brought fresh ideas to the problem.

Having developed a working prototype, Hickman had to turn his invention into a commercially saleable product. In order to protect all his hard work and the 'intellectual property' associated with his invention, he patented his design in 1968 (Patent number GB 1267032-5 and US 3615087). Lacking capital and detailed knowledge and experience of the DIY market, he initially sought to find an existing manufacturer willing to take up his design. He offered his idea to eight companies that manufactured tools and DIY products, hoping that they would take out a licence to manufacture his design by which he would get a royalty payment for each item sold. All rejected it. Typical was the reply he got from Stanley Tools (Figure 1.3).

Unable to interest existing manufacturers, he began to manufacture the product himself, marketing it as the 'Workmate®'. By the early 1970s he was making 14 000 units annually (Van Dulken, 2000: p. 168). Then in 1972, Black & Decker, the US manufacturer of power tools and DIY products and one of the companies that had originally rejected his idea, approached him. They negotiated a licence to manufacture the Workmate®, under which Hickman was to receive 3 per cent of the sales price of every Workmate® as a royalty payment. It took a great deal of effort to persuade Black & Decker's American parent company that the concept was viable.

They need not have worried. The product was a huge success, in part because there was nothing like it on the market. Additional patents were filed over the next decade as improvements to the product were devised, but the basic concept and the main features of the design remained unaltered. Over the next 30 years more than 55 million units were sold and it is still in production and selling well today.

QUESTIONS

1 What kind of innovation does the Workmate® represent?
2 What motivated Hickman to develop the Workmate®?
3 Would you classify Hickman as an inventor or an innovator?
4 At what point did the Workmate® cease to be an invention and become an innovation?
5 Explain what is meant by appropriability?
6 Why was appropriability likely to be an issue (i.e. ensuring that Hickman was rewarded for his innovation)?
7 Why did Hickman initially try to license his invention?
8 What was Black & Decker able to bring to the process of innovation?
9 What does the Workmate® case tell you about the difference between invention and innovation?

Figure 1.3 Rejection Letter from Stanley Tools

STANLEY

STANLEY WORKS (Great Britain) LTD

WOODSIDE · SHEFFIELD 3 · ENGLAND

Directors: G. D. TANNETT (Chairman & Managing)   F. W. BALDREY   W. M. DINGLEY   A. G. JONES   W. R. MILLS   H. G. PEASE (U.S.A.)   J. L. SKIDMORE (& Secretary)

CABLES: STANLYWORX SHEFFIELD 3

TELEX: 54150.   TELEPHONE: SHEFFIELD 78678

Your Ref. RH/EC
Our Ref. NLW/CD

1st April, 1968

a very appropriate date!

R. P. Hickman, Esq.,
Director,
Hickman Designs Limited,
Badgers,
Middle Street,
Nazeing,
Essex.

Dear Mr. Hickman,

Following our telephone message to your wife on Friday, 29th March, this is to confirm that we have completed our investigation and have reluctantly decided that we do not wish to manufacture and market your patented Multiwork-unit.

Our reasons are as follows:

1. We have found your estimates and costings to be realistic, but we calculate that, with overheads and distribution costs, plus the trade discounts applicable to the ironmongery trade, the retail price of the unit would need to be at least £15. As you know, this price will bring your unit very much closer to the price of full size work benches than originally anticipated, and this will restrict demand.

2. Our investigations in Europe produced a very luke-warm reaction. In general, it was felt that the potential could be measured in dozens rather than in hundreds.

3. We believe the marketing suggestions in your letter, dated 26th March, are an essential to achieve success, However, we feel that we would be unable to divert sufficient sales effort from hand tools to devote the necessary time and energy which your project deserves. In essence, we believe that your Multiwork-unit could be best sold by an organisation which could give your project one hundred per cent of its attention.

Sources: Hickman and Roos (1996); Van Dulken (2000)

part i: what is innovation?

QUESTIONS FOR DISCUSSION

1 Why do so many biographies of people associated with successful innovations appear more preoccupied with inventions?

2 Why do there appear to be more accounts of invention compared to accounts of innovation?

3 Explain the difference between invention and innovation. Which do you consider the most important and why?

4 Outline the stages in the process of innovation.

5 Using an innovator of your choice, indicate the impact of his/her innovation.

6 Explain what is meant by an organisational innovation.

7 Explain why today many innovations are more the result of organisational rather than individual activity.

8 What are the benefits to be gained from being first to market an innovation?

9 What is the link between innovation and risk?

10 What personal qualities do you consider are essential for successful innovators?

11 In what ways can an organisation influence, for better or for worse, the rate at which it generates innovations?

12 What impact do process innovations tend to have on an organisation?

13 Why is it that today innovation is often a collective activity?

14 Why do you think successful innovators are often not the best qualified individuals in their field?

15 Ron Hickman, the man who developed the Workmate® workbench has been quoted as saying, 'You cannot invent to order'. What do you think he meant by that statement?

16 Why is it that innovations often come from organisations that are not the established leaders in their field?

17 Take an example of what you consider to be an organisation with a strong record of innovation and explain why you feel it is good at innovation.

ASSIGNMENTS

1 Prepare a briefing document for the Workmate®. This should be no more than six pages in length and should provide the reader with a clear understanding of the nature of the Workmate®, as well as the factors that account for its success as an innovation and the important lessons you feel it holds for would-be innovators. As part of the briefing prepare some slides that will enable you to make a presentation in class.

To carry out this task, you will need to classify the product, identify prospective purchasers/consumers, distinguish product features, identify competitor products, etc. Your aim should be to show what it is about the Workmate® that makes it a good example of an innovation.

You will need to carry out some basic fact-finding research. Possible sources are department stores and catalogues that stock the Workmate® (e.g. Argos catalogue). If you know someone who owns a Workmate®, have a look at it.

2 EITHER prepare a profile on 'Ron Hickman – Innovator', OR prepare a profile of an innovator of your choice.

If you are going to profile Ron Hickman, the case study obviously provides a valuable starting point, particularly the references that appear in the notes at the end of the case. You would be well advised to access the referenced texts in order to find out more about Ron Hickman and his activities as an innovator. You should say something about his background, particularly the expertise and experience that enabled him to innovate (this could well include something about his work prior to the Workmate®). You should aim to analyse and identify the factors that you think explain Hickman's success as an innovator, and the implications for prospective innovators, whether individuals or organisations.

If you are profiling somebody else, you are advised to look at the notes to the case study as a guide to the sort of sources that are likely to be useful. You will probably find that biographies, industry studies and similar sources will be useful. Internet searches may well enable you to find short profiles of the individuals you are interested in. However, be warned that such profiles often lack detail and tend to treat the subject matter in an unsophisticated and uncritical manner.

RESOURCES

Like innovations, valuable resources often come from unlikely sources. Remember too that there is life beyond the Internet!

Printed sources

For an introduction to some of the practicalities of doing innovation, autobiographies and biographies are a valuable resource, because they often provide good accounts of innovation. In particular, they often provide a lot of detail about the process of innovation. One has to be careful because they are prone to being subjective, not unsurprisingly. Hence one sometimes needs to disentangle extraneous detail or the writer's personal views regarding obstacles and hurdles, but they have the great advantage that they usually provide lots of detail.

Good examples of the genre are James Dyson's (1997) *Against the Odds: An Autobiography*, Trevor Baylis's (1999) *Clock this: My Life as an Inventor* and Tim Berners-Lee's (2000) *Weaving the Web: The original design and ultimate destiny of the World Wide Web by its inventor*. While good examples of biographies are Golley's (1996) *The Genesis of the Jet: Frank Whittle and the Invention of the Jet Engine* and if you can get hold of a copy, McElheny' s (1998) *Insisting on the Impossible: The Life of Edwin Land*. Though written from different perspectives, each provides, as well as an account of the innovator's life, a detailed account of the innovation process.

Another possible avenue is to use company histories. These vary enormously, but if a company history exists and that company was active in terms of innovation, it may well be useful. An example would be Pugh's (2001), *The Magic of a Name*, a history of aero engine maker Rolls-Royce which provides an account of the innovation of the wide chord fan that transformed the efficiency of the modern large jet engine. Similarly Reader's (1980) *Fifty years of Unilever 1930–1980* provides an excellent account of the innovation of frozen foods in the 1950s.

Another potential seam of material about innovations comes from books that provide a collection of innovation case studies. Rarely written as case studies, they often provide useful if brief accounts of well-known innovations. One needs to be slightly careful because they often focus on the 'heroic' aspects of invention rather than the less dramatic but often more significant 'nuts and bolts' of innovation. Good examples include, Van Dulken (2000) *Inventing the 20th Century: 100 Inventions that Shaped the World*, Nayak and Ketteringham's (1993) *Breakthroughs!* and, though now rather old, Jewkes *et al.* (1969) *The Sources of Invention*.

Journal articles can have an important part to play too, although, there are fewer than one might expect that deal specifically with innovations. An obvious starting point is to look at journals that deal specifically with innovation such as *R & D Management*, *Research Policy*, *Technovation*, *Technology Analysis and Strategic Management* and *International Journal of Innovation Management*. Mainstream management journals like the *British Journal of Management* and the *Journal of Management Studies* do include studies of innovation, though they focus on innovation issues rather than specific instances of innovation. But do not stop there. Specialist journals can often come up trumps, though they are usually difficult to locate. An example would be Saxenian's (1983) article 'The genesis of Silicon Valley', which appeared in the journal *Built Environment* and provides an account of the origins of Silicon Valley. It is not an account of a particular innovation but it does provide valuable insights into innovation in the mid-twentieth century. Another example would be Cox, Mowatt and Prevezner's (2003) 'New product Development and Product Supply within a Network Setting: The Chilled Ready-Meal Industry in the UK' which appeared in *Industry and Innovation* and provides an account of the innovation of chilled ready-meals in the UK in the 1990s.

Websites

There are a number of websites that provide an introduction to innovation, explaining terminology and providing helpful definitions. Two useful websites are http://www.innovation.gov.uk and http://www.invent.org.uk . The former is maintained by the Department of Trade and Industry and its function is to help individuals and organisations understand innovation. As such it provides a very useful starting point. The latter is maintained by the Institute of Patentees and Inventors, a small non-profit-making organisation that gives advice to inventors. Useful, but be warned that inventors are not necessarily innovators. Another possibility is http://businesslink.gov.uk which covers aspects such as design, intellectual property and R & D as well as innovation.

But be warned, most websites simply do not provide the level of detail or the level of analysis that is expected on an undergraduate course. A truly magnificent exception is http://www.bricklin.com, the official website of the innovator who gave us the spreadsheet, Dan Bricklin. Not only does it provide an excellent first-hand account of the innovation process surrounding the development of the first spreadsheet in 1978, it even includes examples of early advertisements and publicity materials as well as a copy of the first commercial spreadsheet, *VisiCalc*, that you can download and use.

Video and Film

An unlikely source perhaps, but innovators tend to capture the popular imagination and consequently the better-known ones, especially if they are Americans, have not escaped Hollywood's attention. An example of a film that deals with innovation would be *Tucker: The Man and his Dream* (1988), available on video and DVD and starring Jeff Bridges. It is the story of Preston Tucker the man who developed the innovative Tucker car of the 1940s. Similarly videos/DVDs covering the lives of innovators such as Henry Ford and Thomas Edison are available.

REFERENCES

Anderson, I. and R. Kennedy (1986) *Sinclair and the Sunrise Technology*, Penguin Books, Harmondsworth.

Automobile Magazine (1989) *The Man who Designed the Elan*, March.

Baylis, T. (1999) *Clock this: My Life as an Inventor*, Headline Book Publishing, London.

Berners-Lee, T. (2000) *Weaving the Web: The original design and ultimate destiny of the Web by its inventor*, Harper Business, London.

Betje, P. (1998) *Technological Change in the Modern Economy: Basic Topics and New Developments*, Edward Elgar, Cheltenham.

Christensen, C.M. (1997) *The Innovator's Dilemma: When New Technological Cause Great Firms to Fail*, Harvard University Press, Boston, MA.

Cox, H., S. Mowatt and M. Prevezer (2003) 'New Product Development and Product Supply within a Network Setting: The Chilled Ready-Meal Industry in the UK', *Industry and Innovation*, 10 (2), pp. 197–217.

Crombac, G. (2001) *Colin Chapman: The Man and His Cars*, Patrick Stephens Ltd, Sparkford.

DTI (2004) *Succeeding Through Innovation, Creating Competitive Advantage through Innovation: A Guide for Small and Medium Sized Businesses*, Department of Trade and Industry, London.

part i: what is innovation?

Dyson, J. (1997) *Against the Odds*, Orion Business, London.

Freeman, C. and L. Soete, (1997) *The Economics of Industrial Innovation*, 3rd edn, Continuum, London.

Golding, R. (1984) *Mini: Thirty Years On*, 2nd edn, Osprey Publishing, London.

Golley, J. (1996) *The Genesis of the Jet*, Airlife, Marlborough.

Hickman, R.P. and M.J Roos, (1996) 'Workmate', in E. Taylor, (ed.) *Innovation, Design, Environment And Strategy*, T302 Technology, Block 2 Readings and Cases, Open University, Milton Keynes.

Hounshell, D.A. and J.K. Smith, (1988) *Science and Corporate Strategy: Du Pont R&D, 1902-1980*, Cambridge University Press, Cambridge.

Howells, J. (2005) *The Management of Innovation & Technology*, Sage Publications, London.

Jaffé, D. (2003) *Ingenious Women: From Tincture of Saffron to Flying Machines*, Sutton Publishing, Stroud.

Jewkes, J., D. Sawers and R. Stillerman (1969) *The Sources of Invention*, Macmillan, London.

Lee, C.M., W.F. Miller, M.G. Hancock and H.S. Rowen, (2000) *The Silicon Valley Edge: The Habitat for Innovation and Entrepreneurship*, Stanford University Press, Stanford, CA.

McElheny, V.K. (1998) *Insisting on the Impossible: The Life of Edwin Land, Inventor of Instant Photography*, Prospero Books, Reading, MA.

Mowery, D. and N. Rosenberg, (1998) *Paths of Innovation: Technological Change in 20th century America*, Cambridge University Press, Cambridge.

Nahum, A. (2004) *Issigonis and the Mini*, Icon Books, Cambridge.

Nayak, P.R. and J.M. Ketteringham, (1993) *Breakthroughs!*, Mercury Business Books, Didcot.

Parkhouse, S. (2001) *Powerful Women: Dancing the Glass Ceiling*, J. Wiley and Sons, Chichester.

Pugh, P. (2001) *The Magic of a Name: The Power Behind the Jets*, Icon Books, Cambridge.

Reader, W.J. (1980) *Fifty Years of Unilever 1930-1980*, Heinemann, London.

Robson, G. (1999) *Cosworth: The Search for Power*, 4th edn, Haynes Publishing, Sparkford.

Rogers, E.M. (1995) Diffusion of Innovation, 4th edn, The Free Press, NY

Rothwell, R. (1989) 'Small Firms, Innovation and Industrial Change', *Small Business Economics*, **1** (1) pp. 51-64.

Saxenian, A. (1983) 'The Genesis of Silicon Valley', *Built Environment*, **9** (1), pp. 7-12.

Tidd, J. and M. Brocklehurst (1999) 'Routes to Technological Learning and Development: An Assessment of Malaysia's Innovation Policy and Performance', *Technological Forecasting and Social Change*, **62**, pp. 239-257.

Tippler, J. (2002) *Lotus 25 & 33*, Sutton Publishing, Stroud.

Van Dulken, S. (2002) *Inventing the 20th Century: 100 inventions that shaped the world*, British Library, London

CHAPTER 2

types of INNOVATION

OBJECTIVES

When you have completed this chapter you will be able to:

- Distinguish the different forms that innovation can take, such as product, process and service innovation
- Differentiate and distinguish between the different types of innovation, such as radical and incremental innovation
- Describe each type of innovation
- Analyse different types of innovation in terms of their impact on human behaviour, business activity and society as a whole.

INTRODUCTION

The notion that innovation is essentially about the commercialisation of ideas and inventions suggests that it is relatively straightforward and simple. Far from it, not only is the step from invention to commercially successful innovation often a large one that takes much effort and time, but innovations can and do vary enormously. In addition the term 'innovation' is widely used, probably because it frequently has very positive associations, and is often applied to things that really have little to do with innovation, certainly in the sense of technological innovation. The purpose of this chapter is to try and produce some sort of order from the apparent chaos and confusion surrounding innovation.

MAKING SENSE OF INNOVATION

If innovation comes in a variety of shapes and sizes and is used by different people to mean different things then making coherent sense of the subject is not an easy task. Grouping innovations into categories can help. Essentially, putting innovations in groups should make it easier to make sense of innovation as a whole simply because one can then take each group in turn and subject it to detailed scrutiny. If it is easier to make sense of a small group than a large one then we should be on the way to making sense of innovation.

Two kinds of categorisation are attempted. The first centres on different forms of innovation. 'Form' in the sense in which the term is used here applies to the use or application of the innovation. Three applications are considered: product, service and process innovations.

The second categorisation is based on the degree of novelty associated with the innovation and implies that there are different degrees of novelty associated with innovation. As a result, one sometimes finds that things described as innovations actually involve little or no novelty. Take the case of a new wrapper for a chocolate bar. For the people marketing the product, the new wrapper may well appear to be a significant innovation, hence justifying the use of words like 'innovation' and 'innovative' in promotional campaigns; but the reality is that if the same type of wrapper is already in use on other similar products there really is very little innovation. On the other hand, one can have innovations such as television, developed by John Logie Baird (Kamm and Baird, 2002), which not only transformed the nature of leisure time, created a new creative industry and provided employment for thousands, but also went on to transform a whole host of other aspects of our society including politics, advertising, the provision of information and sport. Recognising different degrees of novelty, this categorisation considers four types: radical, architectural, modular and incremental.

FORMS OF INNOVATION

This categorisation is based on the idea of applications or uses for innovation. By this we mean areas or fields where innovations are used. It is possible to differentiate three principal applications for innovation: products, services and processes.

Product Innovation

Product innovations loom large in the public imagination. Products, especially consumer products, are probably the most obvious innovation application. The Dyson bagless vacuum cleaner is an example of a product innovation. James Dyson developed what he terms 'dual-cyclone' technology (Dyson, 1997) and used it to create a new more efficient vacuum cleaner. As a vacuum cleaner it is a consumer product and what makes it an innovation, i.e. what is 'innovative' about it, is that it functions in a quite different way from a conventional vacuum cleaner. It is still a vacuum cleaner and it does what vacuum cleaners have always done, it extracts dust and other items of household debris from carpets and upholstery; but the innovation lies in the way in which it functions. Instead of employing a fan to suck dust into a bag, it dispenses with the bag and uses Dyson's patented dual-cyclone technology to extract dust and place it in a clear plastic container. It is a nice example of a product innovation because it is an everyday household product where you can actually see the innovation at work, a fact that James Dyson, an experienced industrial designer and entrepreneur, no doubt had in mind when he designed his first bagless vacuum cleaner, the Dyson 001.

From a commercial perspective the attraction of product innovations is that the novelty of a new product will persuade consumers to make a purchase. It is no surprise that 'product development' is one of the four business strategies put forward by Ansoff (1988) for the future development of a business. Of course, product innovations do not have to be consumer products, they can just as easily be industrial products such as machinery and equipment.

Service Innovations

Often overlooked but equally important are service innovations, which take the form of new service applications. One reason why service innovations fail to attract as much attention as product innovations is that they are often less spectacular and less eye-catching. This probably has something to do with the fact that, where innovation is concerned, the public imagination has always tended to identify with inventions, rather than innovation as such. Because of their high novelty value, inventions are usually products.

Service innovations typically take the form of a new way of providing a service, often with a novel and very different business model. Occasionally they even take the form of an entirely new service. The creation of the 'Direct Line' telephone insurance business is a good example of the first type of service innovation. For years the insurance business had been transacted via high street outlets, door-to-door, by post or through intermediaries known as insurance brokers. Peter Wood, the creator of the Direct Line telephone insurance business, realised that with appropriate on-line computer services, it would be possible to cut out these expensive and unproductive ways of dealing with the public and deal direct with the customer via the telephone. Developments in computing and telecommunications in recent years have given rise to a whole raft of service innovations very similar to Direct Line where new technologies are used both to provide customers with a better service and to enable service providers to improve their productivity by providing it more cheaply.

MINI-CASE: SOUTH WEST AIRLINES

Founded in the late 1960s by Herb Kellner, it was South West Airlines that started the 'no frills' revolution in air travel.

In Europe in the last 10 years air travel has been transformed by the introduction of low-cost services offered by 'no frills' carriers. The innovation which these carriers introduced has been the provision of easily accessible scheduled short-haul services at fares very much lower than those offered by conventional scheduled airlines. The result has been an enormous increase in both numbers travelling by air and the range of destinations served.

Yet this was not a European innovation. The pioneer of low-cost 'no frills' air transport was South West Airlines based in Texas. Under its charismatic founder, Herb Kellner, South West Airlines had to fight legal battles with local competitors for the first four years of existence just to be allowed to fly. Competitors argued there simply was not enough business to

▶

warrant another airline in the region. When it did finally get airborne it was faced with a price war with Braniff and other airlines as they tried to drive it out of business.

Based at Love Field in downtown Dallas, South West Airlines was able to survive by offering customers a very different package from conventional airlines. The package included low fares (usually 60 per cent below conventional airlines), high frequencies, excellent on-time departure rates and direct sales (i.e. no travel agents). What was not being offered was meals, pre-assigned seats, different classes of seating and connecting flights. This was achieved by means of: a single aircraft type (then and now the Boeing 737), smaller low-cost airports, rapid turnarounds (typically 15–20 minutes), high load factors, and point-to-point services.

The 'no frills' service package diverted some traffic away from existing carriers but, more significantly, it generated a lot of new business, especially leisure and business passengers who could be persuaded to fly rather than drive. As Herb Kellner (Dogannis, 2001: p. 128) put it: 'we are not competing with airlines, we're competing with ground transportation'.

De-regulation of airline services in the US in 1978 meant that South West Airlines was well placed to expand in Texas with this innovation in airline service. Traffic growth proved well above average. South West was able to expand by adding more capacity to its fleet, but instead of adding routes as airlines normally did, Kellner's strategy was to increase flight frequency on existing routes.

It worked. Today South West Airlines is the fifth biggest carrier in the US, and is the most consistently profitable airline in the country, yet it has stuck to its innovative business model. Not only that, the model has been copied with great success in Europe, first by Ryanair (Dogannis, 2001) and then by a host of other airlines including easyJet and BMI Baby to create a low-cost revolution in air travel across the continent.

Source: Procter (1994)

Sometimes one gets innovations that take the form of completely new services. eBay, the Internet auction, and Lastminute.com the clearing house for late bookings on anything from holidays to gifts, would probably come under this heading. So too would Federal Express the brainchild of Frederick W. Smith. Operating in an established industry – parcel delivery – Smith pioneered the idea of overnight delivery using a hub-and-spoke system (Nayak and Ketteringham, 1993). During the day trucks collect parcels and bring them to an airport where they are sorted and then flown overnight to a hub near their destination ready for delivery the next day.

Process Innovations

If service innovations come second behind product innovations, then process innovations almost certainly come a poor third. Yet process innovations often have an even bigger impact on society than either product or service innovations. The early nineteenth-century Luddite movement in and around Nottingham (Chapman, 2002), where stocking knitters who worked on machines in the home took to rioting and breaking the new, more efficient, machines located in factories, because they feared that the new machines would destroy their livelihoods, is testimony to the power of process innovations.

Although generally less well known than product innovations, examples of process innovations, including ones that have had a dramatic impact on society as a whole, abound.

The humble photocopier, developed by Chester Carlson, may not sound like a spectacular innovation, yet it had a big impact on the way in which administrative systems in offices are organised. One has only to look at what happens in an office when the photocopier breaks down to see how reliant we are upon it.

Much less well known, but just as significant in terms of its impact on society, is the 'float glass' process developed by Alistair Pilkington, in which plate glass is manufactured by drawing glass out across a bed of molten tin (Quinn, 1991). Prior to the introduction of this process innovation, plate glass used for shop windows and office windows was expensive and of poor quality, largely because the only way of getting a flat surface was to grind it and polish it. The float glass process at a stroke eliminated the need for time-consuming grinding and polishing it, leading to a dramatic fall in costs. Architects and property developers could now afford to specify large sheets of plate glass when constructing new buildings, where in the past they would have been prevented because of the cost. The result can be seen in building construction in the past thirty years, where everything from office blocks and hotels to airports and shopping malls now employs large expanses of glass.

Assembly Time	Craft Production, 1913 (minutes)	Mass Production, 1914 (minutes)	Reduction in Effort (%)
Engine	594	226	62
Magneto	20	5	75
Axle	150	26.5	83
Components into vehicle	750	93	88

Table 2.1 Craft v. Mass Production at Ford 1913–1914
Source: 'The Machine That Changed the World', by James P. Womack, Daniel T. Jones and Daniel Roos. Rawson Associates/Scribner, an imprint of Simon and Schuster Adult Publishing Group. © 1990 by James P. Womack, Daniel T. Jones, Daniel Roos and Donna Sammons Carpenter

Nor is it only process innovations that affect working practices and the physical infrastructure of towns and cities. Process innovations often have a big impact on the economics of production. As Table 2.1 shows, Henry Ford's introduction of the moving assembly line at his new Highland Park plant in Detroit in 1913 resulted in a dramatic reduction in manufacturing effort. Improved productivity on this scale enabled him dramatically to reduce the price of his Model T car. The price of a Ford Model T, which in 1908 was $850, fell to $600 in 1913 and $360 by 1916 (Freeman and Louçã, 2001: p. 275). As Ford reduced his prices, demand took off and the car, which had hitherto been an ostentatious toy only available to a small wealthy elite, was opened to a broad cross-section of society. Other examples of innovation are given in Table 2.2

Today a similar revolution in production is taking place, but this time the revolution is occurring not on the factory floor but in the office. Business-to-business (B2B)

E-commerce is dramatically reducing the need for paperwork and those who process paper, namely administrators. It is no surprise that all sorts of business organisations from airlines to insurance companies offer a discount for buying online. Buying online means less paper and money spent processing paper. One has only to look at the size of the discounts offered to get an idea of the efficiency gains that firms can make.

Form	Innovation	Innovator	Country
Product	Walkman	Akio Morita/Sony	Japan
	Ballpoint Pen	Laszlo Biro	Hungary
	Television	John Logie Baird	UK
	Spreadsheet	Dan Bricklin	USA
Service	Telephone Insurance	Peter Wood/RBS	UK
	Credit Card	R. Schneider/ F. McNamara	USA
	Internet Bookstore	Jeff Buzos	USA
	'No Frills' Airline	Herb Kelleher/ Rollin King	USA
Process	Moving Assembly Line	Henry Ford	USA
	Float Glass	Alistair Pilkington	UK
	Single Minute Exchange of Dies (SMED)	Shigeo Shingo/Toyota	Japan
	Computerised Airline Reservations (SABRE)	IBM/American Airlines	USA

Table 2.2 Forms of Innovation

TYPES OF INNOVATION

It has long been noted that one can differentiate innovations in terms of the degree of novelty associated with them. Some innovations employ a high degree of novelty, while others involve little more than 'cosmetic' changes to an existing design. This distinction between big-change and small-change innovations has led some to group innovations as either radical or incremental (Freeman, 1982). However, differentiating innovations using just two classes in this way is rather limited and does not bring out the subtle but important differences between innovations. In particular it fails to show where the novelty often lies. To cater for this Henderson and Clark (1990) use a more sophisticated analysis. This incorporates both radical and incremental innovation but within a more wide-ranging analysis that is both robust and meaningful. Henderson and Clark's (1990) analytical framework provides a typology that allows us to analyse more modest innovations and at the same time predict their impact in terms of both competition and the marketplace.

Although this typology focuses primarily on product innovations it can equally be applied to service and process innovations.

At the heart of Henderson and Clark's analytical framework is the recognition that products are actually systems. As systems they are made up of components that fit together in a particular way in order to carry out a given function.

Example

Pen	=	knib + ink storage + stem + cover + ink
System	=	interaction of components

Henderson and Clark (1990) point out that to make a product normally requires two distinct types of knowledge:

- Component knowledge
  i.e. knowledge of each of the components that performs a well-defined function within a broader system that makes up the product. This knowledge forms part of the 'core design concepts' (Henderson and Clark, 1990) embedded in the components.

- System knowledge
  i.e. knowledge about the way the components are integrated and linked together. This is knowledge about how the system works and how the various components are configured and work together. Henderson and Clark (1990) refer to this as 'architectural' knowledge.

MINI-CASE: AUTOMATIC WASHING MACHINE

The modern automatic washing machine has been subjected to a variety of types of innovation. The washing machine is a system for washing clothes. The components comprise: motor, pump, drum, programmer, chassis, door and body. These components are linked together into an overall system. Component knowledge is the knowledge that relates to each of the components. System knowledge, on the other hand, is about the way in which the components interact. The interaction is determined by the way in which the system is configured. Responsible for the design and development of the system, washing machine manufacturers frequently buy in component knowledge by buying components and then assembling them into a finished product.

Washing machines have been affected by both incremental and architectural innovation. Changes in the spin speed are an example of incremental innovation. The spin speed determines how dry the clothes will be when they come out of the machine. Twenty years ago the fastest machines spun at up to about 1000rpm. Gradually spin speeds have risen and today the fastest machines spin at 1600rpm. Although these advances have resulted in improved performance, the system has remained unchanged. However, there have been architectural

►

innovations in the washing machine field. In the 1960s most washing machines were 'twin tubs', where the washer and the spinner were completely separate and placed alongside each other with access via the top of the machine. Clothes had to be manually moved from the washer into the spinner. The automatic washing machine with the washer and spinner combined in a single drum, allowing all the operations to be completed in a single cycle, was an architectural innovation. Similarly when Dyson launched its Contrarotator™ washing machine in November 2000 this was another architectural innovation. This has not just one drum, as on a conventional machine, but two that rotate in opposite directions. This change in the configuration of the system gives an entirely different washing system.

Henderson and Clark (1990) use the distinction between component and system knowledge to differentiate four categories or types of innovation (Figure 2.1). They use a two-dimensional matrix. On one axis are component changes, on the other linkage (i.e. system) changes:

Figure 2.1 Typology of Innovations

| | Components/core concepts | |
System/linkages	Reinforced	Overturned
Unchanged	Incremental innovation	Modular innovation
Changed	Architectural innovation	Radical innovation

Source: 'Architectural Innovation: The Reconfiguration of Existing Product Technologies and the Failure of Established Firms' by: Rebecca M. Henderson, Kim B. Clark, vol 35, pp. 9–30, March 1990. Reproduced with kind permission of Administrative Science Quarterly

In this analysis Incremental and Radical innovation are polarised as being at opposite extremes. However the analysis introduces two intermediate types between these two extremes (Table 2.3): modular innovation and architectural innovation:

Innovation	Components	System
Incremental	Improved	No change
Modular	New	No change
Architectural	Improved	New configuration/architecture
Radical	New	New configuration/architecture

Table 2.3 Changes Associated with Types of Innovation

Incremental Innovation

Incremental innovation refines and improves an existing design through improvements in the components. However, it is important to stress that these are improvements not changes: the components are not radically altered. Christensen (1997) defines incremental innovation as: 'a change that builds on a firm's expertise in component technology within an established architecture'. In the case of the washing machine example used earlier, incremental innovation would be a matter of offering a machine with a more powerful motor to give faster spin speeds.

Incremental innovations are the commonest. Gradual improvements in knowledge and materials lead to most products and services being enhanced over time. However, these enhancements typically take the form of refinements in components rather than changes in the system. Thus, a new model of an existing and established product (perhaps described as a 'mark 2' version) is likely to leave the architecture of the system unchanged and instead involve refinements to particular components. With the system and the linkages between components unchanged and the design of the components reinforced (through refinements and performance improvements) this places such innovations in the top-left-hand quadrant of Figure 2.1, where they are designated 'incremental innovations'.

Radical Innovation

Radical innovation is about much more than improvements to existing designs. A radical innovation calls for a whole new design, ideally using new components configured (i.e. integrated into the design) in a new way. In Henderson and Clark's (1990) terms: 'Radical innovation establishes a new dominant design, and hence a new set of core design concepts embodied in components that are linked together in a new architecture.'

Radical innovations are comparatively rare. Rothwell and Gardner (1989) estimated that at the most about 10 per cent of innovations are radical. Radical innovation is often associated with the introduction of a new technology (Table 2.4). In some cases this will be a transforming technology, perhaps even one associated with the transforming effect of a Kondratiev long wave (see Chapter 3).

Radical Innovation	Technology	Impact on Society
Telephone	Telecommunications	New means of mass communication
Jet Airliner	Jet power	Growth of mass travel, foreign holidays
Television	Television	New leisure activity, entertainment
Personal Computer	Microprocessor	New administrative system, Internet services, e.g. banking

Table 2.4 Radical Innovations

In terms of Henderson and Clark's framework radical innovation is located in the bottom-right-hand quadrant, at the opposite extreme from incremental innovation, as it involves both new components and a new design with a new architecture that links the components together in a different way.

MINI-CASE: NEVER ASK PERMISSION TO INNOVATE

In 1956, a small American company invented a device called the 'Hush-a-Phone'. It was a plastic cup designed to be attached to the microphone end of a telephone handset in order to facilitate telephone conversations in noisy environments - rather like cupping your hand over the phone.

When Hush-a-Phone appeared on the market, AT&T - then the monopolistic supplier of telephone services to the US public - objected, on the grounds that it was a crime to attach to the phone system any device not expressly approved by AT&T. Hush-a-Phone had not been thus approved. The Federal Communications Commission agreed with AT&T. The fact that the device in no way 'connected' with the network was neither here nor there. Hush-a-Phone was history.

A few years later, when Paul Baran proposed the packet-switching technology which eventually underpinned the Internet, AT&T first derided and then blocked its development. One of AT&T's executives eventually said to Baran: 'First, it can't possibly work, and if it did, damned if we are going to allow the creation of a competitor to ourselves.'

Note the verb 'allow'. In a single word it explains why we should never permit the established order to be gatekeepers of innovation.

This is not widely understood by legislatures or governments, and it is particularly not understood by our own dear DTI (aka the Department of Torpor and Indolence), which thinks that the way to encourage innovation is to get all the established players in an industry together and exhort them to do it.

Innovation comes in two forms. The first is incremental - the process of making regular improvements to existing products and services. This is a cosy, familiar business which is easily accommodated by the established industrial order and by its regulatory bodies. It is what governments and corporations have in mind when they declare they are in favour of innovation.

The second kind of innovation is the disruptive variety - defined as developments that upset, supersede or transform established business models, user expectations and government frameworks and create hitherto unimagined possibilities. In other words, change that upsets powerful apple-carts.

This is the kind of innovation that the established order really fears - and often tries hard to squash. Yet, if our societies and economies are to remain vibrant, it is the only kind of innovation that matters. We are thus faced with a dilemma: on the one hand, we need disruptive innovation; on the other, the established order will never make it happen. So what do we do?

This is the central policy issue confronting every modern government. Yet the answer - as a striking new pamphlet by Demos argues - is staring us in the face. It involves learning from the history of the Internet. The reason it spurred such an explosion of disruptive change is that it was an *innovation commons* - an uncontrolled space equally available to all. A whole raft of powerful technologies - for example, the World Wide Web, streaming audio, video conferencing, Internet telephony, instant messaging, peer-to-peer networking, interactive gaming, online auctions, chat - came into being because their inventors had unfettered access to the network. They did not have to ask the permission of AT&T or BT or the DTI to implement their ideas. If the invention was good enough, then it could, and did, conquer the world.

The lesson for the UK – and particularly Ofcom, the new omnipotent communications regulator – is that the preservation of a commons is vital if real innovation is to be nurtured here. This means, for example, that when analogue TV is switched off, some of the liberated spectrum should be retained as an unlicensed commons so that people can experiment and innovate with it.

Like all great ideas, it is simple. The only question is whether it is simple enough for the DTI to get it.

Source: John Naughton, Observer, 2002

Modular Innovation

Modular innovation uses the architecture and configuration associated with the existing system of an established product, but employs new components with different design concepts. In terms of Henderson and Clark's framework, modular innovation is in the top-right quadrant.

MINI-CASE: CLOCKWORK RADIO

An example of this type of innovation would be the clockwork radio, developed by Trevor Baylis. Radios have been around for a very long time. They operate on the basis of electrical energy, normally provided via either an external power supply or batteries. The clockwork radio is an innovation that employs a different form of power supply, one that utilises a spring-based clockwork mechanism. The other components of the radio, such as the speakers, tuner, amplifier, receiver, etc. remain unchanged. As a radio, the clockwork radio operates in the same way as other radios. It employs the same kind of architecture in which the various components that make up the system are configured and linked together in the normal way. However, being clockwork it does not require an external power sources and this is a very valuable feature in those parts of the world which do not benefit from regular uninterrupted power supplies.

Source: Baylis (1999)

As with incremental innovation, modular innovation does not involve a whole new design. Modular innovation does, however, involve new or at least significantly different components. In the case of the clockwork radio it is the power source that is new. The radio operates in much the same way as any other radio.

The use of new or different components is the key feature of modular innovation, especially if the new components embrace a new technology. New technology can transform the way in which one or more components within the overall system can operate, but the system and its configuration/architecture remains unchanged.

Clearly the impact of modular innovation is usually less dramatic than is the case with radical innovation. The clockwork radio illustrates this well. People still listen to the radio in the way they always have; but the fact that it does not need an external power source means that new groups often living in relatively poor countries without access to a stable and reliable supply of electricity living can get the benefit of radio. Clockwork radio has also opened up new markets in affluent countries – for example, hikers who want a radio to keep in touch with the outside world. It has also provided an important 'demonstration' effect as it has led to other products, such as torches, being fitted with this ingenious and environmentally friendly source of power.

Architectural Innovation

With architectural innovation, the components and associated design concepts remain unchanged but the configuration of the system changes as new linkages are instituted. As Henderson and Clark (1990, p. 12) point out. 'the essence of an architectural innovation is the reconfiguration of an established system to link together existing components in a new way.' This is not to say that there will not be some changes to components. Manufacturers may well take the opportunity to refine and improve some components, but essentially the changes will be minor, leaving the components to function as they have in the past but within a new re-designed and re-configured system.

MINI-CASE: SONY WALKMAN

The Sony Walkman provides a good example of architectural innovation. The Walkman when it first came out was a highly innovative new product, but it involved little or no new technology. All the main components that went into the Walkman were tried and tested having been used on a variety of other products. Portable audio tape recorders that could both play and record music had been on the market for many years. Designers at Sony started with an existing, small, audio cassette tape recorder, the Pressman (Henry and Walker, 1991), a small lightweight tape recorder designed for press reporters. They proceeded to remove the recording circuitry and the speakers, and added a small stereo amplifier. A set of lightweight headphones completed the package. Because there were no speakers the new machine needed much less power. The absence of speakers meant it could be made much smaller, while the fact that it needed much less power meant it could use only small batteries making it very much lighter. A very different kind of system with a very different kind of architecture began to emerge. So the Walkman was born. It was new type of audio product. It was a personal stereo, that enabled its young, mobile users to listen to music whenever and wherever they wanted, and without being harassed by older generations concerned about noise.

Source: Sanderson and Uzumeri (1995)

The Walkman was a huge commercial success, selling 1.5 million units in just two years (Sanderson and Uzumeri, 1995). However, the significance of the Walkman is not just that it sold well. It illustrates the power that is sometimes associated with architectural innovations. As well as securing Sony's future as a consumer electronics manufacturer, it had a much wider impact on society. It was soon copied by other manufacturers, but more significantly it changed the behaviour of consumers. Young people found they could combine a healthy lifestyle while continuing to listen to music so that the Walkman may be said to have helped promote a whole range of activities like jogging, walking and use of the gym.

THE VALUE OF AN INNOVATION TYPOLOGY

None of the types of innovation outlined using this framework is entirely watertight. Inevitably there is overlap and there will be many occasions when it is a matter of judgement as to in which category an innovation should be placed. However, this is not really an issue. What matters is the general value that comes from attempting a categorisation of innovations. Categorisation, and in particular this form of categorisation, helps to show that innovations are not homogeneous. Innovations vary. Consequently, any analysis of innovation needs a degree of sophistication that can isolate exactly where the nature of the innovation lies. In the process this should enable the more discerning analysts to cast a more critical eye upon some of the wilder claims surrounding objects that are described using that much over-used adjective: 'innovative'.

Categorising innovations in this way can also help to show that the influence of technology and technological change can vary considerably. Technology works in a variety of ways. However, its impact will differ enormously when applied to whole systems or when, for comparison, it is applied to individual components. Hence, this form of categorisation has a predictive power, such that those who use it can much more effectively evaluate the potential impact of a particular innovation.

Distinguishing four different types of innovation can also help to explain why the responses of firms to the introduction of new technologies will often vary. The analysis means that perhaps we should not be surprised that some firms do not respond positively to some new technologies. If the technology affects components we can expect a rapid take-up of a new technology, because it is likely to reinforce the competitive position of incumbent manufacturers. On the other hand, if the technology leads to system changes and the introduction of new architectures, the incumbents are less likely to be happy about the changes, as their position will be eroded. In Schumpeter's words, we are likely to see 'creative destruction' at work.

This typology can also help in understanding the evolutionary process associated with technological change. When a new technology appears, it frequently leads to a proliferation of competing system designs each with a different architecture. One could see exactly this happening when the first cars were developed – there was a multiplicity of competing architectures, and again when the first video recorders appeared. Eventually through a process of 'shakeout' a common system architecture or 'dominant design' evolved and was adopted by all manufacturers. This kind of evolutionary process is in fact very common and carries major implications for would-be innovators and entrepreneurs. They will need to recognise that, if they enter the industry during its early years, they can expect there to be a period of shakeout eventually. It is even more important that they recognise that the dominant design that will eventually emerge will not always be technically superior to its rivals. The qwerty keyboard is evidence that sometimes technically inferior designs emerge as the dominant design.

CASE STUDY: THE GUTS OF THE NEW MACHINE

(This is an abridged version of an a article from the *New York Times*, 30 November 2003.)

Two years ago this month, Apple Computer released a small, sleek-looking device it called the iPod. A digital music player, it weighed just 6.5 ounces and held about 1000 songs. There were small MP3 players around at the time, and there were players that could hold a lot of music. But if the crucial equation is 'largest number of songs' divided by 'smallest physical space,' the iPod seemed untouchable. Yet the initial reaction was mixed: the thing cost $400, so much more than existing digital players that it prompted one online skeptic to suggest that the name might be an acronym for 'Idiots Price Our Devices'.

Since then, however, about 1.4 million iPods have been sold. For the months of July and August, the iPod claimed the No. 1 spot in the MP3 player market both in terms of unit share (31 percent) and revenue share (56 percent), by Apple's reckoning. It is now Apple's highest-volume product. Whether the iPod achieves truly mass scale – like, say, the cassette-tape Walkman, which sold an astonishing 186 million units in its first 20 years of existence – it certainly qualifies as a hit and as a genuine breakthrough.

So you can say that the iPod is innovative, but it's harder to nail down whether the key is what's inside it, the external appearance or even the way these work together. One approach is to peel your way through the thing, layer by layer.

The Aura

Before you even get to the surface of the iPod, you encounter what could be called its aura. The commercial version of an aura is a brand, and while Apple may be a niche player in the computer market, the fanatical brand loyalty of its customers is legendary. Leander Kahney has even written a book about it, *The Cult of Mac*. As he points out, that base has supported the company with a faith in its will to innovate – even during stretches when it hasn't. Apple is also a giant in the world of industrial design. The candy-colored look of the iMac has been so widely copied that it's now a visual cliché.

But the iPod is making an even bigger impression. Bruce Claxton, who is the current president of the Industrial Designers Society of America and a senior designer at Motorola, calls the device emblematic of a shift toward products that are 'an antidote to the hyper lifestyle', which might be symbolized by hand-held devices that bristle with buttons and controls that seem to promise a million functions if you only had time to figure them all out. 'People are seeking out products that are not just simple to use but a joy to use.' Moby, the recording artist, has been a high-profile iPod booster since the product's debut. 'The kind of insidious revolutionary quality of the iPod', he says, 'is that it's so elegant and logical, it becomes part of your life so quickly that you can't remember what it was like beforehand.'

The idea of innovation, particularly technological innovation, has a kind of aura around it, too. Imagine the lone genius, sheltered from the storm of short-term commercial demands in a research lab somewhere, whose tinkering produces a sudden and momentous breakthrough. Or maybe we think innovation begins with an epiphany, a sudden vision of the future. Either way, we think of that one thing, the lightning bolt that jolted all the other pieces into place. The Walkman came about because a Sony executive wanted a high-quality but small stereo tape player to listen to on long flights. A small recorder was modified, with the recording pieces removed and stereo circuitry added. That was February 1979, and within six months the product was on the market.

The iPod's history is comparatively free of lightning-bolt moments. Apple was not ahead of the curve in recognizing the power of music in digital form. Various portable digital music players were already on the market before the iPod was even an idea. The company had, back in the 1990's, invented a technology called FireWire, which is basically a tool for moving data between digital devices - in large quantities, very quickly. Apple licensed this technology to various Japanese consumer electronics companies (which used it in digital camcorders and players) and eventually started adding FireWire ports to iMacs and creating video editing software. This led to programs called iMovie, then iPhoto and then a conceptual view of the home computer as a 'digital hub' that would complement a range of devices. Finally, in January 2001, iTunes was added to the mix.

And although the next step sounds prosaic - we make software that lets you organize the music on your computer, so maybe we should make one of those things that lets you take it with you - it was also something new. There were companies that made jukebox software, and companies that made portable players, but nobody made both. What this meant is not that the iPod could do more, but that it would do less. This is what led to what Jonathan Ive, Apple's vice president of industrial design, calls the iPod's 'overt simplicity'. And this, perversely, is the most exciting thing about it.

The Surface

The surface of the iPod, white on front and stainless steel behind, is perfectly seamless. It's close to impenetrable. You hook it up to a computer with iTunes, and whatever music you have collected there flows (incredibly fast, thanks to that FireWire cable) into the iPod - again, seamless. Once it's in there, the surface of the iPod is not likely to cause problems for the user, because there's almost nothing on it. Just that wheel, one button in the center, and four beneath the device's LCD screen.

'Steve (Jobs) made some very interesting observations very early on about how this was about navigating content,' Ive says. 'It was about being very focused and not trying to do too much with the device - which would have been its complication and, therefore, its demise. The enabling features aren't obvious and evident, because the key was getting rid of stuff.'

Later he said: 'What's interesting is that out of that simplicity, and almost that unashamed sense of simplicity, and expressing it, came a very different product. But difference wasn't the goal. It's actually very easy to create a different thing. What was exciting is starting to realize that its difference was really a consequence of this quest to make it a very simple thing.'

Only Apple could have developed the iPod. Like the device itself, Apple appears seamless: it has the hardware engineers, the software engineers, the industrial designers, all under one roof and working together. 'As technology becomes more complex, Apple's core strength of knowing how to make very sophisticated technology comprehensible to mere mortals is in even greater demand.' This is why, (Jobs) said, the barrage of devices

▶

made by everyone from Philips to Samsung to Dell that are imitating and will imitate the iPod do not make him nervous. 'The Dells of the world don't spend money' on design innovation, he said. 'They don't think about these things.'

As he described it, the iPod did not begin with a specific technological breakthrough, but with a sense, in early 2001, that Apple could give this market something better than any rival could. So the starting point wasn't a chip or a design; the starting point was the question, What's the user experience? 'Correct', Jobs said. 'And the pieces come together. If you start to work on something, and the time is right, pieces come in from the periphery. It just comes together.'

The Guts

What, then, are the pieces? What are the technical innards of the seamless iPod? What's underneath the surface? A lot of people were interested in knowing what was inside the iPod when it made its debut. One of them was David Carey, who for the past three years has run a business in Austin, Tex., called Portelligent, which tears apart electronic devices and does what might be called guts checks. He tore up his first iPod in early 2002.

Inside was a neat stack of core components. First, the power source: a slim, squarish rechargeable battery made by Sony. Atop that was the hard disk – the thing that holds all the music files. At the time, small hard disks were mostly used in laptops, or as removable data-storage cards for laptops. So-called 2.5-inch hard disks, which are protected by a casing that actually measures about 2 3/4 inches by 4 inches, were fairly commonplace, but Toshiba had come up with an even smaller one. With a protective cover measuring just over 2 inches by 3 inches, 0.2 inches thick and weighing less than two ounces, its 1.8-inch disk could hold five gigabytes of data – or, in practical terms, about a thousand songs. This is what Apple used.

On top of this hard disk was the circuit board. This included components to turn a digitally encoded music file into a conventional audio file, the chip that enables the device to use FireWire both as a pipe for digital data and battery charging and the central processing unit that acts as the sort of taskmaster for the various components. Also here was the ball-bearing construction underlying the scroll wheel.

Exactly how all the pieces came together – there were parts from at least a half-dozen companies in the original iPod – is not something Apple talks about. But one clue can be found in the device itself. Under the Settings menu is a selection called Legal, and there you find not just Apple's copyright but also a note that 'portions' of the device are copyrighted by something called PortalPlayer Inc. That taskmaster central processing unit is a PortalPlayer chip.

Most early MP3 players did not use hard disks because they were physically too large. Rather, they used another type of storage technology (referred to as a 'flash' chip) that took up little space but held less data – that is, fewer songs. PortalPlayer's setup includes both a hard disk and a smaller memory chip, which is actually the thing that's active when you're listening to music; songs are cleverly parceled into this from the hard disk in small groups, a scheme that keeps the energy-hog hard disk from wearing down the battery.

Apple won't comment on any of this, and the nondisclosure agreements it has in place with its suppliers and collaborators are described as unusually restrictive. Presumably this is because the company prefers the image of a product that sprang forth whole from the corporate godhead – which was certainly the impression the iPod created when it seemed to appear out of nowhere two years ago. But the point here is not to undercut Apple's role: the iPod came together in somewhere between six and nine months, from concept to market, and its coherence as a product given the time frame and the number of variables is astonishing. Jobs and company are still correct when they point to that coherence as key to the iPod's appeal; and the reality of technical innovation today is that assembling the right specialists is critical to speed, and speed is critical to success.

Still, in the world of technology products, guts have traditionally mattered quite a bit; the PC boom viewed from one angle was nothing but an endless series of announcements about bits and megahertz and RAM. That 1.8-inch hard disk, and the amount of data storage it offered in such a small space, isn't the only key to the iPod, but it's a big deal. Apple apparently cornered the market for the Toshiba disks for a while. But now there is, inevitably, an alternative. Hitachi now makes a disk that size, and it has at least one major buyer: Dell.

The System

My visit to Cupertino happened to coincide with the publication of a pessimistic install-ment of *The Wall Street Journal's* Heard on the Street column pointing out that Apple's famous online music store generates little profit. About a week later Jobs played host to one of the 'launch' events for which the company is notorious, announcing the availabil-ity of iTunes and access to the company's music store for Windows users. The announcement included a deal with AOL and a huge promotion with Pepsi. The message was obvious: Apple is aiming squarely at the mainstream.

This sounded like a sea change. But while you can run iTunes on Windows and hook it up to an iPod, that iPod does not play songs in the formats used by any other seller of digital music, like Napster or Rhapsody. Nor will music bought through Apple's store play on any rival device. This means Apple is, again, competing against a huge number of players across multiple business segments, who by and large will support one another's products and services. In light of this, says one of those competitors, Rob Glaser, founder and C.E.O. of RealNetworks, 'It's absolutely clear now why five years from now, Apple will have 3 to 5 percent of the player market.'

Jobs, of course, has heard the predictions and has no patience for any of it. Various contenders have come at the iPod for two years, and none have measured up. Nothing has come close to Apple's interface. Even the look-alike products are frauds. 'They're all putting their dumb controls in the shape of a circle, to fool the consumer into thinking it's a wheel like ours', he says. 'We've sort of set the vernacular. They're trying to copy the vernacular without understanding it.' (The one company that did plan a wheel-driven product, Samsung, changed course after Apple reportedly threatened to sue.) 'We don't underestimate people', Jobs said later in the interview. 'We really did believe that people would want something this good, that they'd see the value in it.'

▶

The Core

What I had been hoping to do was catch a glimpse of what's there when you pull back all those layers – when you penetrate the aura, strip off the surface, clear away the guts. What's under there is innovation, but where does it come from? I had given up on getting an answer to this question when I made a jokey observation that before long somebody would probably start making white headphones so that people carrying knockoffs and tape players could fool the world into thinking they had trendy iPods.

Jobs shook his head. 'But then you meet the girl, and she says, "Let me see what's on your iPod." You pull out a tape player, and she walks away.'

Source: © 2003 – The New York Times Magazine

QUESTIONS

1 What is novel about the iPod?
2 What type of innovation would you class the Sony Walkman as and why?
3 What type of innovation would you class the iPod as and why?
4 What does the author mean by 'lightening-bolt' moments?
5 If 'the iPod did not begin with a specific technological breakthrough' as Steve Jobs maintains, how can it still be classed as an innovation?
6 What is licensing and why did Apple choose to license its Firewire technology?
7 Why, according to the author, does innovation especially technological innovation, have an 'aura' around it? Give an example of another product with an aura.
8 What do you think Steve Jobs means when he says that the iPod is about 'navigating content'?
9 Why does Steve Jobs believe that imitators of the iPod like Dell do not pose a threat?
10 By the end of 2004, total sales of iPods had reached 10 million. What does this imply about the prediction of Rob Glaser of RealNetworks for the iPod's market share?

QUESTIONS FOR DISCUSSION

1 **What is the value of being able to categorise innovations?**

2 Why may large established firms be wary of radical (disruptive) innovations?

3 **Why do product innovations tend to attract more public attention than service or process innovations?**

4 Why do process innovations sometimes have wide-ranging consequences for society?

5 **Identify two process innovations which have had a big impact on society.**

6 Differentiate between component knowledge and system knowledge.

7 Choose an example of an everyday household object (e.g. an electric kettle) and identify some of the incremental innovations that have taken place.

8 Why are only a small proportion of innovations typically radical?

9 Why is the Sony Walkman an example of architectural innovation?

10 What type of innovation is Apple's iPod?

ASSIGNMENTS

1 Using any household object of your choice (e.g. vacuum cleaner, hairdryer, etc.) identify and analyse the following:
   - System function
   - Components
   - System linkages
   - Incremental innovation

   Outline what you consider to be the rationale behind ONE recent incremental innovation.

2 Identify a product that has been the subject of modular innovation. Analyse where the innovation has occurred and the impact this has had on the product. Explain why you think this is a case of modular innovation, noting how the system architecture has remained unchanged.

3 What is a system? Take an example of a system and analyse it using a diagram to show the components and the linkages between them. Indicate where there have been examples of a) incremental innovation and b) architectural innovation.

4 What is meant by radical innovation? Take an example of radical innovation and analyse the impact it has had on society. Take care to differentiate between the different groups within society that have been affected.

5 What is meant by the term 'creative destruction'. Explain, using appropriate examples, the link between creative destruction and radical innovation.

RESOURCES

The paper by Henderson and Clark (1990) that appeared in *Administrative Science Quarterly*, provides an excellent starting point for classifying and categorising innovations. It draws on earlier work to provide a fourfold typology.

Two of the categories, radical and incremental innovation, have substantial literatures of their own. Incremental innovation is covered by works such as Nelson and Winter (1982), Ettlie, Bridges and O'Keefe (1984) and Tushman and Anderson (1986). Radical innovation on the other hand overlaps with other aspects of innovation such as technological discontinuities and crops up in sources as diverse as Rothwell (1986) and Christensen (1997).

When it comes to the product, process and service categorisation of innovation, one finds that product innovations are comparatively well served, but service and process innovations have attracted much less attention. The main reason for this is that product innovation falls within the remit of histories of invention while service and process innovations do not. Among the many excellent histories of invention that give valuable insights into the events surrounding the development process associated with product innovations are studies by Van Dulken (2000) and Nayak and Ketteringham (1993) which provide short outlines of many well-known product innovations. In addition there are a number of other texts with a strong American focus including Basalla (1988), Hughes (1989) and Petroski (1992). Basalla (1988) is particularly interesting because its focus is on the evolution of technology and the links between different innovations.

Service innovations are much more poorly served, largely because technology is often less in evidence and because there is no tangible object. While it focuses mainly on product innovations, the study by Nayak and Ketteringham (1993) is one of the few to cover some well-known examples of service innovations, including Federal Express and Club Mediterranée. Similar studies that include examples of service innovations are Davis (1987) and Henry and Walker (1991).

Process innovations are often the hardest to research. They are often well known but acquiring detailed data about them can be difficult. Industry studies, that is to say studies of how an industry has developed over time, can provide valuable, detailed insights. Examples include Womack *et al.*'s (1990) study of the motor industry, Chapman's (2002) study of the knitting and hosiery branches of the industry, and Dogannis' (2001) study of the airline industry. The first of these is the product of a huge study of the motor industry undertaken in the late 1980s. It provides a lot of detail about mass production – one of the most significant process innovations, as well as later developments such as lean manufacturing. A more specialised study that focuses on the regional rather than the international level is Chapman's (2002) study of knitting and textiles. It builds on earlier work stretching back to the 1960s and shows how process innovations transformed the textile industry. Dogannis has written a number of studies of the airline industry. Dogannis (2001) covers a number of process innovations associated with aircraft. Another important study of the same industry is Hanlon (1999) who provides insights into a number of process innovations including computer reservation systems and hub-and-spoke operations.

REFERENCES

Ansoff, I. (1988) *Corporate Strategy*, revd edn, Penguin Books, Harmondsworth.

Basalla, G. (1988) *The Evolution of Technology*, Cambridge University Press, Cambridge.

Baylis, T. (1999) *Clock This: My Life as an Inventor*, Headline Publishing, London.

Cassidy, J. (2002) *Dot.con: the greatest story ever sold*, Penguin Books, Harmondsworth.

Chapman, S.D. (2002) *Hosiery and Knitwear: Four Centuries of Small-Scale Industry in Britain c1589-2000*, Pasold Research Fund/Oxford University Press, Oxford.

Christensen, C.M. (1997) *The Innovator's Dilemma: When New Technologies Cause Great Firms to Fail*, Harvard Business School Press, Boston, Mass.

Davis, W. (1987) *The Innovators*, Ebury Press, London.

Dogannis, R. (2001) *The airline business in the 21st century*, Routledge, London.

Dyson, J. (1997) *Against the odds: An Autobiography*, Orion Business, London.

Ettlie, J.E., W.P. Bridges and R.D. O'Keefe (1984) 'Organizational strategy and structural differences for radical vs. incremental innovation', *Management Science*, **30**, pp. 682–695.

Freeman, C. (1982) *The Economics of Industrial Innovation*, 2nd edn, Pinter, London.

Freeman, C. and F. Louçã (2001) *As Time Goes By: From Industrial Revolutions to Information Revolution*, Oxford University Press, Oxford.

Hanlon, P. (1999) *Global Airlines: Competition in a transnational industry*, Butterworth-Heinemann, Oxford.

Henderson, R.M. and K.B. Clark (1990) 'Architectural Innovation: The Reconfiguration of Existing Product Technologies and the Failure of Established Firms', *Administrative Science Quarterly*, **35**, pp. 9–30.

Henry, J. and D. Walker (1991) *Managing Innovation*, Sage Publications, London.

Howells, J. (2005) *The Management of Innovation and Technology*, Sage Publications, London.

Hughes, T.P. (1989) *American Genesis: A Century of Innovation and Technological Enthusiasm 1870-1970*, Viking, NY.

Kamm, A. and M. Baird (2002) *John Logie Baird*, National Museum of Scotland.

Naughton, J. (2002) 'Never ask permission to innovate', The *Observer*, 3 November 2002, p. 17.

Nayak, P.R. and J.M. Ketteringham (1993) *Breakthroughs!* Mercury Business Books, Didcot.

Nelson, R. and S. Winter (1982) *An Evolutionary Theory of Economic Change*, Cambridge University Press, Cambridge.

Petroski, H. (1992) *The Evolution of Useful Things*, Alfred A. Knopf, NY.

Procter, J. (1994) 'Everyone versus South West', *Airways*, November–December 1994, pp. 22-29.

Quinn, J.B. (1991) *The Strategy Process: Concepts, Contexts, Cases*, Prentice Hall, Englewood Cliffs, NJ.

Rothwell, R. (1986) 'The role of small firms in the emergence of new technologies', in C. Freeman (ed.) *Design, Innovation and Long Cycles in Economic Development*, Frances Pinter, London, pp. 231-248.

Rothwell, R. and D. Gardner (1989) 'The strategic management of re-innovation', *R & D Management*, **19** (2), pp. 147-160.

Sanderson, S. and M. Uzumeri (1995) 'Managing product families: The case of the Sony Walkman', *Research Policy*, **24**, pp. 761-782.

Tushman, M.L. and P. Anderson (1986) 'Technological discontinuities and organisational environments', *Administrative Science Quarterly*, **31**, pp. 439-465.

Van Dulken, S. (2000) *Inventing the 20th Century: 100 Inventions that Shaped the World*, British Library, London.

Walker, R. (2003) 'The Guts of the New Machine', The *New York Times*, 30 November 2003, p. 68.

Womack, J.P., D.T. Jones and J. Roos (1990) *The Machine that Changed the World*, Rawson Associates, NY.

CHAPTER 3

technological change

OBJECTIVES

When you have completed this chapter you will be able to:

- Distinguish between science and technology
- Describe the nature of technology
- Analyse the link between technological change and the long wave cycle
- Describe the phases of the long wave cycle
- Differentiate and distinguish between technological paradigms and technological trajectories.

INTRODUCTION

In the popular imagination innovation is closely linked to technological change. For its part technological change is frequently portrayed as a succession of spectacular breakthroughs that provide us with ever more ingenious products and rapidly renders existing products and even whole industries obsolete. Not only that, technological change is seen as constantly accelerating so that every age perceives itself as embracing a technological revolution.

The truth of this can be gauged from the statement shown below, which is taken from the Communist Manifesto of 1848.

> " Constant revolutionizing of production, uninterrupted disturbances of all social conditions, everlasting uncertainty … all old established national industries have been destroyed or are daily being destroyed. They are dislodged by new industries … whose products are consumed not only at home, but in every quarter of the globe. In place of old wants satisfied by the production of the country, we find new wants … . The individual creativity of nations become common property. " Marx and Engels (1967: p. 18)

In reality technological change is not new. Technologies have been constantly improving throughout time. However the rate of technological change has understandably accelerated since the industrial revolution of the late eighteenth century. Technology has advanced much more rapidly over the last 250 years compared to the previous 250. Thus technological change and all that it brings with it is very much a feature of a modern industrial society!

This chapter aims to explain the nature of technology. Then having classified what it is we mean by technology it proceeds to look at the changes that new technologies bring with them. In particular the chapter focuses on the link between technology and innovation.

THE NATURE OF TECHNOLOGY

The term 'technology' is defined by Simon (1972) as,

> " knowledge that is stored in millions of books, in hundreds of millions or billions of human heads, and, to an important extent in the artifacts themselves. "

This definition probably strikes a chord with most of us, since we generally associate technology with things, especially machines and equipment whether for the direct use of consumers or for use as part of manufacturing processes.

Frequently technology is linked to science, with scientific discoveries and breakthroughs seen as focusing the development of technology. Although there are links between the two, in fact science and technology are distinct. As McGinn (1991: p. 18) notes,

> " Technology is the human activity which is devoted to the production of technics [material products of human making or fabrication] – or technic-related intellectual products – and whose root function is to expand the realm of practical human possibility. "

while,

> " Science is that form of human activity which is devoted to the production of theory-related knowledge of material phenomena whose root function is to attain an enhanced understanding of nature. "

Thus, technology is concerned with practical knowledge of how to do things and how to make things. To develop technology it is not necessary to understand fully the principles behind phenomena. Science on the other hand is all about understanding. Science involves the application of systematic rigorous methods of enquiry in order to develop logical, self-consistent explanations of phenomena (Littler, 1988). A critical aspect is the application of the scientific method involving observation, the development of hypotheses and the systematic collection of data in order to arrive at explanations. The rigorous application of scientific method results in knowledge being recorded and codified as a formal body of knowledge that can be transferred through books and papers.

Technology on the other hand is embedded in 'artefacts' – that is equipment and machines – which form the most obvious examples and readily identifiable forms of technology. However, as Forbes and Wield (2002) note, technology is not only embedded in artefacts, but also in people and organisations. This form of knowledge is proprietary, i.e. firm-specific. Some is explicit: that is to say, it is codified in documents as patents, drawings, manuals, standard operating procedures and databases. On the other hand much of this proprietary knowledge is tacit. That is to say, it is knowledge that resides within the individual, known but extremely difficult or in some cases impossible to articulate or communicate adequately (Newell et al., 2002).

While science and technology are different, they are nonetheless connected. An understanding of phenomena can assist the development of technology. Sometimes scientific breakthroughs lead to advances in technology. Sometimes it is the other way round. However, in recent times not only has the relationship between science and technology become closer, but developments and advances in science have increasingly led to developments in technology.

It is developments in technology that give rise to technological change. Informed, if not led, by advances in science, the development of technology gives rise to innovations in the form both of new products/services and new processes. These innovations form part of technological change. New products/services lead to new patterns of consumption and new behaviours. New processes lead to new ways of working. The outcome is significant changes in the economic and social facets of human existence – precisely the things that typically go hand-in-hand with technological change.

LONG WAVE CYCLE AND TECHNOLOGICAL CHANGE

One has only to reflect on the course of technological change to appreciate that it is not constant, being sometimes rapid, as in the early years of the Internet, and sometimes relatively sluggish. Nor are the Internet boom years of the late 1990s different from other periods. The 1870s and 1880s saw a flurry of innovations surrounding electricity. This same period also saw innovations surrounding the internal combustion engine. Further evidence of the cyclical nature of innovation and the technological change that accompanies it can be found in the cyclical nature of academic work. It is no coincidence that the rise of the Internet in the early 1990s was accompanied by a renewed interest on the part of academics in technological change, manifested in an increase in academic output of journals and books.

The cyclical pattern of innovation is not a short cycle of 5-10 years like the business cycle, but is much longer. The idea of a long cycle or long wave was analysed by Nicholai Kondratiev (1892-1938), a Russian economist who founded and directed the Institute of Conjuncture in Moscow in the 1920s (Freeman and Louçã, 2001).

Kondratiev was not the first to put forward the idea of a long cycle. Economists had pointed to long term cyclical changes in prices, interest rates and trade before the First World War. However, Kondratiev's analysis in the 1920s did much to bring the idea to a wider audience. As a result, he is particularly associated with the notion of a 50-year cycle of economic activity stretching from a phase of depression through recovery and boom back to depression again. Such was his contribution that this long cycle, or 'long wave' as it is sometimes termed, is often referred to as the 'Kondratiev cycle'.

Starting with the industrial revolution of the late eighteenth century, some four long waves have been completed since then, and we are today well into the fifth cycle or fifth Kondratiev (Table 3.1) (Hall, 1981).

Date	Cycle/Wave	Technology
1780–1830	First	Cotton, Iron, Water Power
1830–1880	Second	Railways, Steam Power, Steamship
1880–1930	Third	Electricity, Chemicals, Steel
1930–1980	Fourth	Cars, Electronics, Oil, Aerospace
1980–	Fifth	Computers, Telecommunications and Biotechnology

Table 3.1 Long Wave Cycles

The idea of a long wave cycle was taken up by Joseph Schumpeter, who came across Kondratiev's work in Germany before he moved to Harvard. Schumpeter used the long wave cycle as a feature of his work on business cycles. Schumpeter (1939) put forward the idea that each long wave represented the application of a new group of technologies, each of which had a very powerful transforming effect on the economy, effectively bringing about another industrial revolution. In Schumpeter's analysis each revolution was based on a major technological change that, just like the first industrial revolution, brought major shifts in productivity, consumption and the organisation of productive activity. The technological change of the first long wave centred on the development of iron smelting using coal, pioneered by Abraham Darby, the application of water power and above all the mechanization of the textile industry especially cotton in Lancashire. Associated with these new technologies was an organisational innovation – the development of the factory system, initiated by Richard Arkwright with his mill at Cromford in Derbyshire.

The second long wave ran from the 1830s to the 1880s and involved technological change again based on a group of new transforming technologies principally in the form of steam power and the introduction of the railways (Freeman and Louçã, 2001). The tendency of major innovations to group together or 'swarm' in this way is a feature of the long wave cycle.

The introduction of the new technology of the railway led to a speculative boom. The 'railway mania' of the 1840s which saw the flotation of a large number of railway companies aiming to profit from the application of the new technology of railways and the anticipated growth in traffic (Figure 3.1). Like most financial booms it led eventually to a financial crash.

Figure 3.1 Passenger Traffic in Britain 1840-1870

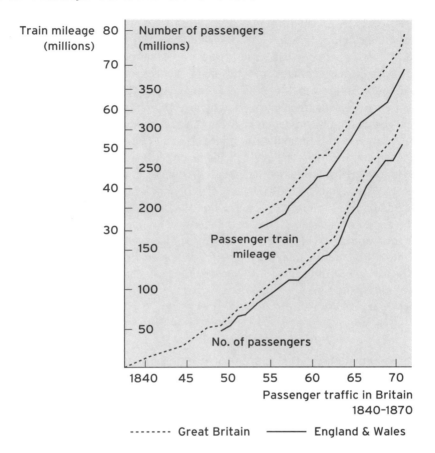

Source: Freeman and Louçã (2001 : p. 189), 'As Time Goes By: From Industrial Revolutions, to Information Revolution', Oxford University Press

MINI-CASE: THE RAILWAY MANIA OF THE 1840S

Railway technology developed piecemeal during the early years of the nineteenth century as a means of moving bulky commodities such as a coal over short distances. However, the opening of the Liverpool and Manchester railway in September 1830 not only brought the new technologies together, it had a very powerful demonstrative effect. It was the first railway constructed between two major centres of population, the first to employ steam-powered locomotives throughout its length, and the first to carry passengers. It was also, relatively speaking, a high-speed line with locomotives capable of speeds of up to 30mph compared to the 12mph achieved by the fastest stagecoach. For the first time a scheduled passenger service base on a railway timetable was available. The Liverpool and Manchester railway proved a huge success.

▶

Over the next decade the railway system was extended gradually so that by 1842 Manchester could be reached from London. In 1844 49 Acts of Parliament authorized 805 miles of new railway at a cost of £20 million. Until this point, however, railways had largely been the concern of what Smiles (1862: p. 236) termed 'the commercial classes'. The rising price of railway shares from this point onwards meant that railway promotion was no longer confined to the Stock Exchange. In 1845 no less than 94 Acts of Parliament were passed authorising an additional 2700 miles of railway. The scale of this activity can be gauged by the fact that the authorised capital of the companies covered by these Acts came to £59 million, a figure directly comparable with the country's national income at this time. Nor was this the height of the railway mania, for in the following year 219 Acts of Parliament covering 4538 miles of railway costing £133 million were passed. Finally, in 1847, the first signs of economic depression brought a decline in rail travel, and an end to railway promotion on the dramatic scale of the previous four years. The end of the speculative boom associated with the new technology of railways also brought about the downfall of George Hudson, the so-called 'railway king' (Beaumont, 2002) and the man most closely associated with railway company promotion. However, by 1852, more than 7500 miles of railway had been completed linking all the major cities in Britain.

Source: Dyos and Aldcroft (1974).

Figure 3.2 Penetration of Steel, USA, 1860–1950 (percentage of all iron and steel products)

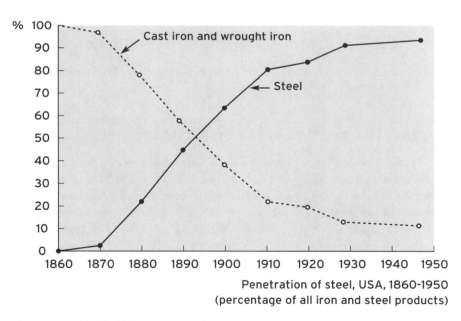

Penetration of steel, USA, 1860-1950 (percentage of all iron and steel products)

Source: Freeman and Louçã (2001: p. 233), 'As Time Goes By: From Industrial Revolutions, to Information Revolution', Oxford University Press

The third long wave, based on the new technologies of electricity, chemicals and steel (Freeman and Louçã, 2001) was from the 1880s to the 1930s and was largely complete by the time Schumpeter was writing. It was accompanied by managerial innovations such as scientific management and the rise of the first large corporations (Freeman and Louçã, 2001). The new transforming technologies themselves found widespread applications in products and services. With a new and much more flexible source of power new industries arose to manufacture the equipment to distribute the power, as well as new industries to make new machines, appliances, instruments and tools. Similarly, the availability of cheap, high-quality steel helped transform existing industries (Figure 3.2). The railways switched from iron rails to steel as did shipbuilding. Along with the new technologies themselves came improvements in productivity derived from the use of scientific management.

The fourth long wave from the 1930s to the 1980s brought new technologies associated with electronics, cars, oil and aerospace. The car industry not only grew dramatically, but also had a wide-ranging impact on the economy. Increased use of cars led to major investments in infrastructure, while the introduction of Fordist mass production led to dramatic changes in working practices. Changes in electronics led to the manufacture of consumer electrical products, most notably radio and television. The fourth Kondratiev was truly the period of the mass market.

The fifth long wave as we shall see in the next section was based on yet another raft of transforming technologies, this time in the form of computers, telecommunications and biotechnology.

Among the most significant features of the long wave is that it follows a regular course. A long climb up from depression involving recovery and prosperity phases leads to a maturity phase where a shallow decline leads to a steep fall in the depression phase (Figure 3.3). Each of these phases has implications for the pattern of innovation.

Figure 3.3 The Long Wave Cycle

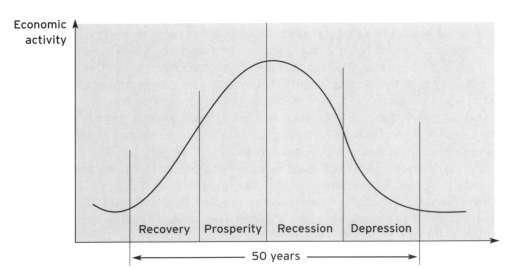

In the recovery phase the discoveries of scientists and investors are developed into new innovations that create entirely new opportunities for investment, growth and employment. Often these new opportunities will be created by newcomers (i.e. new firms) who view technology in a new way. At this point in the cycle there is often a high degree of uncertainty that produces a variety of competing product configurations. Offering significant improvements in performance and a high novelty value, these innovations command a price premium. Under these conditions there will often not be particularly strong pressures for production efficiency. As yet these innovations are finding only a specialist market. Higher prices mean higher profits from these innovations that act as a decisive impulse for new surges of growth which in turn act as a signal for imitators (Freeman, 1986) to enter the fray.

In the prosperity phase the innovations begin to diffuse to a wider range of applications through finding a broader market. As the innovations and their associated new technologies reach a wider market so they become better known and frequently imitations appear. They may well catch the mood of the popular imagination. Often there will be a 'bandwagon' effect as others try to cash in on the new technology. The combination of appropriate financial conditions and a large number of would-be imitators can easily lead to a speculative boom (e.g. the 'Railway mania' of the 1840s and the 'Dot.com' boom of the 1990s) as investors try to cash in on the technological advances. Over-ambitious and unrealistic plans combined with the ever-increasing cost of capital lead to the inevitable financial crash. Such a crash typically heralds the onset of the third phase.

In the third phase, with surplus capacity and diminishing returns as the limits of technological advance become evident, price competition becomes intense. It is at this point that the focus of innovation shifts. The new technology which has hitherto been applied to create new products now begins to spill over into process applications. This may well be where the transformative capacity of the new technology is at its greatest. New production processes can sweep away old working practices, leading to dramatic improvements in productivity.

Finally market saturation leads to ever-greater price competition and declining profitability which are the features of the depression phase of the long wave cycle. So too are decreasing returns that begin to set in as the technological advance reaches its limits. The result is mergers and acquisitions in pursuit of greater efficiency. This is the 'shake out' phase. Despite the depressed and difficult trading conditions, this phase is also the point at which the discoveries, breakthroughs and inventions that will form the basis of the next long wave begin to take place. In Schumpeter's analysis the innovations that occur in the recovery phase will tend to cluster. Hence, the early stages of each long wave are associated with a 'swarm' of new innovations.

The same phenomena are to be found in the work of Mensch (1979). He showed that, as Schumpeter predicted, the rate of innovation over time tends to vary. According to Mensch, innovation is subject to a 'wagon-train' effect (Hall, 1981: p. 534). Thus, while inventions and discoveries can occur at almost any time, innovations tend to bunch together at the end of one long wave and the beginning of another. This is one reason why the transforming effect of technological change can be so dramatic.

THE INTERNET: OUR VERY OWN LONG WAVE

Freeman and Louçã (2001) argue that we are currently witnessing a fifth Kondratiev long wave. Beginning in the 1980s, this long wave is associated with new transforming technologies, comprising computers, telecommunications and the Internet together with developments in biotechnology. These technologies have between them begun to transform a variety of aspects of our daily lives. According to Freeman and Louçã (2001) the 'Dot.com bubble' of the late 1990s shares many of the characteristics of similar bubbles seen in earlier long waves, such as the railway mania of the 1840s and the Wall Street crash of 1929.

As with previous long waves the origins of the many transforming technologies lie in the breakthroughs and inventions that occurred in the downward phase of the fourth Kondratiev. Nonetheless, each technology has itself been transformed by more recent breakthroughs which have gained additional momentum as a result of technological convergence.

Developments in electronics underpin developments in all three of the transforming technologies. Jack Kilby's idea that, instead of transistors being linked together and mounted on a circuit board, they could simply be manufactured from a single piece of silicon, led to the production of the first integrated circuit (IC) as far back as 1958 (Campbell-Kelly, 2003a). ICs were initially very expensive and used for specialist defence applications such as the guidance system of the Minuteman missile. However, Moore's Law, that the capacity of ICs would double every year, led in time to a dramatic reduction in the price of a core technological input. The development of the first microprocessor by Intel in 1971 was another decisive event. Though it still required successive incremental improvements to deliver appropriate performance, these advances in electronics paved the way for big changes in other technologies.

One of these technologies was computing, where the availability of ICs led to the development of the first personal computers. Though they were at first lacking in power and flexibility, within a decade the PC had begun to challenge computing orthodoxy based on large mainframe applications. As the PC grew more capable through advances in electronics, so it multiplied.

Developments in electronics also fed through to telecommunications. Transistors were first used for switching in the 1960s. With the introduction of digital technology in the form of pulse code modulation (PCM), the scope for extending the application of electronics grew still further. Developments in fibre optics and the introduction of packet switching transformed the capacity and versatility of the system.

In the 1990s the three technologies converged to give rise to the Internet. Computers and telecommunications are important transforming technologies in their own right, but it is the Internet that exhibits the greatest transforming potential. Initially giving rise to business-to-consumer (B2C) applications in the upward phase of the long wave, as the cycle appears to be reaching its apex we are now seeing the technology of the Internet diffuse into process applications through business-to-business (B2B) E-commerce.

As with previous long waves, the transformation does not stop at the technology itself. The transformation is changing working lives, business models, leisure patterns and the structure and shape of business organizations themselves.

THE IMPLICATIONS OF THE LONG WAVE CYCLE

The notion of the long cycle has a number of important implications for technological change. First, the idea of technological change being cyclical helps to break away from the popular notion that the rate of technological change is simply accelerating over time. If technological change is cyclical, one can expect different effects at different points in time. The cycle predicts that the distinct phases will be: a hesitant start, fast growth, subsequent saturation followed by decline and stagnation as the limits of technology are reduced. This point has been taken up by Abernathy and Utterback (1978) in their analysis of the life cycle of technologies and industries. This helps to explain a second point – why innovation can vary over time. Product innovations that occur early in the long wave cycle give way eventually to process innovations as a technology diffuses through the economy. This was emphasised by Schumpeter (1939) who argued that the diffusion of innovations was inherently uneven. In the early stages of the recovery phase only a few firms and individuals will be far-sighted enough to bring forward innovations. In the prosperity phase many firms will follow in the wake of successful pioneers. In the maturity phase, diffusion shifts from products to processes.

Third, the long wave cycle reinforces the notion of different types of innovation. In the early phases of the long wave, radical innovations are more likely, while in later phases one can expect to find incremental innovations.

Fourth, the long wave cycle shows how some technologies can have a much bigger impact than others. In particular one can differentiate transforming technologies. These innovations are so wide-ranging in their impact that they cause major perturbations in the economic and social system (Freeman, 1986). Steam power, electricity, oil and latterly the Internet all come within this category. These technologies are not merely important in themselves, they are also important because of the impact they have on a wide range of industries. As a result their transforming power is much greater than other technologies.

Fifth, the long wave cycle shows how technologies often go hand-in-hand with institutional changes (Freeman, 1986). These institutional changes include:

- education and training
- industrial relations
- corporate structures
- systems of management
- capital markets
- the legal framework

Each of the five long waves identified in Table 3.1 was associated with significant institutional changes. In the first long wave it was the introduction of factory production. This not only facilitated the application of new technologies and innovations in textile production, it also brought with it big increases in productivity. These in turn led to lower prices, thereby stimulating demand and major shifts in consumption. In the second wave the institutional change was the joint stock company that facilitated access to greater supplies of capital and permitted larger enterprises. In the third wave it was scientific management and the rise of the large corporation, while in the fourth it was mass production and the large, vertically integrated corporation. Finally, the fifth long wave has been accompanied by further institutional changes, most notably network structures for organisations, alliances and joint ventures and outsourcing.

The significance of these institutional changes lies not just in the impact that they have had on the economic and social life of nations but also in the way in which the long wave cycle shows that they are linked to technological change. This in turn shows that technology is not merely important for its own sake, but for the things that it brings with it. This is where the power of the long wave cycle lies, in its ability to link technology to economic and social aspects of life. The new technology of the internal combustion engine combined with the introduction of mass production did not merely create a new class of product in the form of the car. It changed the nature of work for millions of industrial workers, it brought massive improvements in productivity and, as Freeman and Louçã (2001: p. 263) show, it permitted an unprecedented degree of personal mobility that in turn induced massive growth in infrastructure and supply industries. Dramatic as these changes were at the time, the long wave cycle helps to show that this was neither the first time that technological change had occurred on this scale nor – as we are currently experiencing – was it the last.

Finally, as Perez (1983) describes it, the long wave cycle highlights the impact of transforming technologies and costs. The railways, for instance, were not only the leading industrial sector in terms of investment and employment generation in the Victorian boom of the 1850s and 1860s, they also had a big impact on the costs of many industrial sectors. The ability to transport coal cheaply reduced dramatically the cost of energy for the whole economy. Similarly, in the Internet age developments in information technology have dramatically reduced the cost of storing, processing and transferring information. This in turn has facilitated knowledge acquisition and knowledge transfer, significantly altering the cost base of a wide range of industries.

TECHNOLOGICAL PARADIGMS AND TRAJECTORIES

Another attempt to distinguish differences in the pattern of technological change is embodied in Dosi's (1982) notion of technological paradigms and trajectories.

MINI-CASE: THE JET ENGINE

When it was first introduced, the jet engine formed a new technological paradigm. Over time the jet engine has been developed and improved. A series of technological advances has steadily improved its performance. These advances have included the axial compressor, the twin spool configuration and the turbofan. Initially there was a single technological trajectory as military and civil applications of the jet engine used the same engine. Rolls-Royce's Spey engine, for instance, powered not only a range of commercial airlines including the Trident, the Fokker F28 and the BAC I-II, but also a number of military jets as well including the Phantom and the Buccaneer. However, over the past 20 years jet engines have followed two different technological trajectories. Military engines have increasingly focused on providing improved performance in terms of power output and size. Hence the EJ200 engine (Gunston, 1995) that powers the Eurofighter Typhoon

▶

is smaller, simpler and more compact than the previous generation of military fighter engines and in terms of power delivers some 20 000lbs of thrust. Civil engines in contrast have followed a different trajectory and have become vastly more powerful, with the largest version of the Rolls-Royce Trent engine producing 100 000lbs of thrust, making it five times more powerful than the EJ200 engine. Not only that, civil engines are now physically much bigger too, with the Trent having an overall diameter of some 3 metres. Like other comparable civil engines it is also more fuel efficient, much easier to control and subject to very much less maintenance than a military engine. And yet both types of engine though they have followed different technological trajectories, still fall within the same technological paradigm.

The term 'technological paradigm' borrows heavily from science, in particular the notion of scientific paradigms (Kuhn, 1970) used to identify particular scientific schools of thought. According to Dosi (1982) technological paradigms represent a general area or field of technology in which the search for innovation is conducted by a significant group of innovators, within a particular historical context. As examples of technological paradigms Dosi (1982: p. 152) cites nuclear technologies, semiconductor technologies and organic chemistry technologies. As such, a technological paradigm sets what Von Tunzelmann (1995) describes as 'the technological domain within which technologies evolve'.

A particular technological paradigm therefore effectively delimits the field of enquiry within which innovation is pursued. By setting the boundaries of knowledge in this way, as far as innovation is concerned it confines the search process in terms of the direction of enquiry and the prescriptions sought. A technological paradigm is likely to be based on a selected set of principles. These principles in turn are likely to confine the innovation process in terms of:

- the field of enquiry
- the problems to be solved
- the procedures used
- the generic tasks to which it is applied
- the properties it exploits
- the materials technology it uses

Hence a technological paradigm plays a big part in setting limits to the field of enquiry by defining 'the rules of the game', even though this may be quite unintentional. Indeed Dosi (1982: p. 153) notes that technological paradigms tend to have a powerful 'exclusion effect' that confines the efforts and technological imagination of engineers and whole organisations, making them 'blind' to other technological possibilities.

When a new technological paradigm emerges, it represents a major discontinuity, or shift in thinking. A paradigm shift is likely to be associated with some form of radical innovation that ushers in a new technology. As an example of a paradigm shift Dosi (1982) refers to the switch in electronics from thermionic valves to semiconductors. This involved new principles of operation, new materials and a whole new set of tasks. Similarly, in the field of aerospace the switch from piston to jet engines was another example of a change of technological paradigm. The jet engine demanded new materials in the form of high temperature alloys, new scientific principles and new control systems, and meant confronting an entirely different set of problems.

When a paradigm shift occurs, it can be very difficult for existing or incumbent firms who have made big investments in irreversible capabilities in production, skills, marketing and product support, as well as production capacity and reputation, to make the transition to the new technology. When the new technology of electronics impacted on the cash register industry, it brought major problems for incumbent manufacturers such as NCR who were firmly wedded to the older mechanical technology. NCR's investment in plant capacity, R & D, patents and intellectual property and service networks were rendered obsolete. Even more problematic was the need to rethink the nature of the product. NCR, for instance, saw the new technology of electronics as a way of building cash registers that could add numbers more quickly. Unlike the new entrants into the industry they were unable to re-conceptualise the product as a means by which their customers could manage inventories and supplier relations more effectively (Afuah, 2003).

Similarly, Christensen (1997) shows how the mechanical excavator industry found it hard to re-conceptualise their product when cable technology began to be challenged by the new technology of hydraulics in the 1960s. Failure to rethink the product led to new entrants, such as Joseph Bamford with his 'JCB' hydraulic excavator, taking over the market.

Within any given technological paradigm there will be a number of different technological trajectories. Each technological trajectory represents a specific subset of the technological field defined by a given technological paradigm.

A technological trajectory forms a development path traced out as a given technology develops. However, because a technological trajectory is a subset of a paradigm, it is possible to have a cluster of technological trajectories or directions whose outer boundaries will be defined by the paradigm.

Technological trajectories by their nature are cumulative and typically involve incremental innovations. Development along a particular trajectory involves improving the trade-off between technological variables set by the paradigm. There is therefore an essential continuity about technological trajectories as technological advance builds upon prior advances.

CASE STUDY: HOWEVER THE INTERNET DEVELOPS, IT IS THE CONSUMER WHO WILL BENEFIT

YES, the Internet is helping to crunch down prices, but by how much – and what else is it doing to the whole production process?

This week the UK debate about interest rates has been given a new twist by Sushil Wadhwani, a member of the Bank of England monetary committee. He has just written a paper that looks at the lessons from recent US experience and argues that after a lag, similar effects should become apparent here.

Crucially, productivity growth may be faster than expected as a result of the Net. The reason that much of an increase in productivity has not been evident outside America may be partly because the costs of adjusting to the new technology have obscured the

▶

gains being made. In any case he believes that business-to-business E-commerce will result in significant savings in costs.

This being Britain, the argument leads into a debate about interest rates. Put crudely, we do not need higher interest rates to hold down inflation, because the combination of greater competition and rising productivity will do it for us.

That spin is understandable. People care about their mortgages, companies care about their borrowing costs, and both care about the level of sterling – though private individuals rather like the strong pound because it makes foreign holidays so affordable, while companies dislike it because it makes exporting tougher.

But seeing the impact of the Internet through the prism of British interest rate policy is to focus on one tiny effect of a great global phenomenon. What it does to the structure of industry, to the nature of competitiveness, to the ability of developed economies to continue delivering productivity gains – all this is vastly more important.

And we are still guessing wildly. At most we have three or four years' experience of the impact of the Internet, in practice more like 18 months. In normal economic terms this is nothing, for you cannot really assess the implications of any new technology, or indeed any new economic event, until you have tracked it through a full economic cycle. Once we have come through the next downswing and seen how the Internet has altered the response of the economy to a fall-off in demand then we can start to make some better guesses. But even then they will still be guesses.

But we can, thanks to the couple of years of US experience, see some of the issues that matter, and it might be helpful here to identify two: the impact of business-to-business E-commerce and the fragility, or otherwise, of current ratings of technology stocks.

The first matters because it will determine the outcome of the key micro-economic issue facing the world: whether the business community really can deliver a sharply improved economic performance. The second matters because it will determine the key macro-economic issue: might a crash of share values be the thing that plunges the world into the next downswing?

The starting point with E-commerce is to realise that the bit we all write about, what businesses are doing to use the Net to sell services to the consumer, is much less important than the bit we do not write about, what businesses are doing to use the Net to streamline their production chain. Growth in business-to-consumer E-commerce is tiny compared with business-to-business and the gap is likely to widen. Quite suddenly, companies are figuring out ways of using the Net to cut out human beings and paper from the complex business of ordering components, managing stocks, maintaining equipment, and so on.

Humans make mistakes and paperwork costs money and time. You know the story about an automated factory that was run by a man and a dog? Why a man and a dog? The man was there to feed the dog and the dog was there to bite the man if he tried to touch any of the machinery. The Internet makes it possible to apply this principle to the entire production process. But it takes time to learn how to do it, and many factories are in the transition process, spending money on new kit and making it work, without yet obtaining the savings in labour and the improvement in quality and time that the technology promises.

The first assessment of the impact of business-to-business investment from people like Goldman Sachs suggests that it will boost long-term productivity growth and reduce the sustainable rate of unemployment and thereby boost economic growth of developed countries by one percentage point a year for five years.

There is an offset to this over the next five years from a decline in the growth of the labour force, but the very long-term effect will still be to boost economic activity by about the 5 per cent figure.

My own guess, for what it is worth, is that this estimate is both too high and too low. In the short-term it is too high. One percentage point a year across the entire economies of the developed countries requires an enormous shift in the way every business operates. It is a tall order. On the other hand, the effect of this new technology will not just last five or ten years. This will affect the way we live for a generation or more. It will take a full generation refining the technology and figuring out how to use it before we can be sure that we have exploited its full potential. In the long term the 5 per cent figure is therefore too low.

The markets, of course, have bought the hi-tech story, up-rating these stocks again and again. As my colleague Jeremy Warner has pointed out on several occasions, were it not for a couple of hi-tech sectors we would, in Britain at least, be in a bear market. Have the markets pushed too far, demanding with these ratings improvements in growth that cannot be achieved?

Goldman here argues that it is 'a tough call, but not necessarily fatal'. Interestingly, it also believes that the trigger for a downgrading of equities will not be the aggressive valuations put on technology stocks. If there is a trigger, it will most probably be a reassessment of the cost of capital (higher interest rates) or a reassessment of the earnings outlook. They do not see either happening soon and remain reasonably confident of the hi-tech sector.

We will see. The big point here surely is that we are seeing something akin to the railways, the steamship, the telegraph, the radio, the car, maybe even the moving production line of Henry Ford. New technologies fuel economic growth. They also inspire booms. Some of the protagonists get immensely rich. Others fail. The only thing we can be really sure of is that ultimately the certain beneficiaries are the consumers.

Source: The Independent, 25.02.02 by McRae

QUESTIONS

1 What is productivity and how can improvements (gains) in productivity enable developed nations to continue to grow?

2 Why does McRae think we are still guessing about the likely impact of the internet?

3 What is the difference between business-to-consumer (B2C) and business-to-business (B2B) E-commerce?

4 Why, according to McRae, is business-to-consumer less important than business-to-business E-commerce?

5 Where would you expect to find (i) business-to-business E-commerce and (ii) business-to-business E-commerce in the phases of the long wave?

6 What does the long wave cycle predict in terms of the sorts of organisations likely to be important in the development of the Internet?

7 Why does McRae regard the Internet as comparable to the railways, the steamship, radio and the car?

8 Why is business-to-business E-commerce likely to lead to big productivity gains over the long term?

QUESTIONS FOR DISCUSSION

1 **Distinguish between technological paradigms and technological trajectories.**

2 Why do process innovations tend to occur during the later phases of the long wave cycle?

3 **Where does the 'Dot.com' bubble fit within the long wave cycle? To what extent do you consider it to have been predictable?**

4 What is technology? Use examples to illustrate your answer.

5 **Distinguish between science and technology.**

6 What is tacit knowledge and why is it an important aspect of technology?

7 **What is the Kondratiev long wave cycle?**

8 How can the Kondratiev long wave cycle contribute to our understanding of innovation?

9 **What is meant by the term 'diffusion' and why is it an important aspect of innovation?**

10 How did Joseph Schumpeter link together innovation and the long wave cycle?

ASSIGNMENTS

1 What is technology and what is the link between technology and economic development?

2 Using a period of your choice, show how the long wave cycle is associated with the introduction of one or more new technologies

3 What is a speculative boom? How would you account for the Dot.com boom of the late 1990s?

4 Explain what is meant by the term 'technology trajectory'. Use a technology of your choice to illustrate your answer.

5 Why does the emergence of a new technological paradigm often seem to create problems for existing manufacturers?

RESOURCES

Books

The resources available for technological change are limited. There is comparatively little on either the long wave cycle or Nicholai Kondratiev. Fortunately the field covered by technological trajectories and paradigms is rather better covered.

Many books on innovation mention the long wave cycle (Goffin and Mitchell, 2005), but the treatment is often little more than a brief introduction. Detailed coverage of the long wave cycle is provided by Freeman and Soete (1997), Tylecote (1992) and a comparatively recent offering from Freeman and Louçã (2001). This last-named text is a really excellent source. Not only does it explain the theory of the long wave cycle, but it provides detailed accounts of each long wave cycle, with particular emphasis upon the current fifth Kondratiev.

The features of the fifth Kondratiev are also covered in a variety of texts detailing recent developments in computing and the Internet (Campbell-Kelly, 2003a; Naughton, 1999). Similarly studies of the Dot.com bubble also provide valuable insights. One of the best examples of such texts is Cassidy (2002) which provides an in-depth analysis of the bubble and the factors that led to it.

Technological paradigms and trajectories are well covered by the prolific output of Dosi which really starts with Dosi (1982). Other notable contributions have come from Nelson and Winter (1977) and Mowery and Rosenberg (1989). Tidd, Bessant and Pavitt (2001) contains an excellent chapter on technological trajectories and paradigms.

REFERENCES

Abernathy, W.J. and J. Utterback (1978) 'Patterns of industrial innovation', in M.L. Tushman and W.L. Moore (eds) *Readings in the Management of Innovation*, pp. 97–108, Harper Collins, New York.

Afuah, A. (2003) *Innovation Management: Strategies, Implementation and Profit*, 2nd edn, Oxford University Press, New York.

Beamont, R. (2002) *The Railway King: A Biography of George Hudson*, Review Books, London.

Campbell-Kelly, M. (2003a) *From Airline Reservations to Sonic the Hedgehog: A History of the Software Industry*, The MIT Press, Cambridge, MA.

Campbell-Kelly, M. (2003b) 'The rise and rise of the spreadsheet', in M. Campbell-Kelly, M. Croarken, R. Flood and M. Robson (eds) *The History of Mathematical Tables: From Sumer to Spreadsheet*, Oxford University Press, Oxford.

Cassidy, J.C. (2002) *Dot.con: The greatest story ever sold*, Penguin Books, Harmondsworth,

Christensen, C.M. (1997) *The Innovator's Dilemma: When New Technologies Cause Great Firms to Fail*, Harvard Business School Press, Boston, MA.

Dosi, G. (1982) 'Technological paradigms and technological trajectories', *Research Policy*, **11** pp. 147-162.

Dyos, H.J. and D.H. Aldcroft (1974) *British Transport: An Economic Survey from the Seventeenth Century to the Twentieth*, Pelican Books, Harmondsworth.

Forbes, N. and D. Wield (2002) *From Followers to Leaders: Managing technology & innovation*, Routledge, London.

Freeman, C. (1986) 'The role of technical change in national economic development', in A. Amin and J.B. Goddard (eds) *Technical Change, Industrial Restructuring and Regional Development*, Unwin & Hyman, London.

Freeman, C. and F. Louçã (2001) *As Time Goes By: From Industrial Revolutions to Information Revolution*, Oxford University Press, Oxford.

Freeman, C. and L. Soete (1997) *The Economics of Industrial Innovation*, 3rd edn, Continuum, London.

Goffin, K. and R. Mitchell (2005) *Innovation Management: Strategy and Implementation Using the Pentathlon Framework*, Palgrave, Basingstoke.

Gunston, B. (1995) *The Development of Jet and Turbine Aero Engines*, Patrick Stephens Ltd, Sparkford.

Hall, P. (1981) 'The geography of the fifth Kondratieff cycle', *New Society*, 25 March 1981, pp. 535-537.

Kuhn, T.S. (1970) *The Structure of Scientific Revolutions*, University of Chicago Press, Chicago.

Littler, D. (1988) *Technological Development*, Philip Alan, Oxford.

Marx , K. and F. Engels (1967) *The Communist Manifesto,* Penguin Books, Harmondsworth.

McGinn, R.E. (1991) *Science, Technology and Society*, Prentice Hall, Englewood Cliffs, NJ.

McRae, H. (2000) 'However the internet develops, it is the consumer who will benefit', The *Independent*, 25 February 2000, p. 23.

Mensch, G. (1979) *Stalemate in Technology: innovations overcome the depression*, Ballenger, NY.

Mowery, D. and N. Rosenberg (1989) *Technology and the Pursuit of Economic Growth*, Cambridge University Press, Cambridge.

Naughton, J. (1999) *A Brief History of the Future,* Phoenix, London.

Nelson, R. and S. Winter (1977) 'In search of a useful theory of innovation', *Research Policy*, **6**, pp. 36-76.

Newell, S., J. Robertson, H. Scarborough and J. Swan (2002) *Managing knowledge work*, Palgrave, Basingstoke.

Perez, C. (1983) 'Structural Change and the Assimilation of New Technologies in the Economic and Social System', *Futures*, **15**, pp. 357-375.

Schumpeter, J.A. (1939) *Business Cycles,* McGraw-Hill, New York.

Simon, H.A. (1972) 'Technology and Environment', *Management Science,* **19** (10) pp. 1110-1121.

Smiles, S. (1862) *The lives of George and Robert Stephenson*, John Murray, London.

Tidd, J., J. Bessant and K. Pavitt (2001) *Managing Innovation: Integrating Technological, Market and Organizational Change*, 2nd edn, J. Wiley, Chichester.

Tylecote, A. (1992) *The Long Wave in the World Economy: The Present Crisis in Historical Perspective*, Routledge and Kegan Paul, London.

Von Tunzelmann, N. (1995) *Technology and Industrial Progress: The Foundations of Economic Growth*, Edward Elgar, Cheltenham.

Womack, J.P., D.T. Jones and D. Roos (1990) *The Machine that Changed the World*, Rawson Associates, NY.

WHAT DOES INNOVATION INVOLVE?

CHAPTER 4

THEORIES of INNOVATION

OBJECTIVES

When you have completed this chapter you will be able to:

- Distinguish the principal theories associated with technological innovation
- Apply theories to the analysis of innovations
- Demonstrate the predictive capability of innovation theories
- Evaluate the most appropriate theories for the analysis of innovations
- Demonstrate practical benefits that can be derived from the application of innovation theories.

INTRODUCTION

Innovation is not inevitable. We have already seen that there are different types of innovation. It can occur in different ways and take place at different rates. Similarly, the impact of innovation varies according to the type. This is where theory comes in. Theories help us to make sense of innovation. In particular theories enable us to identify patterns of innovation. These patterns will be linked to different types of innovation. At the same time innovation theories possess a predictive capability that allows us to indicate the likely course of innovation and the impact that it will have. Given that innovation theories allow us to make sense of innovation, they provide powerful tools of analysis. These tools mean that the analysis of innovation is not just a matter of describing what happened. Using theories one can identify patterns of innovation, make comparisons between innovations and predict possible outcomes from the process of innovation.

This chapter aims to introduce a number of theories of innovation. They come from a variety of sources, though most are closely linked to technological aspects of innovation. In addition the chapter shows how these theories can be used as tools for analysing innovations and provides scope for gaining practical experience of doing this. This should enable the reader to come to a clearer appreciation of the nature of innovation.

WHO NEEDS THEORY?

Innovation is a practical business. We have already seen that it is concerned with the 'commercialisation' of discoveries and inventions. Inevitably this has a lot do with practicalities. For instance, finding a market for the product or working out how the product can be manufactured relatively cheaply and easily. Despite this, it is actually very important to be able to relate innovations to a body of theory. Why should this be?

Innovation has a number of features, but three in particular stand out as being ones where theory in one form or another can contribute to our understanding. These features are:

■ Complexity
■ Populism
■ Interest in success/failure

Innovation is a complex phenomenon and one that embraces a number of academic disciplines. Essentially scientific and technological by nature, as a phenomenon it has important economic and social consequences. This overlap makes it all the more necessary to have a body of theory that can aid and assist in analysing the phenomenon. Otherwise, we are in danger of trivialising innovation because there is no single easily identifiable body of knowledge upon which we can readily and easily draw.

Probably because it is new and exciting, innovation attracts a lot of column centimetres in the press. As a result much that is written about innovation tends to be journalistic. While this helps to popularise innovation and bring it to the attention of a large number of people, it can sometimes make analysing and accounting for innovation difficult. In particular it tends to mean that a good deal of what is written about innovation tends to highlight personalities, particular events and success stories, rather than the range of activities, collective effort and organisational support that are a vital part of innovation. It also means that accounts of innovation often include a lot of extraneous material and lack hard data.

Finally, much that is written about innovation tends to focus on success or failure. The fact that an innovation is successful is what makes it interesting. Accounts of innovation therefore tend to be written for a variety of purposes including entertainment, personal aggrandisement, money and the like. All too often the reasons for the success (or for that matter the failure) are ignored, with attention instead focusing on the extent of the success. Yet, if we are to understand innovation, what matters is to understand the reasons for success, and theory can help in making sense of events so that one comes to identify and evaluate these reasons.

THEORIES OF INNOVATION

There are many theories associated with innovation, of which four are presented here. The selection of four theories is fairly arbitrary. One could very easily have included many more. However, the four selected here offer scope for using a number of different theories in different contexts without unduly complicating the picture. There are other texts that can provide the reader with more theories should he or she so wish (Ettlie, 2000).

The four theories are:

- Technology S-curve
- Punctuated equilibrium
- Dominant design
- Absorptive capacity

These theories are associated primarily with technological innovation. Despite the emphasis on technology within the process of innovation, they all provide adequate scope for analysing innovations in general.

TECHNOLOGY S-CURVE

One of the central ideas behind the theory of the Technology S-curve is the notion of a technology life cycle. This implies that over time the capability of a technology to deliver improved performance will vary. Early in the life cycle the potential for a given investment of engineering effort to deliver improved performance will be high. Successive amounts of additional engineering effort produce ever greater improvements in performance. This is the well-known 'learning curve' effect, which results from a new technology becoming better understood, better controlled and more widely diffused. Eventually this will begin to lessen in terms of the relationship between inputs and performance until a point is reached where increasing engineering effort produces diminishing returns in terms of performance improvement. This implies that a given technology eventually reaches some kind of 'natural limit' as it matures (Figure 4.1).

Figure 4.1 Technology S-curve

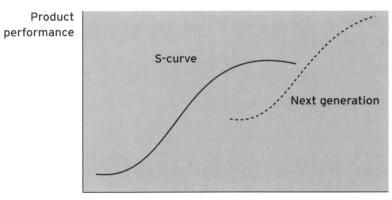

Source: Foster (1986)

Why does a technology mature?

Foster (1986) argues that technologies simply have physical limits. He cites the technology of sailing ships whose speed was limited by the physics of wind and water. The tea clippers that traded between China and Europe in the last years of the nineteenth century were amongst the most efficient sailing ships ever built, representing the high point of sailing ship technology. They were eclipsed by steamships which substituted the new technology of steam power for wind. As a technology matures so performance improvements get less and less and it requires a radical innovation associated with a new technology. The piston-engined aircraft of the Second World War represented the physical limit of propeller technology in terms of speed. To go any faster required a new technology. That technology took the form of the jet engine pioneered by Whittle in the UK (Golley, 1996) and Von Ohain in Germany (Conner, 2001). The jet engine worked in a completely different way from the piston engine, requiring new skills, new knowledge and new materials. Yet it had the potential to power aircraft at much higher speeds.

Sahal (1981) suggests that technological maturity is also a matter of scale and complexity. Scale is a matter of the product getting impossibly large in the pursuit of improved performance, while complexity is a matter of there being more and more components. In order to escape the problems caused by scale and complexity a radical innovation is required to escape the limits of the technology.

Christensen (1997: p. 34) suggests that the technology S-curve can be useful not only in terms of its descriptive value (i.e. the extent to which it represents what happens in real life) but also in terms of its predictive value. By this is meant the ability of the Technology S-curve to predict the course or developmental path of innovation. The predictive power of the Technology S-curve is that the point of inflection shows the point at which the existing technology has reached maturity and is going into relative decline. The technology may still be making an effective contribution, but the point has been reached where one might expect to find a successor technology (i.e. another Technology S-curve) to arise. The challenge is to identify and develop the successor technology. The inability to anticipate new technologies has been cited as the reason why incumbent firms often fail and as the source of advantage for new entrants.

MINI-CASE: DISK DRIVES

A hard disk drive is a storage device used to store data on a computer. The first ones were developed by IBM in the 1950s and today most modern PCs have a hard disk drive. Indeed adverts for PCs typically stress the capacity of the hard disk drive (e.g. 200 Giga Byte hard disk) as a feature. The hard disk drive consists of a read-write head mounted on an arm that swings over the surface of a rotating solid disk (much like the operation of a record-player deck for playing vinyl records). The first hard disk drives used on large mainframe computer installations were 14 inches in diameter. In the 1970s smaller 8-in. drives were developed for a new generation of microcomputers produced by firms such as Data General and Hewlett-Packard. These smaller drives were produced by new entrant firms such as Shugart and Quantum. Gradually the capacity of the smaller 8-in. hard disk drives rose and they began to invade the mainframe computer market where their lower cost and greater reliability began to tell. Then in the 1980s Seagate Technology introduced a 5.25-in. hard disk drive. These were of no interest to microcomputer manufacturers, so Seagate pioneered new applications. They turned to the makers of the new personal computers such as Apple and IBM. At this point it was by no means certain that PC users wanted a hard disk drive – data storage was by so-called 'floppy disks'.

By the early 1990s the first 3.5-in. hard disk drive had appeared developed by a small Scottish company – Rodine. Again there was little interest from existing PC manufacturers. But the idea was taken up by a new company Connor Peripherals a spin-off from 5.25-in. hard drive manufacturer, Seagate Technology. The 3.5-in. drive differed significantly from its 5.25-in. counterpart. Many of the mechanical parts were replaced with electronics, making the drive significantly lighter and more rugged. These attributes were of no particular concern for existing PC manufacturers, so again firms like Connor Peripherals sought new applications such as the growing market for portable and laptop PCs where weight, ruggedness and power consumption were critical issues. In time the 3.5-in. hard disk drive became the industry standard.

Source: Christensen (1997)

PUNCTUATED EQUILIBRIUM

Abernathy and Utterback (1978) used the distinction between incremental and radical innovation to show how established industries go through periods of stability with changes confined to incremental innovations. They recognised that eventually the stability is broken by a radical innovation that is highly disruptive, bringing the period of stability and equilibrium to an end. According to Loch and Huberman (1999) this is caused by initial uncertainty surrounding the new technology, leading to experimentation and instability. As the technology becomes better understood the changes become incremental and stability and equilibrium return.

Punctuated equilibrium is an evolutionary theory. As a theory of innovation, technology is a central force shaping the pattern of innovation. Tushman and Anderson (1986) note that it is the evolution of technology that gives rise to the distinctive pattern of innovation associated with punctuated equilibrium. Technology evolves not on a smooth continuous basis, but via a succession of fits and starts. Major technological breakthroughs are relatively rare. Tushman and Anderson (1986) cite the cases of Chester Carlson and the development of xerography (photocopying) and Alistair Pilkington and the development of the float glass process for manufacturing plate glass, as examples of discontinuities. They observe (Tushman and Anderson, 1986: p. 41) that technological change is: 'a bit-by-bit cumulative process until it is punctuated by a major advance'.

Both xerography and float glass were major advances. Each in its own way was highly disruptive, but it was followed by a period of relative stability. This is what occurred in both instances. Innovations followed the major advances but they were incremental innovations that resulted in modest product improvements rather than significant changes.

The discontinuities that punctuate periods of equilibrium are linked to major technological innovations. These represent technical changes that are step changes. Tushman and Anderson (1986: p. 44) describe them as: 'so significant that no increase in scale, efficiency or design can make older technologies competitive with the new technology'. The new technology represents a step change which is why no amount of incremental change for the existing technology will render it competitive.

A key feature of technological discontinuities is that they require new skills, new abilities and new knowledge in both the development and the manufacture of the product. As a result such innovations can be 'competence destroying'. This means that existing firms are unable to use the knowledge and experience they have accumulated during the period of equilibrium. Given that the existing knowledge represents a big investment made over a long period of time, they are likely to want to make use of this knowledge. Hence they are more likely to opt for incremental innovations.

Nor is it just that 'incumbent' firms possess knowledge and expertise linked to the old technology that gives rise to inertia; other factors include:

- Traditions
- Sunk costs
- Internal political constraints
- Commitment to outmoded technology

If working practices have changed very little in a long time, they may have become so widely accepted and so deeply engrained in the organisation that they have become traditions. As traditions, the rationale and reasoning behind the working practices in question may have long since disappeared, but because this is the way things have been done for a long time, traditions are very hard to break. Nor are traditions confined to working practices: they can cover all aspects of a business.

Sunk costs are costs associated with prior investments. These could cover equipment, buildings, systems or even training. Firms make investments all the time, but where the investment is technology-specific the cost of the investment is a sunk cost. The key aspect is that the investment cannot be transferred to the new technology – it is sunk in the old technology. If the organisation has invested heavily, then it may be expecting to spread the cost over future output. As a result the organisation may be reluctant to break from the old technology.

Internal political constraints can arise for all sorts of reasons. If managers have a strong commitment to the old technology – perhaps by virtue of their training or their knowledge – they may well be reluctant to embrace a new technology they know little about. Not only will this lead to a reluctance to innovate on their part, it may even stop others pursuing innovation.

All of these factors serve to constrain or limit the responsiveness of existing firms. This helps to account for the existence of equilibrium. Under these conditions existing firms may confine themselves to incremental innovations, thereby prolonging the period of equilibrium. Eventually, however, radical innovations lead to discontinuities that punctuate the equilibrium. A period of 'ferment' then ensues, when technological uncertainty leads to a number of competing product architectures or product designs. This can extend not only to technical aspects such as product configuration and product performance, but also to marketing aspects such as product pricing and market boundaries.

MINI-CASE: THE PERSONAL COMPUTER (PC)

The arrival of the personal computer represented a technological discontinuity. Previously computers had been large, specialised pieces of equipment operated by computer specialists. The PC changed all that. Suddenly there was a computer anyone could use. But to begin with there were large numbers of competing designs. They had very different configurations in terms of storage devices, video display units and input devices. For example, some PCs had twin disk drives, others single disk drives, some used 5½-in. disks, others 3½-in. disks, some had external disk drives others internal disk drives. Apple Computer went some way to creating a common product configuration with their Apple II machine, but it was the arrival of the IBM PC that created a configuration that was copied by others and effectively created the industry standard.

These circumstances, with a high level of technological and market uncertainty, offer opportunities for new entrants. During a period of equilibrium, new entrants would normally find they were at a disadvantage to incumbents, but when technological discontinuities arise and a process of ferment occurs, the tables may be turned. While incumbents may be stuck with 'legacy' problems such as sunk costs, unwanted skills and obsolete plant, new entrants can respond more effectively to the new conditions precisely because they are unencumbered by the baggage of an old technology, old ways of doing things and an outdated view of the world.

MINI-CASE: THE JET AGE

The evolution of the modern airliner exemplifies punctuated equilibrium, because there have been a small number of path-breaking designs that have changed both the nature and the economics of air travel. The Douglas DC3 was the first modern airliner. One of the first twin-engined monoplane designs, its introduction in the late 1930s made the provision of a modern scheduled passenger service feasible. It was followed over the next three decades by a succession of improved but essentially similar designs. Some were a bit bigger, some were a bit faster but the nature of the airline business remained unchanged. It was still a minority activity with a high proportion of long-distance travel being undertaken by sea. The introduction of the De Havilland Comet in the 1950s was a technological discontinuity. It heralded the start of the so-called 'jet age'. It was much faster, and more comfortable than the piston-engined airliners then in use. Though commercially unsuccessful the Comet demonstrated the advantages of jet travel. It was followed by a series of competing designs mostly with different design configurations. The Comet was relatively small carrying only 60 passengers and with its engines set into the wing. The French Caravelle airliner had its engines at the rear and could carry nearly 100 passengers.

▶

However, it was the Boeing 707 that set the standard. It was the first commercially successful jet airliner, capable of carrying 150 passengers with its engines mounted in pods under the wing. The next technological discontinuity was the introduction of the Boeing 747 jumbo jet in the 1970s. It was not only much more economical but quieter and capable of a much longer range. Its arrival marked the beginning of mass air travel.

Source: Heppenheimer (1995)

As a theory that helps to explain the pattern of innovation exhibited in real life, punctuated equilibrium has its limitations. It is a theory in which technology plays a central role. Some might argue that it is a theory of technological evolution. Similarly it is an external theory in the sense that it tells us little about how innovation is carried out inside the firm. However despite this, punctuated equilibrium is of value as an innovation theory. It helps to explain inertia, in particular the reluctance of existing firms to adopt a new technology, which in turn explains why some innovators find it hard getting existing firms to take an interest in new ideas. The case of James Dyson (Dyson, 1997) and his attempts to get established vacuum cleaner manufacturers to embrace his new more efficient dual-cyclone technology which eliminated the need for a dust bag, is a good example. What was happening was that incumbent firms were comfortable with the equilibrium that existed in the vacuum cleaner industry. Dyson's ideas posed a technological discontinuity that threatened their investments in know-how and manufacturing capability. Consequently, they were not keen to embrace the new technology. Another merit of punctuated equilibrium is that it integrates the typology of innovation that distinguishes radical and incremental forms of innovation. Finally, punctuated equilibrium provides a good fit with reality where technological discontinuities can be very disruptive.

DOMINANT DESIGN

MINI-CASE: THE SPREADSHEET

Microsoft's Excel is familiar to all, so much so that few of us use any other spreadsheet and many people would be hard pressed to name another. Today Excel represents a dominant design where spreadsheets are concerned, yet it was a comparatively late entrant. The first spreadsheet was VisiCalc which appeared about 1980. Despite an early lead, within two years VisiCalc had more than twenty competitors including Supercalc, Lotus 1-2-3, Multiplan (Microsoft's offering later re-named Excel) and Quattro. Each competitor was able, as a follower and imitator, to incorporate additional features into its spreadsheet software. Quattro introduced a WYSIWYG (What You See Is What You Get) facility, while Lotus 1-2-3 introduced integrated presentation graphics and a small database into the spreadsheet. Whereas VisiCalc had been designed to work on the Apple II computer, the newer spreadsheets were designed to operate on newer PCs as they became available. As the IBM PC replaced the Apple II as the most widely used personal computer and became the desktop computer used not only by small companies but large ones as well, spreadsheets designed from the outset for this machine, such as Lotus 1-2-3, began to get the upper hand in the marketplace. VisiCalc, though it had been the innovator as the first spreadsheet to go on the market, was soon eclipsed. Lotus 1-2-3, very rapidly took over the market. In time Microsoft revamped its own spreadsheet, Excel. With GUI facilities, Microsoft's Excel built up a strong position in the Apple Macintosh market but was completely overshadowed by Lotus 1-2-3 in the IBM PC market. However, in the late 1980s Microsoft released Excel for Windows. At this time the Windows operating system was something of a novelty, but as it came to dominate the IBM PC market in the early 1990s, Lotus was caught without a Windows version of its product. Users wanting to take advantage of the new operating system had to buy Microsoft Excel, because it was the only spreadsheet for Windows then available (Campbell-Kelly, 2003b). Within a very short space of time Microsoft Excel completely dominated the market.

Source: Campbell-Kelly (2003a; 2003b)

A dominant design is a design or product configuration that comprises 'the one that wins the allegiance of the marketplace, the one that competitors and innovators must adhere to if they hope to command a significant market following' (Nordström and Biström, 2002: p. 713). Quite literally it is a configuration that all or most firms eventually follow.

The theory of a 'dominant design' is linked to ideas about the evolutionary development (Teece, 1986) of science. In science ideas are constantly evolving. The process of evolution comprises two stages, a pre-paradigmatic phase, when many ideas are circulating and no one explanation of a phenomenon holds sway, and a paradigmatic phase, when a single explanation or theory becomes widely accepted. The switch to the latter phase and the emergence of a dominant paradigm signals scientific maturity. This paradigm remains the accepted view until perhaps even it is eventually overturned by another paradigm, as in the seventeenth century when Copernicus's theories of astronomy overturned those of Ptolemy (Teece, 1986).

Dosi (1982) provides a perspective of technological innovation that evokes this evolutionary principle and parallels notions of scientific evolution. According to Dosi the early stages of technological innovation are characterised by a state of 'flux' or 'ferment' (Anderson and Tushman, 1990), where product designs and configurations are fluid. There will be large numbers of competitors, with rivalry based on competing designs, each of which is markedly different. In this pre-paradigmatic phase no one design or configuration stands out. The evolution of the bicycle (Rosen, 2002) which generated a proliferation of competing designs or forms in the late nineteenth century, including the famous 'penny-farthing', illustrates this pre-paradigmatic period of flux.

Eventually an evolutionary process involving variation, selection and retention (Basalla, 1988) leads to the emergence of a single design that forms the dominant design (see Figure 4.2). Teece (1986) likens this to a process akin to a game of 'musical chairs' where competing designs gradually fade away to leave a dominant design. Others have used the term 'shake-out' to describe the way in which most of the early designs fall away. When a dominant design emerges, competition then shifts away from design and towards other variables such as branding, promotion and price. Under these new circumstances new factors, such as scale and learning, become more important and specialised capital assets begin to replace general-purpose capital assets as firms seek lower unit costs through economies of scale and learning. In the case of the bicycle, the dominant design was the 'safety bicycle' that emerged at the end of the nineteenth century from the proliferation of different forms, to include a variety of features we would all recognise today, including a triangular all-metal frame, bearings, chain drive to the rear wheel and pneumatic tyres (Anderson and Tushman, 1990).

Figure 4.2 Dominant Design and the Technology Cycle

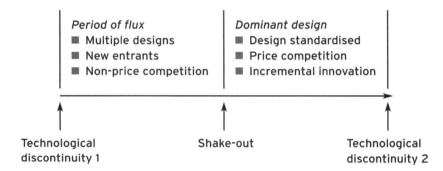

Source: Adapted from Anderson and Tushman (1990)

Utterback (1993) has shown that dominant designs are more likely to appear in mass markets, such as typewriters, bicycles, sewing machines, televisions and cars (Freeman and Louçã, 2001), which can justify the investment in specialised capital assets. This is borne out by well-known examples of dominant designs that include the qwerty typewriter keyboard, Ford's Model T, Douglas's DC3 airliner, JVC's VHS video recorder and Microsoft's Windows operating system.

How do dominant designs arise? Abernathy and Utterback (1978) suggest three possible factors that can give rise to dominant designs. First, consumer preference, where a particular package of factors present in one design finds favour with consumers in meeting their requirements. These factors will not all be technical. In fact, dominant designs are rarely technically superior to rival designs, but the particular package of features appeals to consumers. Sometimes it is the market power of a dominant producer that is the key factor. Abernathy and Utterback (1978) cite the case of IBM and its 360 series mainframe computer which became the de facto industry standard. Finally regulation, either by government or some form of industry body, may be instrumental in a dominant design appearing.

The theory of dominant design contributes to our understanding of innovation in a number of ways. It highlights the importance of the user, who may be less interested in technical features and more interested in usability. It also highlights the importance of standards particularly where compatibility is an issue for the user. Finally it shows the importance of business strategy within innovation. In the spreadsheet case, for instance,

Lotus, like many software companies, lost out to Microsoft because its business strategy anticipated OS/2 succeeding MS-DOS as the principal operating system for PCs (Campbell-Kelly, 2003a: p. 253), and invested in software products to reflect this.

MINI-CASE: THE VIDEO CASSETTE RECORDER (VCR)

The first video recorders were produced by Ampex, an American electronics company. Through licensing and joint ventures RCA and Toshiba of Japan emerged as competitors. However, the market was at this stage confined to professional and industrial users such as TV studios. Despite the early technological lead of US companies, it was Japanese companies who were the first to master the essentials of consumer-oriented product design and mass production. In the race to produce a home video cassette recorder (VCR), Sony and JVC came up with the idea of a ½-in. tape cassette (in the professional market ¾-in. and 1-in. tapes were used). Sony introduced the Betamax system in 1975 and JVC brought out the rival Video Home System (VHS) a year later. Although both systems used ½-in. tape they were incompatible formats because they used different tape-handling mechanisms and cassette sizes, as well as different systems for decoding the signals.

Sony's Betamax was the first compact, inexpensive, reliable and easy-to-use VCR. Its early lead meant that it accounted for the majority of VCR production in 1975–77 and enjoyed steadily increasing sales until 1985. However, it fell behind JVC's rival VHS format during 1978 and steadily lost share thereafter. From 1980 onwards VHS had become the dominant design – a position it held until DVD appeared 20 years later.

How did VHS become a dominant design? This is a particularly pertinent question when one considers that JVC was one of the smaller Japanese consumer electronics companies and compared to Sony had limited manufacturing and distribution capabilities. When Betamax and VHS appeared in the mid-1970s, many other companies were experimenting with video recording systems and a variety of formats including Ampex, RCA, Matsushita, Toshiba, Sanyo and Philips. Hence this was a period of considerable 'flux' with a variety of product configurations and formats under development.

The leading contenders were remarkably evenly matched in terms of cost and performance. However, as the nascent market for home video recording developed, JVC proved the more successful in terms of its ability to create effective alliances and licensing agreements. In particular, JVC recognised the importance of European manufacturers. It pursued a deliberate strategy of trying to win over European firms such as Thorn EMI, Blaupunkt and Thomson. JVC was also more active in providing technical assistance. For its part Sony was in some respects hampered by its earlier success. Its strong reputation as a highly successful innovator meant that it had less experience of working with other firms and manufacturing products for them. As a result, despite an early lead on the part of Betamax, VHS became the industry standard as more and more firms adopted it, thereby making it the dominant design.

Source: Cusumano, Mylonadis and Rosenbloom (1992)

ABSORPTIVE CAPACITY

The theory of absorptive capacity differs from previous theories by virtue of the fact that it integrates both the external dimension of innovation, which is concerned with the evolution of technology, and the internal dimension, that is concerned with learning and the knowledge transfer process within the innovating organisation. The emphasis within absorptive capacity upon learning marks it out as providing a very different analytical framework from the S-curve or punctuated equilibrium. This is reflected in the words of Cohen and Levinthal (1990: p. 128), in a seminal article that first expounded the notion of absorptive capacity, when they identified it as being concerned with: 'the ability of a firm to recognise the value of new, external information, assimilate it and apply it to commercial ends'.

Figure 4.3 provides a schema in which one can see recognition, assimilation and application at work. In this schema the external environment where technological evolution takes place is at the top and the internal environment of the firm is at the bottom. The process of recognising external trends and technological opportunities is represented by the arrow linked to the box on the left. This box represents the part of an organisation's absorptive capacity that focuses on assimilation. Hence the link from the external environment to this box represents a conduit or channel through which external ideas and opportunities are fed into the organisation. The capacity of the conduit is a crucial feature of organisations with a strong absorptive capacity. If it is effective, the organisation will be good at recognising external ideas.

While Cohen and Levinthal (1990) acknowledge that 'external influences' are vital for innovation, recognition of such influences is only one feature of absorptive capacity. For effective innovation there has also to be a capability to assimilate ideas within the

Figure 4.3 Absorptive Capacity

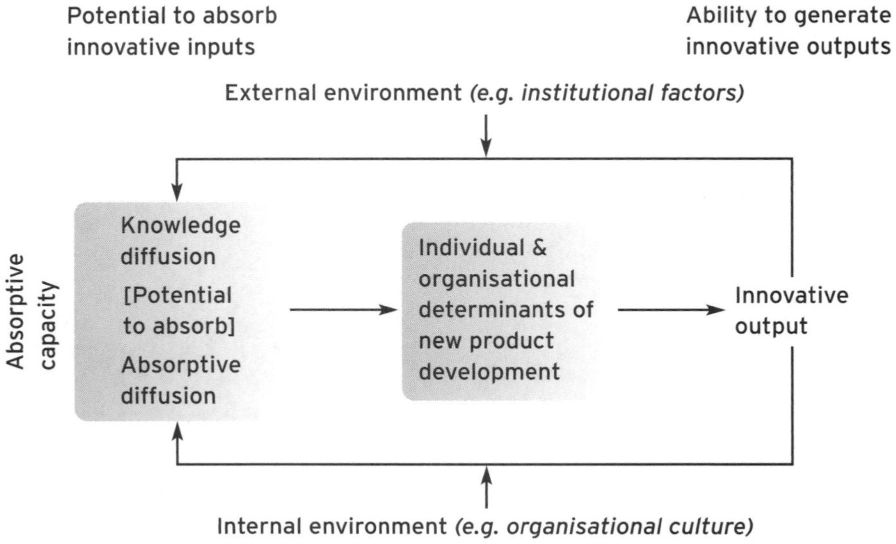

Source: Ettlie (2000: p. 83) from 'Managing Technological Innovation', John Wiley, 2000

organisation. As Figure 4.3 shows, assimilation is dependent upon an ability to bring external ideas in and to absorb such ideas within an organisation. Termed 'knowledge diffusion' this has to extend right across an organisation and to all levels within it. The 'silo' mentality has no place within the theory of absorptive capacity. Accordingly, the theory of absorptive capacity sets considerable store by internal communication systems that effectively transfer knowledge across the different parts of the organisation. In this context Cohen and Levinthal note that shared knowledge and expertise is necessary for good communication.

Nor is this the end of the story, since, for effective innovation, the ideas once absorbed within an organisation have to be applied. An organisation's capability to apply ideas it has absorbed is represented by the box in the centre of Figure 4.3. What determines an organisation's ability to recognise, absorb and apply new ideas so as to achieve effective innovation? Ettlie (2000) reminds us that there is a constant tension between inward-looking (the bottom channel in Figure 4.3) and outward-looking (the top channel in Figure 4.3) absorptive capacity. Effective absorptive capacity requires the organisation to maintain a balance between the two if it is to lead to effective innovation.

The theory of absorptive capacity places a great deal of emphasis on an organisation's ability to learn. Three factors are identified by Cohen and Levinthal (1990) as being critical in developing and extending an organisation's capacity to learn and thence its ability to assimilate and apply new ideas:

- Exposure to relevant knowledge
- Presence of prior related knowledge
- Diversity of experience

Exposure to relevant knowledge means that the organisation and its staff need to utilise appropriate networks in order to ensure that they keep abreast of developments in the field. The importance attached to prior related knowledge is linked to the assimilation process. Cohen and Levinthal (1990) argue that the ability to recognise the value of new knowledge and assimilate it into the organisation is a function of the accumulated prior knowledge within the organisation This in turn emphasises the importance of knowledge within this theoretical model. Assimilation requires that knowledge be evaluated, which in turn requires prior knowledge. It is because of the learning process that Cohen and Levinthal also stress diversity of experience. The greater the range of experience within the organisation, the greater the scope for recognising external ideas and stimuli.

The case study of the McLaren MP4 racing car with its carbon fibre chassis that appears at the end of this chapter provides a powerful illustration of the theory of absorptive capacity (see also Chapter 5). Carbon fibre had been known about for some time (Henry, 1988). Designers had used small amounts of carbon fibre in non-structural applications such as wings, but it took a relative outsider to the Formula 1 community, like John Barnard, to attempt to build an entire chassis out of carbon fibre. Significantly it was his external contacts, particularly in the aerospace industry and the US, that gave the McLaren team access to the necessary knowledge and expertise.

In this context Bruce and Moger (1999) observe that there is a danger that organisations that place excessive emphasis on increasing efficiency and cost reductions through the pursuit of increased specialisation using mass-production-type activities risk impairing the organisation's absorptive capacity. This is because the division of labour aimed at producing repetitive, mass-production-type activities reduces the diversity of experience of those working within the organisation, and as a result the scope to build up absorptive capacity, which Cohen and Levinthal argue is a cumulative process, is limited.

Absorptive capacity is more sophisticated than some of the other theories of innovation. It highlights the importance of external knowledge as a critical component in innovation. It helps to explain why some organisations, even where they are exposed to external knowledge, may be poor innovators, because they cannot absorb and make use of the knowledge, and it serves to show why networks and networking can be so important to innovation. However, probably the greatest strength of absorptive capacity and the reason why it has been widely used by those researching the field is that it integrates and brings together a number of ideas. These include ideas about technological evolution, the learning process and networking. Absorptive capacity offers a synthesis that draws these different strands together. In the process it offers a powerful tool for analysing innovation.

HOW DOES THEORY HELP THE INNOVATOR?

Theories of innovation, like any other form of theory, have three main contributions to offer when it comes to analysing innovation. These contributions are:

- Descriptive
- Analytical
- Predictive

The descriptive contribution includes the identification of key events, the description of the course of events and how the events are linked so as to provide an accurate account of the process of innovation based on relevant evidence. Most people when asked about innovation will describe a process. This is not surprising, because innovation typically involves a linear process in which activities follow one from the other. While describing the course of innovation in this way often makes a good story – hence the large number of books about innovations (especially biographies which recount how the innovation came about) – unfortunately it does not provide a great deal in the way of explanation.

The analytical contribution is all about uncovering and explaining why the innovation occurred. In order to be able to offer an explanation of why one innovation occurred in the way that it did, or why another innovation was particularly successful, one needs to be able to identify patterns or points of commonality between innovations. This permits the identification of causal relationships so that the success or failure of an innovation is not seen as merely a succession of accidents. At the same time it helps to be able to draw comparisons between different innovations. Undertaking these sorts of activities requires some tools of analysis. Innovation theories can provide these analytical tools.

Finally, the predictive element allows us to understand why particular innovations succeeded or failed so that others can learn and in future take actions that will enhance the likelihood of success. Why should this be? Prediction can offer a number of benefits:

- The course of events associated with innovation can be anticipated.
- Problems and difficulties can be predicted in advance.
- There is scope for planning so that resources can be used more effectively.
- There is a greater likelihood of a successful outcome to the innovation.

Consequently, while theories of innovation may appear to be abstract and unrelated to the practical problems associated with innovation, in fact they can offer some very practical benefits, all the more so when one considers that there is often a high level of uncertainty associated with innovation.

CASE STUDY: McLAREN MP4

Despite being highly successful for most of the previous decade, by 1980 the McLaren Formula One team had slipped to the back of the grid, having been uncompetitive for the last couple of seasons. This lack of success led to the team being taken over by Ron Dennis, who had previously run the Project 4 team in Formula Two, and being re-formed as McLaren International. As an outsider who was not part of the established Formula One scene, Dennis looked not to existing teams for a designer but instead picked John Barnard who had previously worked in Indycar racing in the US.

At this time all Formula One cars were of monocoque construction with the body and the chassis forming an integral unit made from sheets of aluminium riveted together. This type of chassis construction had been pioneered by Colin Chapman of Lotus some 20 years earlier. Borrowed from the aerospace industry, aluminium monocoque chassis construction combined torsional stiffness with light weight and strength (Tippler, 2001). This type of chassis superseded the spaceframe chassis made of lengths of tubular steel welded together to create a structure that was relatively light and yet strong. Spaceframe construction had been popular in the immediate postwar years because it was cheap, easy to use and employed readily available materials (Tippler, 2001). These had been important considerations when there was little money in motor racing. However, the lightness and rigidity of the aluminium monocoque chassis led to it being rapidly adopted in the 1960s when the maximum size of engine was reduced from 2.5 to 1.5 litres. With much less power available Formula One designers like Colin Chapman turned their attention to ways of improving handling, so that if racing cars were not going to go as fast on the straights they would at least go round the corners more quickly. However, by the time John Barnard came to design his first Formula One racing car for McLaren International in 1980 a different set of design parameters had come into play.

John Barnard felt that the key to producing a design that would be competitive on racing circuits across the world lay in designing a chassis that was narrower than the ones other teams were using. This was at a time when the teams were increasingly experimenting with aerodynamics, particularly the use of 'ground effects' to give cars better road-holding capabilities. Ground effects involved fitting a special type of wing underneath the car. In order to fit the biggest possible underwing, Barnard wanted to use a narrower chassis than the other teams were using. Unfortunately a narrower chassis would also be one with less torsional stiffness, meaning that it would flex going round corners, thus upsetting the handling and eliminating the gain from a bigger underwing. In order to maintain torsional stiffness, Barnard reasoned that he would have to consider

▶

building the chassis from a material other than the conventional aluminium sheeting used by other teams. Steel would be stiffer but it would also be heavier. On the other hand carbon fibre offered not only lightness, but stiffness and great strength.

However, no one had ever made a chassis out of carbon fibre: in fact the material had never been used for any large structure before. Carbon fibre applications had been confined to individual components. In the aerospace industry carbon fibre had been used for compressor blades and engine cowl panels on aero engines (Spinardi, 2002), both applications where lightness and strength were at a premium. The problem was that Formula 1 teams were not familiar with carbon fibre technology.

Undaunted, Barnard designed the new McLaren MP4 with a carbon fibre chassis. How though was the team to acquire carbon fibre technology? Rather than acquiring the technology directly, McLaren opted to gain access to carbon fibre technology by subcontracting manufacture of the chassis. However, Barnard was unable to find a facility in Britain capable of doing carbon fibre work on this scale; nobody was interested. Then an American friend of Barnard's who had worked in aerospace, suggested Hercules Aerospace, an American company that built missile components and parts for the AV8B jump-jet and the F18 fighter using carbon fibre (Spinardi, 2002). Hercules had a department that undertook specialised one-off jobs. They agreed to build the chassis. Although it took McLaren's drivers a little while to get used to the new car, team leader John Watson was soon putting in competitive times with the new MP4. In consecutive races, Watson finished third in Spain, second in France and then won the British Grand Prix at Silverstone.

Source: Cooper (1999)

QUESTIONS

1 What type of innovation does the McLaren MP4 represent and why?
2 Using an appropriate diagram show how the Technology S-curve can be applied to explain innovations in Formula One chassis design.
3 How might the theory of absorptive capacity be used to explain this particular case of innovation?
4 What does the McLaren MP4 case tell us about mechanisms that enable organisations to acquire external knowledge?
5 Why do you think the McLaren International team in its new form was able to assimilate external knowledge?
6 To what extent might the introduction of the carbon fibre chassis be termed a technological discontinuity and if so what would be the implications for a new team?
7 To what extent does the theory of punctuated equilibrium explain innovations in Formula 1 chassis design since the Second World War?

QUESTIONS FOR DISCUSSION

1 **Which of the four theories outlined makes the greatest contribution to our understanding of innovation and why?**

2 How can a theory of innovation have a predictive capability (use appropriate examples to support your case)?

3 **Many potential innovators who have developed and patented an innovation have found they could attract little interest when they sought to persuade established companies to take out a licence. To what extent do any of the theories of innovation provide an explanation for this behaviour?**

4 Why do innovations often come from outside the industry?

5 **Explain what is meant by a technological discontinuity and show how such discontinuities are linked to theories of innovation.**

6 Use the theory of the Technology S-curve to distinguish between radical and incremental types of innovation.

7 **Draw the Technology S-curve for the washing machine. How does this theory help to explain innovations in the field?**

8 According to the theory of punctuated equilibrium – why is the rate of innovation not constant?

9 **What is the significance of knowledge transfer within the theory of absorptive capacity?**

10 How might outsourcing reduce a firm's absorptive capacity?

ASSIGNMENTS

1 Take an account of innovation (this could be from a biography of an innovator, or a television programme or a film) and use any *one* theory of innovation to explain why and how innovation occurred.

2 Provide a detailed critique of any one of the theories of innovation.

3 What is the connection between notions of evolution, particularly technological evolution, and theories of innovation?

4 What is the value of theories of innovation for (a) would-be innovators and (b) policy-makers?

part II: what does innovation involve?

RESOURCES

There is no comprehensive account of theories of innovation. Ettlie (2000) provides an overview that includes a range of different theories, but each theory is only mentioned briefly. Tushman and Anderson's (2004) collection of readings on innovation is useful. It includes some important papers on innovation theory including a number on the theory of dominant design. The textbook by Burgelman, Maidique and Wheelwright (2001) also includes a number of the same papers, but it is more wide ranging and includes Cohen and Levinthal's (1990) seminal paper on absorptive capacity and Teece's (1986) paper on dominant design.

Without a comprehensive treatment of the theories of innovation the reader has little choice but to rely on the key texts for each particular theory. These include Foster (1986) and Christensen (1993) for the Technology S-curve, Cohen and Levinthal (1990) for absorptive capacity, Tushman and Anderson (1986) for punctuated equilibrium and Basalla (1988), Teece (1986) and Anderson and Tushman (1990) for dominant design.

Several of these papers also provide excellent case studies that use a particular theory to explain and account for innovation. Notable examples include: Tushman and Anderson's (1986) study of technological discontinuities in airliners and computers; Anderson and Tushman's (1990) study of process innovations in the plate glass industry; Christensen's application of the Technology S-curve to innovations in the hard disk drive (1993) and mechanical excavator (1997) industries.

There are plenty of other case studies. Among the more interesting are Pinch and Bijker's (1987) study of dominant design in bicycles, Heppenheimer's (1995) study of dominant design in jet engines, David's (1985) study of dominant design in typewriters, Campbell-Kelly's (2003a: 2003b) study of spreadsheets.

REFERENCES

Abernathy, W.J. and J.M. Utterback (1978) 'Patterns of Industrial Innovation', *Technology Review*, **80** (7), pp. 40-47.

Anderson, P. and M.L. Tushman (1990) 'Technological Discontinuities and Dominant Designs: A Cyclical Model of Technological Change', *Administrative Science Quarterly*, **35**, pp. 604-633.

Basalla, G. (1988) *The Evolution of Technology*, Cambridge University Press, Cambridge.

Bruce, M. and S.T. Moger (1999) 'Dangerous Liaisons: An Application of Supply Chain Modelling for Studying Innovation within the UK Clothing Industry', *Technology Analysis and Strategic Management*, **11** (1), pp. 113-125.

Burgelman, R.A., M.A. Maidique and S.C. Wheelwright (2001) *Strategic Management of Technology and Innovation*, 3rd edn, McGraw-Hill/Irwin, NY.

Campbell-Kelly, M. (2003a) *From Airline Reservations to Sonic the Hedgehog: A History of the Software Industry*, The MIT Press, Cambridge, MA.

Campbell-Kelly, M. (2003b) 'The rise and rise of the spreadsheet', in M. Campbell-Kelly, M. Croarken R. Flood and M. Robson (eds) *The History of Mathematical Tables: From Sumer to Spreadsheet*, Oxford University Press, Oxford.

Christensen, C.M. (1992) 'Exploring the Limits of the Technology S-curve: Part 1 Component Technologies', *Production and Operations Management*, **1** (4), p. 334.

Christensen, C.M. (1993) 'The Rigid Disk Drive Industry: A History of Commercial and Technological Turbulence', *Business History Review*, **67**, pp. 531–588.

Christensen, C.M. (1997) *The Innovator's Dilemma: Why New Technologies Cause Great Companies to Fail*, Harvard Business School Press, Boston, MA.

Cohen, W.M. and D.A. Levinthal (1990) 'Absorptive Capacity: A New Perspective on Learning and Innovation', *Administrative Science Quarterly*, **35**, pp. 128–152.

Conner, M. (2001) *Hans Von Ohain: Elegance in Flight*, American Institute of Aeronautics and Astronautics, Renton, VA.

Cooper, A. (1999) 'The Material Advantage', *Motor Sport*, **LXXV** (3), pp. 32–37.

Cusumano, M.A., Y. Mylonadis and R.S. Rosenbloom (1992) 'Strategic Maneuvering and Mass-Market Dynamics: The Triumph of VHS over Beta', *Business History Review*, **66**, pp. 51–94.

David, P. (1985) 'Clio and the economics of QWERTY', *Economic History*, **75**, pp. 332–357.

Dosi, G. (1982) 'Technological Paradigms and Technological Trajectories', *Research Policy*, **11**, pp. 147–163.

Dyson, J. (1997) *Against the Odds: An Autobiography*, Orion Business, London.

Ettlie, J. E. (2000) *Managing Technological Innovation*, John Wiley & Sons, NY.

Fiol, C.M. (1996) 'Squeezing Harder Doesn't Always Work: Continuing the Search for Consistency in Innovation Research', *Academy of Management Review*, **24** (4) p. 1012.

Foster, R.J. (1986) *Innovation: The Attacker's Advantage*, Summit Books, NY.

Freeman, C. and F. Louçã (2001) *As Time Goes By: From Industrial Revolutions to Information Revolution*, Oxford University Press, Oxford.

Golley, J. (1996) *Genesis of the Jet: Frank Whittle and the Invention of the Jet Engine*, Airlife, Marlborough.

Henry, A. (1988) *Grand Prix Design and Technology in the 1980s*, Hazleton Publishing, Richmond.

Heppenheimer, T.A. (1995) *Turbulent Skies: The History of Commercial Aviation*, J. Wiley and Sons, NY.

Loch, C.H. and B.A. Huberman (1999) 'A Punctuated Equilibrium Model of Technology Diffusion', *Management Science*, **45** (2), p. 160.

Nordström, K. and M. Biström (2002) 'Emergence of a dominant design in probiotic functional food development', *British Food Journal*, **104** (9), pp. 713–723.

Pinch, T. and W. Bijker (1987) 'The social construction of "artefacts"' in W. Bijker, T. Hughes and T. Pinch (eds) *The Social Construction of Technological Systems: New Directions in the Sociology and History of Technology*, MIT Press, Cambridge, MA.

Rosen, P. (2002) *Framing Production: Technology, Culture, and Change in the British Bicycle Industry*, The MIT Press, Cambridge, MA.

Rosenbloom, R.S. and M. Cusumano (1987) 'Technological Pioneering and Competitive Advantage', *California Management Review*, **29** (4), pp. 51–76.

Sahal, D. (1981) *Patterns of Technological Innovation*, Addison-Wesley, Reading, MA.

Spinardi, G. (2002) 'Industrial Exploitation of Carbon Fibres in the UK, USA and Japan', *Technology Analysis and Strategic Management*, **14** (2), pp. 381–398.

Teece, D.J. (1986) 'Profiting from Technological Innovation: Implications for Integration, Collaboration, Licensing and Public Policy', *Research Policy*, **15**, pp. 285–305.

Tippler, J. (2001) *Lotus 25 and 33*, Sutton Publishing, Stroud.

Tushman, M.L. and Anderson, P. (1986) 'Technological Discontinuities and Organizational Environments', *Administrative Science Quarterly*, **31**, pp. 439–465.

Tushman, M.L. and P. Anderson (2004) *Managing Strategic Innovation and Change*, 2nd edn, Oxford University Press, New York.

Utterback, J.H. (1993) *Mastering the Dynamics of Innovation*, Harvard University Press, Boston, MA.

CHAPTER 5

SOURCES of INNOVATION

OBJECTIVES

When you have completed this chapter you will be able to:

- Review the innovation process
- Distinguish the different ways in which the innovation process can commence
- Analyse the diverse sources of innovation
- Evaluate the relative importance of different sources of innovation.

INTRODUCTION

This chapter focuses on the invention stage of the innovation process. It is concerned with where new ideas for innovation come from. In a sense it is concerned with the 'eureka' moment, when an individual has an idea for a new product or service or figures out how a production process could be dramatically improved. In reality of course innovation rarely starts in that way. Similarly, there are those convinced that innovation is really all about research effort and, providing enough is spent on research and development (R & D), innovations will follow. While R & D effort can lead to innovation, the example of many large corporations such as Xerox with its PARC laboratory in Silicon Valley shows that there is no guaranteed connection between the two. One of the primary reasons for this is that there is a diversity of sources of innovation (see Figure 5.1). This chapter aims to explore this diversity by showing how and when factors like user knowledge, serendipity and an external perspective can be the starting point for the innovation process.

INSIGHT - A FLASH OF GENIUS

Models of the innovation process, especially those that portray the process as linear and sequential, place idea generation as the starting point of the process. Within a very structured new product development process this might well involve the application of a variety of creativity techniques (e.g. brainstorming) designed to throw up a number of possible ideas that can then be developed and evaluated. However, innovation is generally less structured and idea generation is more likely to take the form of a sudden 'insight' that follows months or even years of painstaking research. It is this insight that really lies at the heart of the invention phase of the innovation process. Insight is the starting point that leads to invention and thence to innovation.

Figure 5.1 Insights and Sources of Innovation

Insight is the moment of inspiration when an idea occurs to an individual and forms the basis of an idea. According to Usher (1954) insight is the point where a mental act goes beyond the skill normally expected of some one trained in the field. How does the insight arise? Is it simply a sudden moment of inspiration or are there any patterns one can observe?

There appear to be at least some patterns that form the basis of insight:

- Association
- Adaptation
- Analogy
- Serendipity/Chance

Association

Association involves the bringing together of two apparently unconnected ideas. With association, insights in one field provide solutions in another. There might be two different fields but essentially the same activity can be used to solve a problem. The insight arises as a result of bringing work from two different fields together. Bette Nesmith Graham provides an example of insight arising through association. She was working as a secretary in a Texas bank when she devised a method of removing typing errors. She had noticed that signwriters painted over errors in order to remove them and decided to take a small paintbrush and some white paint to work to see if they would enable her to correct her typing mistakes. The results were sufficiently encouraging for her to try and develop the idea further. Working on a trial-and-error basis in her kitchen, with occasional technical help from her son's chemistry teacher, after five years she had perfected the idea to the point where it was a practical product. When IBM, then the largest manufacturer of electric typewriters, showed no interest in her invention, she set up production on her own in her garage. The business prospered and in time her correcting fluid became a mass-market product sold and used all over the world. Amazingly it even survived the coming of the word processor which should have rendered it obsolete.

Adaptation

Adaptation on the other hand involves taking an existing solution to a problem and adapting it so that it can be used for a different purpose in another field. James Dyson's dual-cyclone bagless vacuum cleaner provides an example of this form of insight (Dyson, 1997). Dust extraction using a cyclone was well established for industrial applications, and it was while working on an industrial dust extraction system designed to remove fine particles of paint from the atmosphere that the idea occurred to him that the cyclone principle could be applied to vacuum cleaners. The amount of adaptation required was considerable. Indeed, the originality of Dyson's system was that it employed not one but two cyclones.

Analogy

Analogy is exemplified where a principle used in one situation is used for a different purpose in another. Georges de Mestral's invention of the Velcro fastener provides an example of analogy lending to an insight. De Mestral was a Swiss engineer who began to think about alternatives to conventional zip fasteners (Van Dulken, 2000). While walking through the woods with his dog one night he noticed how burs stuck to his dog's fur and his own clothing. Curious, he examined the burs under a microscope and discovered that the surface of the bur contained large numbers of tiny hooks. It was these tiny hooks that caught in the strands of fur and the loops of wool on the clothing. After some eight years of effort De Mestral finally developed a method of reproducing the hooks and loops in woven nylon. He called the new product Velcro which was short for *velours croché*, meaning hooked velvet.

Serendipity/Chance

Finally there is chance, or as it is sometimes termed, 'serendipity', where supposedly random occurrences giving rise to a new insight form the basis of an idea that ultimately leads to an innovation. Taylor (1996) gives the example of Alexander Graham Bell and the invention of the telephone. It was in June 1875 while working on improvements to the telegraph system, which enabled messages to be transmitted over telegraph wires using morse code, that one of his assistants by chance over-tightened a clamping screw. When the telegraph was used Bell heard a sound and began to realise how sound waves in the air could be made to vary the strength of electrical current in a wire. It was almost a year later that Bell, having filed a patent, was able to transmit the first words. Nonetheless, the idea that formed the basis of the innovation was the result of the earlier chance event.

Whatever the reason for the insight – whether it is association, adaptation, analogy or even serendipity – a common reaction of most people is to think how obvious the idea is. This high-lights a very important feature of insights: that is, no matter how they arise, the skill, indeed some might say the genius, is being able to recognise their potential value. In Bell's case it was his prior knowledge of the field that enabled him to interpret the event in such a way that he was able to gain an insight. Had others witnessed it, lacking Bell's knowledge and experience, it is highly unlikely that they would have had an insight in the same way. Clearly Bell was fortunate that the chance event occurred in his presence, but one should not under-estimate that what is required is a combination of chance, prior knowledge and experience; only then is there any likelihood of what some might describe as a 'flash of genius'.

SOURCES OF INNOVATION

If the notion of 'insight' is central to the inspiration that lies at the heart of an innovation, it then raises the question of whose insight is it? In other words, can we exercise a degree of precision with regard to the person or organisation that forms the source of the inno-vation? Many potential sources of innovation have been identified. However if one narrows the sources down to mainly persons or organisations, the list of potential sources becomes more manageable:

- Individuals
- Corporate Undertakings
- Users
- Outsiders
- Spillovers
- Process Needs

INDIVIDUALS

In the popular media the individual inventor still reigns supreme. Not only is the typical invention portrayed as something produced by an individual, but the whole exercise is typically portrayed in heroic terms as the individual versus the system.

In fact, a study by Jewkes *et al*. (1969) showed that the individual inventor was alive and well. In a study that covered some 70 important innovations that occurred during the twentieth century, Jewkes *et al*. (1969) found that in around half the cases the source of the innovation was a single person either working on his/her own or at least independent of a corporate undertaking. Only one-third of the innovations had as their source the research laboratory of a corporate undertaking, the remainder being simply difficult to classify. More recent studies (Amesse *et al*., 1991) support the general pattern identified by Jewkes *et al*. (1969).

The resilience of the individual inventor is linked to a number of factors. First, there is the growth in the small-firms sector that has taken place during recent years. Second, a variety of organisational devices, such as strategic alliances, have enabled small firms to work with large firms. Third, innovation is associated with applications of technology and while large firms may be proficient where the development of technology is concerned, small firms will often have greater knowledge of applications. A further factor has been the increased popularity of spin-off companies. Not only do these provide a means for individual inventors to leave the corporate sector and set up on their own in order to develop an innovation, they also, as in the case of Silicon Valley, provide a very powerful role model for would-be innovators. In this context the growth of the venture capital industry over the last three decades has provided a powerful force to support this kind of trend. Finally, as Christensen (1997) has pointed out, the emergence of disruptive new technologies is something to which the corporate sector often finds it difficult to adapt. Consequently, some of the new technologies of the last quarter of the twentieth century have helped promote the cause of the individual inventor/innovator. Not least of these have been some of the technologies associated with computing. The examples of Steve Jobs and Steve Wozniak and the personal computer and Dan Bricklin and the spreadsheet stand as testimony to the success of the individual inventor/innovator in this context.

CORPORATE UNDERTAKINGS

In the twentieth century there appeared to be good reasons for believing that large firms were becoming an increasingly important source of innovation. As we saw in Chapter 1, this was a time when the corporate R & D laboratory became an important source of innovation, reducing the importance of the individual inventor as a source of patents (Mowery and Rosenberg, 1998).

Nor did it apply merely to those innovations industries where the government was an important customer, as in the case of aerospace. In many industries the growth of the mass market put increasing emphasis on the size of the undertaking. Only large firms had the capability to integrate a number of different business functions, or the resources to ensure that increasingly stringent safety requirements were met or the marketing and distribution facilities required to meet the needs of the mass market.

The primacy of large firms as a source of innovation does seem to be borne out in some cases. There are large firms such as 3M or Merck or Intel that not only spend a great deal on R & D as a proportion of turnover, but also have a strong record on innovation. Similarly, there are some industries, such as aerospace and pharmaceuticals where large firms have been the main source of innovation. However, it is important to note that there is no necessary connection between the size of a firm and its record on innovation and similarly there have been large firms that commit large amounts of resources to R & D, which nonetheless exhibit a very poor record of successful innovations.

MINI-CASE: WIDE CHORD FAN

One of the most significant components in a large jet engine is the fan fitted to the front of the engine. Not only does the fan provide much of the engine's power, it also contributes significantly towards its efficiency and hence fuel economy. Rolls-Royce has pioneered innovations in fan technology. It produced the world's first jet engine fitted with a fan, the Rolls-Royce Conway engine, used on both the British VC10 and the American Boeing 707 airliners. In the past the fan comprised a large number of long thin blades. The

▶

blades had to be thin in order to keep the weight down, thereby ensuring that in the event of blade failure it would be contained within the engine. However, thin blades were prone to flexing and had to be braced by means of a 'snubber' or reinforcing ring attached to the mid-point of each blade. This arrangement was effective but had an adverse effect on engine efficiency because the snubber reduced the aerodynamic efficiency of the fan thereby lowering the overall fuel efficiency of the engine. Rolls-Royce's chief designer Adrian Lombard recognised that the use of blades with a 'wide chord' (i.e. substantially wider blades than the thin ones then in use) would be more rigid and therefore eliminate the need for the snubber. This would make the engine more fuel efficient, a feature that offered a significant competitive advantage when it came to persuading airlines to buy the engine. The problem was to find a material that was both light and rigid. The company developed a carbon fibre blade but unfortunately this proved unable to withstand the impact of bird strikes and had to be replaced with a thin solid titanium blade.

Despite this setback Rolls-Royce's scientists and engineers persisted with their research into wide-chord blades. During the 1980s after many years of research they came up with an innovative hollow titanium wide-chord blade. It was made from two sheets of titanium with a honeycomb section in between and proved both light and strong. Introduced on the Rolls-Royce 535E4 engine that powered the Boeing 757, the wide-chord blade not only had a more efficient aerodynamic profile, but was sufficiently rigid to dispense with the need for a snubber. The wide-chord fan fitted to the 535E4 engine was 92 per cent efficient compared to the fan fitted to earlier versions of the engine which was only 88 per cent efficient. As a result the 535E4 engine gave major savings in fuel over other similar engines.

Rolls-Royce then applied the same wide-chord fan blade technology derived from its research to other engines in its product portfolio. Over the years the company extended its research into wide-chord fan blades to develop wider and more efficient blades, culminating in the 'swept fan'-type blade for the Rolls-Royce Trent 900 engine fitted to the Airbus A380 that flew for the first time in April 2005.

Source: Smith and Rogers (2004)

USERS

The idea that users can be an important source of innovations is particularly associated with the pioneering work of Von Hippel (1976). He was among the first to show that in certain industry sectors, users play a critical role not only in generating ideas for innovations, but also in their subsequent development. Von Hippel's work focused on sectors producing scientific equipment, specifically instruments used for gas chromatography, nuclear magnetic resonance spectrometry, ultraviolet absorption spectrometry and transmission electronic microscopy. Von Hippel was able to show that in each case the idea that formed the basis of the instrument came from, and was initially developed by, a user who was a member of the scientific community. Only when the idea had been developed into a working prototype was it transferred to a manufacturing company for commercial production.

It is significant that Von Hippel focused on scientific equipment as an industry sector. This sort of equipment is widely used in scientific research and as Rothwell (1986) points out, in this field scientific researchers form the focal point of state-of-the-art expertise. In addition the nature of their work – research – means that they often have to construct new kinds of equipment in order to allow them to move forward the frontiers of knowledge. Experimentation by its nature requires monitoring and measuring equipment, and new forms of experimentation may well require new forms of equipment. Similarly, in chemistry, chemists working in government laboratories and universities often need to devise new forms of analytical equipment in order to further their research. As Von Hippel points out, manufacturers of scientific instruments are simply not sufficiently closely involved in scientific research to perceive or predict the new requirements in the field which would enable them to make the initial invention. Similarly, users do not possess the capability to make scientific instruments so that once they have developed a working prototype they then turn to manufacturers to produce the equipment in quantity.

Work by Shaw (1985, 1998) showed a similar pattern of activity in the medical equipment sector in the UK. Shaw utilised a sample of 34 medical equipment innovations undertaken by 11 companies. In this case the users were clinicians or physicians usually working as consultants within particular medical specialisms who typically used medical equipment to facilitate improved diagnosis or therapy for patients. The innovations included such things as neonatal oxygen monitoring systems, portable, battery-operated, variable-speed syringe drivers and respiratory recording and monitoring systems (Shaw, 1998).

Figure 5.2 Medical Equipment Innovation Network

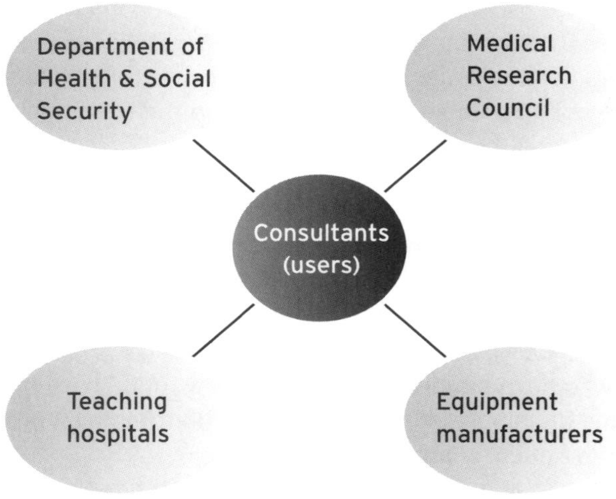

Source: Shaw (1998)

Shaw found that in the medical equipment field the innovation process involved a network of actors (see Figure 5.2), including consultants, small medical equipment manufacturing companies, the Medical Research Council (MRC), the Department of

Health and Social Security (DHSS) and teaching hospitals. However the 'prime actor' in the network was the user, that is to say, the consultant. According to Shaw (1998) the innovation process typically began with the consultant specifying a need. He or she would then outline the conceptual framework of a piece of equipment that could meet this need. Then, working with technicians from the hospital where the consultant was based, a hand-built prototype would be devised. Funding for the development of a prototype typically came from the Department of Health and Social Security or the Medical Research Council. The prototype would then be tested and evaluated by the consultant in the hospital. If the prototype showed promise, medical equipment manufacturers would become involved with a view to putting the prototype into production. At this point there would also be some form of market assessment. Consultants typically played a crucial role in this by seeking the views of fellow consultants in their field who were lead-users and as such had the power to specify the equipment needs of hospitals, especially teaching hospitals. In this context Shaw (1998: p. 440) noted that 'consultants acted as both market and technological gatekeepers'. When the medical equipment manufacturers had produced the first examples of the new equipment, it was then sent out for operational testing in hospitals. The use of teaching hospitals, especially ones that were Medical Research Council-designated centres, played an important part in ensuring the credibility of the equipment, since assessments of quality were often based not just on functional utility but on 'the stature and networking ties of the consultant who had carried out the tests and published the results'. Co-operation with consultants (i.e. users) also helped the medical equipment manufacturers to understand the constraints of DHSS budgets which could be critical in terms of their timing of the launch of new equipment.

Thus, users in the form of consultants not only formed the source of innovation, they also contributed to other aspects of the innovation process through using their professional networks to access rich sources of information. What remains clear however is that in this field the users are the primary source of innovation. Manufacturers are insufficiently close to developments in medical science to be the source of innovation. Rather consultants, especially those based in designated centres of excellence which are working on state-of-the-art techniques are the prime actors in innovation and the ones who act as the source of innovation.

OUTSIDERS

A consistent feature of innovation over many years has been the substantial proportion derived not from those working in a given field, be it an industry or a particular sector, but from outsiders who have hitherto had little to do with it. There is a case for arguing that outsiders provide an important source of innovation. Table 5.1 provides some examples of innovations developed by outsiders.

To what extent were these individuals outsiders? Chester Carlson, the inventor of the photocopier, worked for an electrical company analysing patents (Van Dulken, 2000). Steve Jobs and Steve Wozniak the pioneering Apple II computer innovators (Campbell-Kelly, 2003a) were college drop-outs and although Wozniak worked for Hewlett-Packard, he was in calculators not computers. John Barnard the designer of the McLaren MP4, the world's first carbon-fibre racing car, was new to Formula One, having previously worked in the US on Indycars (Cooper, 1999). Finally Jeff Bezos, who pioneered Internet-based

Innovation	Company	Innovator	Date
Photocopier	Haloid Corporation	Chester Carlson	1938
Personal computer	Apple Computer	Steve Jobs & Steve Wozniak	1977
Carbon fibre F1 racing car	McLaren-International	John Barnard	1981
Internet bookstore	Amazon.com	Jeff Bezos	1995

Table 5.1 Outsiders as Innovators

retailing through the creation of Amazon.com, was a fund manager in the financial services industry (Cassidy, 2002).

None of these individuals worked in the field where they were to achieve success as innovators. They were not part of the community in which their innovation was based and in that sense they were outsiders.

What do outsiders have that industry insiders lack? In analysing why outsiders are an important source of innovation it is worth exploring the role of industry insiders. Within any industry there will tend to be what Galbraith (1958: p. 35) describes as the 'conventional wisdom', which comprises 'the ideas which are esteemed for their acceptability', and in the context of innovation is likely to extend to accepted ways of approaching and picturing issues and problems. In short, assumptions may well be deeply embedded and as such go unquestioned. Where people work in a group of like-minded specialists or belong to such groups, the phenomenon may be even more pronounced. Similarly, groups can be insulated from the world around them with a collective perspective that leads all too easily to assumptions and ideas going unquestioned. The insularity may be worse if firms have close relationships with their customers. It was noted by Christensen (1997) that, in some circumstances, paying close attention to customers and customer needs may actually be counterproductive for innovation, in so far as it leads to greater insularity as the firm's outlook becomes more specialised and ignores some of the wider trends in technology and potential customers.

Outsiders may be able to avoid at least some of these pitfalls. Since they are not part of an established community, they are likely to have fewer inhibitions when it comes to challenging accepted ideas. Similarly, outsiders may be more willing to try unorthodox ideas precisely because they are not familiar with the 'conventional wisdom'. The case of Chester Carlson illustrates this well. When trying to develop a means of copying he found that the conventional wisdom prescribed chemical methods for reproducing photographs, a field closely related to his experimental work on document copying. Lacking expertise in chemistry Carlson was obliged to pursue a completely different direction involving the use of an electrical method which he later called 'electrophotography' (Van Dulken, 2000). As well as being willing to try unorthodox ideas and approaches, outsiders may also be more willing to try simple ideas.

However, one of the biggest advantages enjoyed by outsiders is that they often have external contacts in fields which may be unrelated but nonetheless prove useful. A feature of these contacts is likely to be their diversity, enabling the innovator to draw from a relatively wide knowledge base. We saw in Chapter 4 that this was true of John Barnard when he designed the McLaren MP4 racing car. The established practice was for the chassis of a racing car to be constructed from sheets of aluminium riveted together to form a tub. This was light, strong and relatively easy to build. While carbon fibre had been used on racing cars, it had only been used for single components such as the aerodynamic wing at the rear of the car. The idea of building the whole chassis (which was the main part of the car) from carbon fibre was revolutionary. However, Barnard had a friend who worked for British Aerospace in Weybridge where the Harrier jet was built. The friend explained how carbon fibre was being used in the aerospace industry. In this way Barnard was able to confirm that his idea was feasible. Unfortunately, it still proved impossible to get any one in the UK to undertake the construction of anything as big as a car chassis from carbon fibre. At this point Barnard was again able to call on one of his contacts outside Formula One, in this case a former colleague from the US, who suggested that Hercules Aerospace of Salt Lake City, Utah, who built guided-missile components using carbon fibre, might be able to help (Henry, 1988). Hercules Aerospace not only had the expertise and the facilities, they also had an R & D section willing to undertake one-off jobs.

Thus outsiders may possess a range of advantages over industry-insiders. Not only are they likely to be more open to new approaches and willing to challenge existing ideas, but the range of external contacts they can draw on means their absorptive capacity, as far as external linkages are concerned, is likely to be greater too.

SPILLOVERS

Spillovers typically occur when a firm benefits from another firm's investment in R & D. The nature of the spillovers can vary, but they might for instance result from one firm making an investment in R & D that leads to a scientific discovery or the development of a new product that other firms are able to imitate or copy. Alternatively, if the firm that has developed the new product chooses not to commercialise it, it might license it to others. Either way a firm other than the one that made the initial investment is able to bring an innovative new product to market.

Two examples illustrate this source of innovation. Dan Bricklin and his company Software Arts developed VisiCalc, the world's first spreadsheet in the late 1970s. However, the spreadsheet idea was soon copied by other firms including Lotus with its spreadsheet 1-2-3, Borland with Quattro and Microsoft with Multiplan (later re-named Excel). Although Software Arts was the first mover (being the first to get a product to market) it was ultimately overtaken first by Lotus. Quite literally after Software Arts had developed the spreadsheet and proved the concept in the marketplace, the idea then spilled over and became public knowledge to be taken up by others. (At the time software could not be patented in the US, unlike the situation today.) Similarly when Du Pont developed Teflon in the late 1950s, they chose not to become heavily involved in applications, on the grounds that this was not their area of expertise. Instead it was left to Bill Gore, a researcher at Du Pont, to go it alone and develop fabrics using Teflon, in particular the high-performance, weatherproof fabric Gore-Tex.

Spillovers are likely to occur in situations where it is difficult to prevent others from appropriating the benefits from an invention. Intellectual property rights are the means by which inventors normally endeavour to prevent others appropriating benefits. Success is dependent on being able to engage a tight appropriability regime. Sometimes despite best endeavours this proves difficult. Software Arts was hampered by the fact that, at the time it developed VisiCalc, software could not be patented and the Software Arts' founders had to rely on copyright. A similar thing happened with EMI's CAT scanner where the principle was relatively easily understood and complementary assets such as training, product support and servicing proved key features of competitive advantage (Teece, 1986). Under these circumstances, even though patents were employed, it was not possible to prevent knowledge spilling over into the public domain for other firms to take up. Of course, as we saw in the previous chapter, some firms have a deliberate policy of innovating through the use of spillovers. Such firms tend to allow others to pioneer a new product and then, employing a 'follower' strategy, they move in to undertake successful innovation.

MINI-CASE: MOSAIC – THE USER FRIENDLY WEB BROWSER

'From: Marc Andreessen (marca@ncsa.uiuc.edu)
Sat, 23 Jan 93 07:21:17-0800
By the power vested in me by nobody in particular, alpha/beta versions of 0.5 of NCSA's Motif-based information systems and World Wide Web browser, X Mosaic, is hereby released,
Cheers
Marc'

With this message on 23 January 1993, a young undergraduate computer science student at the University of Illinois at Urbana-Champaign, announced the launch of the world's first web browser for non-professional computer users (Naughton, 1999: p. 245). It was the brainchild of Marc Andreessen, an undergraduate student working part time at the National Center for Supercomputing Applications (NCSA), and as an innovation it was to prove the 'Killer App' of all time.

While earlier browsers were line-mode browsers that required a considerable knowledge of computing to operate, Mosaic was graphics-based, displayed Web pages inside windows, employed scroll bars, buttons and menus, and had 'Back' and 'Forward' buttons to facilitate movement between pages. Unlike earlier browsers it was easy to use, easy to install, and able to operate on a variety of operating systems including Microsoft Windows. It was friendly and it was free. Mosaic spread like wildfire, and within months there were hundreds of thousands of users. Mosaic not only proved the concept of a user-friendly browser, it helped immensely in popularising the World Wide Web. Before Mosaic, navigating the Internet was nearly impossible for non-professional computer users. Significantly, professional users appeared to have little interest in changing this situation. The arrival of Mosaic meant that any one could surf the net. As Table 5.2 shows, just about any one did. Within two years the volume of Internet traffic comprising web pages went from virtually nothing to a quarter of the total. As Naughton (1999: p. 247) puts it, 'after Mosiac appeared the Web went ballistic'.

Date	Ftp %	Telnet %	Netnews %	Irc %	Gopher %	Email %	Web %
May 1993	42.9	5.6	9.3	1.1	1.6	6.4	0.5
Dec 1993	40.9	5.3	9.7	1.3	3.0	6.0	2.2
June 1994	35.2	4.8	10.9	1.3	3.7	6.4	6.1
Dec 1994	31.7	3.9	10.9	1.4	3.6	5.6	16.0
Mar 1995	24.2	2.9	8.3	1.3	2.5	4.9	23.9

Table 5.2 Internet Traffic by Protocol 1993–95
Source: 'A Brief History of the Future: The Origins of the Internet', by John Naughton, Weidenfeld & Nicolson, a division of The Orion Publishing Group

Ironically, while the University of Illinois at Urbana-Champaign made this possible, the university obtained comparatively little direct benefit. When a commercial browser, called Netscape Navigator and owing much to Mosaic, was launched by a company set up by Jim Clark the founder of Silicon Graphics and employing Marc Andreessen and other members of the Mosiac team, the university received $1 million in compensation, but when the Internet took off, Netscape Communications, the company involved, like many Dot.com companies, was soon worth more than $1 billion.

Sources: Cassidy (2002); Naughton (1999)

PROCESS NEEDS

Sometimes the demands of a manufacturing process will act as a stimulus to innovation. This is probably more likely to be true of mature industries where the pressure of competition will be a powerful force for efficiency savings. Manufacturing firms that compete on price (i.e. pursuing a 'cost leadership' strategy) will always be keen to find new ways of lowering the cost of production.

Two well-known examples illustrate this process at work, in particular the way that the requirements of the process can be a source of innovation. As noted in Chapter 2 Henry Ford's development of the moving assembly line at his manufacturing plant at Highland Park in Detroit came as part of his drive to lower costs and boost sales (Womack *et al.*, 1990). Alistair Pilkington's development of the 'float glass' process in the 1960s and 1970s fits a similar pattern. At that time the manufacture of plate glass, which was increasingly being used in construction, was expensive and time consuming because it required sheets of glass to be subjected to grinding and polishing in order to obtain a flat service. The float glass process developed by Pilkington did away with these stages in the production process. Instead glass was drawn directly from the furnace over a bed of molten tin. Although it took several years to perfect this innovation, it did eventually result in plate glass being produced much more quickly and at much lower cost. So great was the improvement that Pilkingtons were able to license the process to other glassmakers and it is still widely used today. The significance of float glass as an innovation is that it is a clear case of a process need, in this case effectively a production bottleneck, being the source of the innovation. As is often the case it was the existence of a bottleneck that provided the stimulus to innovate.

CHAPTER 5: SOURCES OF INNOVATION

WHOSE INNOVATION IS IT ANYWAY?

This chapter has outlined a diverse range of possible sources of innovation. Judgements about the relative importance of these sources are difficult. There can be little doubt that in the late nineteenth and early twentieth centuries innovation was very much the province of individuals. Individuals had ideas for new products and by and large they were the ones who turned them into successful innovations, often creating and building a successful company in the process.

By the mid-twentieth century things had changed. Increasingly the corporation was the source of innovation. As innovation became more technical and relied more and more on breakthroughs in science, so in the 'modern' era large corporations, able to fund research laboratories, were the source of innovation.

In the last quarter of the twentieth century further change could be observed in some industry sectors. The example of Silicon Valley and other clusters of high-technology activity increasingly points to small firms founded by technical entrepreneurs as an important source of innovation.

While it does seem there has been a revival of interest in small firms, it would be wrong to see this as 'putting the clock back'. Significantly, the context, particularly the institutional context, has changed. The new institutional context increasingly makes it possible for organisations to externalise innovation. Consequently, one increasingly sees outsiders/external parties as a source of innovation. However, outsiders innovate less and less in isolation. Institutional changes have brought forth a range of co-operative arrangements including: alliances, joint ventures, technology agreements, sub-contracting and licensing. These have created the conditions where it is feasible for outsiders – including individuals and small firms – to engage in innovation in collaboration with large organisations.

Hence the message of this chapter should perhaps be that the relative importance of sources of innovation is constantly changing, and in any event is likely to vary according to the industry or industry sector involved.

CASE STUDY: VISICALC - THE FIRST ELECTRONIC SPREADSHEET

Computing in the 1970s

Once upon a time there were no computers. Well, there were computers but you never saw them. Back in the 1970s computers were big machines housed in special rooms with a small army of programmers and systems analysts to look after them. They were used for routine standardised tasks, such as payroll – calculating each employee's wages and making appropriate deductions. It was a task well suited to the computer's capability to do large numbers of calculations in a short space of time. Payroll software was usually bespoke, that is to say tailored to the specific requirements of each application, which was why they required support staff. Then, towards the end of the decade the arrival of the microprocessor, an integrated circuit which contained all the essential parts of a computer's central processing unit contained on a single chip, led to a dramatic change in computing technology as the first so-called 'microcomputers' began to appear.

The Microcomputer

The first microprocessor-based computer was the Altair 8800 manufactured by Micro Instrumentation Telemetry Systems. It was sold in kit form for assembly by computer hobbyists, and was featured on the cover of Popular Electronics in January 1975, when the magazine observed that it was now possible to buy a computer for $400 (Campbell-Kelly, 2003a). The launch of the Apple II in April 1977, the product of the embryonic Apple Computer firm formed the previous year by computer hobbyists Steve Jobs and Steve Wozniak, marked a significant step towards what we would now recognise as a personal computer. The Apple II looked like a computer terminal and consisted of a keyboard, CRT display screen and a CPU all in one package. It was soon followed by a host of imitators. Significantly, the existing computer manufacturers like IBM, Honeywell and DEC, who had their sights still firmly set on the needs of their existing corporate customers, did not enter the microcomputer field at this time.

One of the great attractions of these early machines was that, unlike their much larger counterparts, they operated in real time. Most large machines at that time operated on the basis of batch processing where your task for the computer was not entered via a keyboard but by punched cards or paper tape and joined a queue of other jobs, so that the results were not available until perhaps several hours later. This was fine for standardised routine tasks such as payroll, but immensely time consuming otherwise. Being able to cut out the queuing associated with batch processing and work in real time was immensely attractive for those doing non-routine, non-standardised tasks such as programming or scientific work.

The Software Bottleneck

The new microcomputers were not without their limitations. As an article by Ben Rosen in the *Electronics Newsletter* observed at the time, hardware developments had outpaced developments in software. Most hobbyists did their own programming and were not too bothered, but for the professional microcomputer user, such as someone running a small business, the absence of anything like general-purpose software applications, meant the new computers were of limited value even if, like the Apple II, they were accessible, available and relatively cheap. Unlike hobbyists, potential business and professional users of the microcomputer did not want to write software, they wanted to use software written by some one else which they could purchase and start using straight away without having to do any programming themselves.

If the appearance of the Altair 8800 on the cover of *Popular Electronics* was one of the defining moments in the birth of the personal computer, so too was the arrival of a software innovation – VisiCalc, the spreadsheet, described by Campbell-Kelly (2003a: p. 212) as, 'one of the heroic episodes in the history of the personal computer'. In 1978 an American, Dan Bricklin, first came up with what eventually became VisiCalc, the first spreadsheet. Interestingly, Bricklin was not a computer specialist, nor was he employed by a computer company. He graduated in engineering from Massachusetts Institute of Technology (MIT) in 1973. Upon graduation he went to work for the computer firm DEC. Four years later he went back to university enrolling at the Harvard Business School in order to re-train for a career in the financial services industry. It was as a student that he conceived the idea of what was to become VisiCalc. At Harvard he came across old-fashioned pencil-and-paper spreadsheets, which were large sheets of paper divided into columns for financial modelling. Using these sheets was laborious and time consuming because any changes in the parameters of a model meant rubbing out figures, recalculating them and writing in new ones. Bricklin's initial idea was simple. The sheet would be displayed on-screen and the user would be able to enter data directly, just as if writing figures on a sheet of paper. Having done that, the user had merely to highlight the requisite figures with the cursor to get the sum (or whatever other mathematical operation they wanted). Then, any changes in the parameters of a financial model could be made just by typing in new numbers with recalculation taking place automatically. In the ensuing weeks Bricklin sketched out his ideas as designs on paper. These early designs were mostly to show how the various instructions to the computer would be executed in programming terms and what the computer screen would look like. Not until he was reasonably clear about the operations required did he start programming.

The first prototype was developed by Bricklin over a weekend in the autumn of 1978, using an Apple II computer borrowed from Dan Fylstra (who was later to publish the software). The software was written in Apple BASIC. As a prototype it did not have a scroll facility, but most of the other recognisable features of what we would today recognise as a spreadsheet, such as columns and rows and the capability to do arithmetic, were present. Instead of a mouse the new software used the games paddle on the computer (a legacy of the Apple II's appeal to hobbyists) to move the cursor around the screen. To test the prototype Bricklin used it to model a bank account where payments in and out over time result in variations in the account balance.

VisiCalc – The First Electronic Spreadsheet
Having proved the concept with this first prototype, Bricklin teamed up with a friend, Bob Frankston to develop a commercially viable product. On 2 January 1979 they formed a company, Software Arts Inc. Software Arts then struck a deal with Dan Fylstra, the founding editor of the well-known computing magazine *BYTE* for his publishing company, Personal Software, to publish the new software when it was finally ready for sale. As part of this deal Bricklin and Frankston, as the authors of the software, were to get a royalty of 35.7 per cent on gross sales. As someone with an MBA, Fylstra could see the potential of the new software for financial modelling. It was Fylstra and Frankston who between them came up with the name for the new software: VisiCalc. In the meantime Bricklin, who was still a student at Harvard, continued with development work on the new software.

The first application of the VisiCalc spreadsheet came when Bricklin used it for one of his class assignments, the Pepsi-Cola case study. Bricklin used VisiCalc to do 5-year financial projections that tested a variety of different strategies. Unfortunately, this early version of VisiCalc did not yet have a capability for printing the results, so Bricklin was forced to copy the results off the screen by hand. Despite this his professor was very impressed. Had he used a calculator it would have taken Bricklin much longer to produce the various different scenarios that arose from the different strategies.

Product Launch

The first advertisement for VisiCalc appeared in the computer magazine *BYTE* in May 1979. The advertisement said nothing about VisiCalc. There was no description of the product and nothing to say what it could be used for or what it would do. Instead the advertisement merely posed a question: VisiCalc – How did you ever do without it?

VisiCalc was announced to the public at the National Computer conference in New York in June 1979, though it was to be August before it went on the market. Bob Frankston delivered a paper at the personal computer part of the conference. This was very much a sideshow housed in a hotel near the main conference hall. Most of the 20-strong audience were friends and family. The paper attracted very little attention mainly because it was not about computer hardware or aspects of advanced programming. At the conference Bricklin and Frankston met Bill Gates whom Bricklin described as, 'a young kid well known for his version of BASIC and speeding tickets'. Two days after the conference Bricklin graduated from Harvard.

Another person that Bricklin and Frankston met at the New York conference was Ben Rosen an analyst at Morgan Stanley. He was not slow to spot the potential of VisiCalc and the spreadsheet concept. In a short article in the *Electronics Newsletter* he described VisiCalc as an 'electronic blackboard' where you write out what you want to do, press a key and the software automatically does all the calculations and displays the results. Similarly updates and changes could be made quickly and easily. However, Rosen noted that probably the most outstanding feature of the software was that you did not need to know anything about computers to use it effectively. Prophetically he wrote, 'at $100 VisiCalc could emerge as one of the bargains of our time', adding, 'VisiCalc could some day become the software tail that wags (and sells) the personal computer dog'.

Using some of the early royalty prepayments from Personal Software, Bricklin and Frankston were able to rent a small office and with money borrowed from the bank bought a Prime minicomputer on which they could run software development tools that enabled them to finish the first commercial version of VisiCalc. The first commercial version of VisiCalc, version 1.35, was delivered to the first few customers in the late summer of 1979. Bricklin hand typed the labels. The first real release was version 1.37 delivered in mid-October 1979. At last Bricklin and Frankston had a proper product. Like a book, VisiCalc came in a brown vinyl binding holding the manual, complete with a $5\frac{1}{4}$-inch disk, reference card and registration card. Most of the early versions of VisiCalc were for the Apple computer. VisiCalc retailed initially for $100 and in the first few months sales at 500 copies a month were strong. However, favourable press comments combined with word-of-mouth recommendations propelled sales to 12 000 per month in the second half of 1980 despite ratcheting up the price to $250 (Campbell-Kelly, 2003a: p. 214).

The Personal Computer is Born

VisiCalc made it possible for the first time for ordinary people, especially people with no knowledge of computers or programming, to use a computer to do complete business-related tasks such as drawing up a budget or a business plan without the need for computer specialists. In many respects it was VisiCalc that made the Apple an outstanding success as the first personal computer. Apple sales went from $2.7 million in 1977 to $200 million in 1980 and more then one-third of a billion dollars the following year. It was so successful that eventually even IBM got into the business of personal computers when in 1981 it brought out its IBM Personal Computer.

Product	Application	Publisher	Units sold (cumulative)	Retail price	Retail value
VisiCalc	spreadsheet	Personal Software Inc	700 000	$250	$175m
Wordstar	word processor	MicroPro	650 000	$495	$325m
SuperCalc	spreadsheet	Computer Associates	350 000	$195	$75m
PFS: File	database	Software Publishing	250 000	$140	$35m
Dbase II	database	Ashton-Tate	150 000	$695	$105m
1-2-3	spreadsheet	Lotus Development	100 000	$495	$50m
EasyWriter	word processor	Information Unlimited	55 000	$35	$19m

Table 5.3 Best-selling Software Applications 1983
Source: Martin Campbell-Kelly, 'From Airline Reservations to Sonic the Hedgehog: A History of the Software Industry', MIT Press, p. 215

Although VisiCalc was itself a big success, difficulties surrounding the protection of intellectual property, meant VisiCalc soon had many imitators (see Table 5.3). In time the imitators became better known. One of the most successful was Lotus 1-2-3, produced by the Lotus Development Corporation which eventually bought out Software Arts, and in 1985 Lotus quietly decided not to continue publishing VisiCalc. In time Lotus 1-2-3 gave way to Microsoft's version of the spreadsheet, Excel. It offered the advantage that it was compatible with and linked to Microsoft's best-selling, word-processing package, Word. Ironically, Software Arts' inability to protect its intellectual property and prevent imitators meant that the diffusion of the spreadsheet as an innovation occurred more rapidly. What Dan Bricklin lost in terms of his ability to appropriate the benefits from his intellectual property, the rest of us gained in terms of the speed of diffusion of this important innovation.

Source: Campbell-Kelly (2003a; 2003b); Bricklin (2005)

QUESTIONS
1 Why didn't existing computer firms rush in to provide solutions to the software bottleneck?
2 Why did Ben Rosen's observation that, 'VisiCalc could some day become the software tail that wags the personal computer dog', prove prophetic?
3 Why did the spreadsheet attract very little interest at the National Computer conference in 1979?
4 Why do you think the first advert for VisicCalc carried nothing more than the slogan 'VisiCalc – How did you ever do without it?'

5 Which of the sources of innovation presented in this chapter most appropriately identifies the origins of this innovation and why?

6 What was Bricklin's relationship with the computer industry when he came up with his idea?

7 To what extent was Bricklin helped by his previous experience or lack of it?

8 Why are outsiders sometimes better placed than those within an industry to come up with innovations?

QUESTIONS FOR DISCUSSION

1 **Which of the four forms of insight do you consider is the most important and why?**

2 What factors can you identify that have led to a resurgence of innovation by individuals?

3 **How can the diffusion of an innovation be affected by its source?**

4 In what sorts of industries is one more likely to find corporate innovation and why?

5 **Which of the four theories of innovation gives prominence to outsiders as a source of innovation?**

6 Why are industry insiders sometimes inhibited from engaging in innovation?

7 **What reasons did Christensen put forward for industry insiders being less innovative?**

8 Give an example of a recent innovation undertaken by a network of individuals/organisations?

9 **What has been von Hippel's contribution to our ideas about innovation?**

10 What do we mean by a spillover and how can spillovers contribute to innovation? Use an example to illustrate your answer.

11 **If the sources of innovation are diverse what are the implications of this for companies keen to innovate?**

12 How significant is serendipity/chance as a source of insight that can lead to innovation?

ASSIGNMENTS

1 Take two examples of innovation and compare and contrast the sources of innovation in each case.

2 Using an example of your choice, prepare a short case study of an individual innovator, showing how the initial idea emerged and analysing how this eventually became a successful innovation.

3 Compare and contrast the approach to innovation taken by two large corporations. Account for the differences in approach that you identify.

4 Compare and contrast the 'modern' and 'postmodern' approaches to innovation.

5 Why are outsiders an important source of innovation?

RESOURCES

Books

There are few books that deal specifically with sources of innovation. Two that do are Von Hippel (1998) which focuses on functional sources of innovation such as users, manufacturers, and suppliers, and Leonard (1995) which focuses on external sources.

Other books that deal with sources of innovation indirectly are those that offer a 'compendium' approach to innovation, that is to say they provide detailed accounts of a number of innovations. Among the most useful is Van Dulken (2000) which covers 100 innovations made during the course of the twentieth century and includes a range of well-known household objects like the Rawlplug, Formica, and the safety razor as well as innovations that are more of an industrial nature. Similarly Jewkes *et al.* (1969) give details of a large number of well-known innovations. The book by Davis (1987) is similar but the range of innovations covered is international and includes several comparatively recent innovations. Nayak and Ketteringham (1986) covers rather less in the way of innovations but has the advantage that each of the innovations that it does cover is dealt with in substantially more detail. This is particularly useful as it provides greater scope for analysing the source of the innovation in each case.

Websites

The website*: http://www.bricklin.com/* includes many photos not only of the personalities involved but also the products and the places. If you would like to have a go with VisiCalc the website even includes a downloadable copy of an early IBM PC version of VisiCalc from early 1981. It runs on most modern PCs under MS-DOS in Windows. In addition many university libraries still have a variety of manuals and user guides to VisiCalc! They make interesting reading. So too do journals/trade publications with a long history such as *BYTE*.

REFERENCES

Amesse, F., C. Denranleau, H., Etemad, Y. Fortier and L. Seguin-Dulude (1991) 'The individual inventor and the role of entrepreneurship: A survey of the Canadian experience', *Research Policy*, **20**, pp. 13-27.

Beije, P. (1998) *Technological Change in the Modern Economy: Basic Topics and New Developments*, Edward Elgar, Cheltenham.

Bricklin, D. (2005) Dan Bricklin's website, available online at www.bricklin.com (accessed 5 June 2005).

Campbell-Kelly, M. (2003a) *From Airline Reservations to Sonic the Hedgehog: A History of the Software Industry*, MIT Press, Cambridge, MA.

Campbell-Kelly, M. (2003b) 'The rise and fall of the spreadsheet', in M. Campbell-Kelly, R. Croarken, E. Flood and M. Robson (eds) *The History of Mathematical Tables: From Sumer to Spreadsheets*, Oxford University Press, Oxford.

Cassidy, J. (2002) *Dot.con: the greatest story ever told*, Penguin Books, Harmondsworth.

Christensen, C.M. (1997) *The Innovator's Dilemma: When New Technologies Cause Great Firms to Fail*, Harvard Business School Press, Cambridge, MA.

Cooper, A. (1999) 'The Material Advantage', *Motor Sport*, **LXXV** (3), March 1999, pp. 32-37.

Davis, W. (1987) *The Innovators*, Ebury Press, London.

Dyson, J. (1997) *Against the Odds*, Orion Books.

Galbraith, J.K. (1958) *The Affluent Society*, Penguin Books, Harmondsworth.

Henry, A. (1988) *Grand Prix Car Design and Technology in the 1980s*, Hazleton Publishing, Richmond.

Henry, J. and D. Walker (1991) *Managing Innovation*, Sage Publications, London.

Jewkes, J.D., D. Sawers and R. Stillerman (1969) *The Sources of Innovation*, Macmillan, London.

Leonard, D. (1995) *Wellsprings of Knowledge: Building and Sustaining the Sources of Innovation*, Harvard Business School Press, Boston, MA.

Mowery, D.C. and N. Rosenberg (1998) *Paths of Innovation: Technological Change in 20th century America*, Cambridge University Press, Cambridge.

Naughton, J. (1999) *A Brief History of the Future: The origins of the internet*, Phoenix, London.

Nayak, P.R. and J.M. Ketteringham (1986) *Breakthroughs! How Leadership and Drive Create Commercial Innovations That Sweep the World*, Mercury Business Books, Didcot.

Rothwell, R. (1986) 'Innovation and Re-innovation', *Journal of Marketing Management*, **2** (2), pp. 109-123.

Shaw, B. (1985) 'The Role of the Interaction Between User and the Manufacturer in Medical Equipment Innovation', *R & D Management*, **15** (14) pp. 283-292.

Shaw, B. (1998) 'Innovation and new product development in the UK medical equipment industry', *International Journal of Technology Management*, **15** (3-5) pp. 433-45.

Smith, D.J. and M.F. Rogers (2004) 'Technology Strategy and Innovation: The Use of Derivative Strategies in the Aerospace Industry', *Technology Analysis and Strategic Management*, **16** (4) pp. 509-527

CHAPTER 5: SOURCES OF INNOVATION

Taylor, E. (1996) *An Introduction to Innovation:* Block 1, T302: Innovation, Design, Environment And Strategy, Open University, Milton Keynes.

Teece, D.J. (1986) 'Profiting from technological innovation: Implications for integration, collaboration, licensing and public policy', *Research Policy*, **15**, pp. 285–305.

Usher, A.P. (1954) *A History of Mechanical Inventions*, rev. edn, Harvard College, Boston, MA.

Van Dulken, S. (2000) *Inventing the 20th Century: 100 Inventions that Shaped the World*, British Library.

Von Hippel, E. (1976) 'The dominant role of users in the scientific instruments innovation process', *Research Policy*, **5** (3).

Von Hippel, E. (1988) *The Sources of Innovation*, Oxford University Press, Oxford.

Womack, J.P., D.T. Jones and D. Roos (1990) *The Machine that Changed the World*, Rawson Associates, NY.

the process of innovation

OBJECTIVES

When you have completed this chapter you will be able to:

- Differentiate commercialisation from invention
- Describe the various sources of innovation
- Distinguish the steps in the innovation process
- Differentiate and distinguish the different activities associated with the process of innovation
- Evaluate the techniques available to facilitate the process of innovation
- Differentiate and evaluate the five models of the innovation process.

INTRODUCTION

Inventions and inventors attract a high public profile. There are for instance a large number of popular books about inventions (Brown, 1996; Dyson and Uhlig, 2001; Van Dulken, 2000) and many inventors, such as James Watt and Thomas Edison, are well-known figures. Yet many inventions are commercial failures. For a variety of reasons they fail to make it in the marketplace. Some, like the Sinclair C5 electric car or the De Havilland Comet jet airliner, or the Ford Edsel, are well known, but most have long since disappeared into obscurity.

That many inventions fail commercially serves to highlight the fact that for an invention to be a success, those behind it have to undertake a lengthy and difficult process of innovation.

The innovation process is concerned with the various activities that have to be undertaken in order to turn an invention into a commercial product or service, which consumers, be they individuals or firms, will purchase in large numbers.

This chapter explores the nature of this process. In particular the various activities involved with the exploitation of inventions to make them into commercially viable products and services are examined and explored. In this way the chapter explains what is involved in carrying out innovation. One might say it looks at how firms innovate.

As well as looking at the activities associated with innovation the chapter also looks at different ways of organising these activities to create a process. Several different models of the innovation process are examined. The existence of a number of models of the process reflects the fact that there are distinct and different ways in which firms approach or carry out innovation. It also reflects the changing nature of innovation, especially the increased importance of knowledge and the various different ways in which that knowledge can be channelled into innovation.

THE STEPS IN THE INNOVATION PROCESS

To get from an idea to a product that is on the market and available for consumers to purchase involves a number of steps. Just how many steps and how they are linked together can vary enormously depending on the nature of the product or service. Despite this, one can identify the major steps and the activities involved in the generic innovation process.

This generic innovation process is portrayed in Figure 6.1. It highlights the main steps that have to be undertaken. They are shown in a particular sequence starting with the generation of an idea or research leading to new discovery at one end, and the finished product going onto the market at the other. This is an *idealised* model designed to highlight the activities that have to be undertaken. In real life innovation is not usually as neatly packaged. The steps themselves will not always be as clearly differentiated as shown in the model, nor will they necessarily come in quite this sequence. Development, for instance, often takes place as much after the design stage as before. Effective innovation actually requires that they should not be so clearly defined. Overlap between the steps is desirable, and in some instances many of the steps will be undertaken concurrently. However, for the purposes of explaining what has to be done and the nature of the various activities involved in innovation, it is helpful to portray the innovation process as consisting of a number of well-defined steps or stages.

Figure 6.1 distinguishes a total of seven steps in the innovation process. All seven are associated with innovation and can be said to be part of the innovation process. However,

Figure 6.1 A Generic Model of the Innovation Process

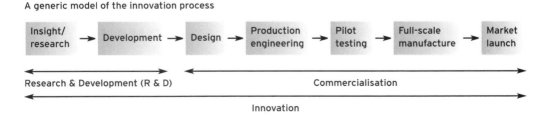

107

as the diagram makes clear, the first two steps, insight/research and development, are particularly associated with the invention. Most accounts of inventions focus on these two steps. The first step is associated with a breakthrough of some sort and the second is about getting it to work. The remaining five steps turn an invention into an innovation. Labelled 'commercialisation' they are the activities involved in transforming an invention into a commercially viable product. Whereas an invention is usually produced as a one-off, it requires innovation to transform it into something that can be produced in quantity and with standards of reliability that consumers have come to expect.

Insight/Research

This generic model of the innovation process begins with an insight that gives rise to an idea, or a new discovery as the result of research. How is it possible to have dual sourcing in this way? Essentially some innovations, particularly technological ones, are the product of a big investment in research, while others are more the result of individual human ingenuity. The former are likely to be associated with scientific discoveries and technological breakthroughs, the latter with the application of new technologies in order to devise better products.

As we saw in the previous chapter, ideas for innovation arise in all sorts of ways. Sometimes the insight that provides the basis of the idea that leads to the innovation comes from association – quite literally something triggers an idea. Sometimes the idea is an adaptation of an existing product. Sometimes it is by analogy that a way of solving the problem is discovered. And sometimes it just happens, more or less by chance.

MINI-CASE: JAMES DYSON AND THE DUAL-CYCLONE VACUUM CLEANER

The innovation in the dual-cyclone vacuum cleaner came from applying industrial dust extract technology to domestic vacuum cleaners. James Dyson had the idea when he installed an industrial dust extraction system incorporating a large, 30-foot-high cyclone dust extractor to remove excess powder from the atmosphere in a powder-coating plant. As James Dyson himself describes it: 'it occurred to me at that moment that there was really no reason why it shouldn't work in miniature – using a cyclone about the size of, say, a Perrier bottle.'
Source: Dyson (1997)

MINI-CASE: GEORGE DE MESTRAL AND VELCRO

George de Mestral was a Swiss engineer. Returning from a walk in the forest one day, he noticed cockleburs sticking to his dog's coat. Intrigued, he put a cocklebur under a microscope and noticed that the surface consisted of thousands of tiny hooks that readily stuck to tiny loops in his clothing. Having noticed that zip fasteners often had a habit of sticking he wondered if the principle of tiny hooks and loops could be used to develop a new type of fastener for clothing. It took him eight years and resulted in 'Velcro' (derived from 'velours croché' meaning hooked velvet), which is now widely used throughout the world on clothing and footwear.
Source: Van Dulken (2002)

MINI-CASE: RON HICKMAN AND THE 'WORKMATE®'

Ron Hickman was a designer for the sports car manufacturer Lotus. He was also a keen do-it-yourself enthusiast. When his DIY activities led to him damaging items of furniture, it gave him the idea of building a light, portable workbench that could be folded away when not in use. Hitherto workbenches had been wooden. They were unsuitable for DIY work because they were heavy and took up space, something that was in short supply in modern homes. Hickman used his experience of designing suspension systems for cars to devise a workbench that was light and strong and yet could be folded away when not in use.

Source: Hickman and Roos (1996)

In all three of the mini-cases shown above, the idea that formed the basis of the innovation was not the product of detailed scientific enquiry but of human ingenuity. Individuals going about their daily business observed a phenomenon in such a way that it gave them a new insight that in turn led them to conceive of a new and much more effective product. In all three instances the innovation was incremental or possibly modular, in that it resulted in a better product rather than an entirely new product. Similarly, all three innovators had scientific or industrial backgrounds, though their innovations were in all three cases in fields that were new to them.

Not all innovations arise from individuals having fresh insight into something. As we saw in the previous chapter, much innovation, especially in technological fields, arises as a result of extensive scientific research carried out over many years often by large teams of people. This sort of research is typically carried out in research laboratories. Sometimes the laboratories are located in universities, in some cases they are government laboratories and in others they are the laboratories of large business corporations.

MINI-CASE: BELL LABS

The Bell Laboratories (or Bell Labs as they are usually known) of the American telecommunications company AT&T have a quite outstanding record of innovation. Since they were founded in 1925, Bell Labs has registered more than 25 000 patents. Nor is it just that they have come up with a large number of innovations; many of them have been highly successful and have had a huge impact on the way we live our lives. Amongst the most significant innovations developed at Bell Labs over the years have been: the transistor, electrical sound recording, the solar cell, digital switching and the laser. These innovations were not the product of individual human ingenuity but rather of sustained collective scientific endeavour, for at their height Bell Labs employed a staff of 25 000.

MINI-CASE: XEROX PARC

Xerox PARC is the Palo Alto Research Centre of the Xerox company. Located in Silicon Valley in California, Xerox PARC came up with a suite of innovations that were to transform the computer industry. Until the 1980s computers were operated by typing in commands and instructions on a line-by-line basis. Even PCs received their instructions in this way. It was scientists at Xerox PARC who came up with the innovations that form the graphical user interface (GUI) now used by personal computers worldwide, including:

- Overlapping windows
- Pull-down menus
- Point-and-click selection by mouse

Source: Campbell-Kelly (2003)

In both the mini-cases shown above, innovations were the result of an intense research effort on the part of the laboratories concerned. These laboratories were run by private companies and their remit was to engage in research in pursuit of technological breakthroughs. Having made the breakthroughs, the technology was then passed on to other parts of the company in order to initiate the rest of the innovation process. This contrasts sharply with insights/ideas of individuals in the sense that the breakthroughs were not associated with specific products or specific customers. In the case of Bell Labs and the laser, for instance, it was to be many years before the technology was incorporated into products.

Although the innovations coming via these two different routes – insight of individuals and the research effort of corporate undertakings – can be very different, often there is a degree of overlap. Take the case of the web browser, Mosaic, featured in the previous chapter. This was the product of a team working in a large, publicly funded research laboratory based in a university in the US. Yet Mosaic was also the product of an individual – undergraduate computer science student, Marc Andreessen. It was his insight of an accessible web browser that lay at the heart of Mosaic, and the exponential growth of the Internet that it unleashed.

Development

Development is about turning ideas and technologies into products. The product that results from the development stage will not be ready to sell to consumers, but it will have many of the operational characteristics of the final product. In short, the product will work, and so will demonstrate the feasibility of placing it on the market, even if there is still a great deal more to do before it is produced in the volumes and to the standards of reliability demanded by consumers.

A feature of the development process is testing. Indeed, development is all about testing, modifying and improving the product. Testing has to take place to ensure that the product works in the way intended, in particular that it will work consistently and with the kind of performance that consumers are likely to demand. This means that development is usually a slow and laborious process. Accounts of new inventions frequently dwell on just how slow and painstaking the development stage can be. James Dyson's account of the development stage of his dual-cyclone vacuum cleaner is typical:

CHAPTER 6: THE PROCESS OF INNOVATION

❝ This is what development is all about. Empirical testing demands that you only ever make one change at a time. It is the Edison principle, and it is bloody slow. It is a thing that takes me ages to explain to my graduate employees at Dyson Appliances, but it is important. They tends to leap in to tests, making dozens of radical changes and then stepping back to test their new masterpiece. How do they know which change has improved it and which hasn't? **❞**
Dyson (1997: p. 124)

Figure 6.2 Models and Prototypes

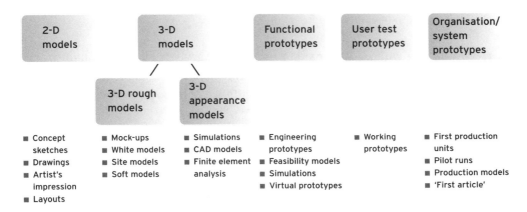

Source: Adapted from Leonard-Barton (1991)

Central to the development stage of the innovation process is the construction of models and prototypes. Figure 6.2, which is adapted from work by Leonard-Barton (1991) provides a typology of models and prototypes that helps to distinguish between the two. The purpose of models is to convey the form, style and 'feel' of an object (Leonard-Barton, 1991). Models serve to communicate the appearance of the proposed product. They are typically used to give an impression of what the product will actually look like. 3D examples, whether they are mock-ups, white models or computer-aided design (CAD) models, enable people to visualise the form of the product. Models are normally non-functional and as such cannot be operated.

Sabbagh (1996) explains how Boeing used mock-ups in developing new airliners. What were termed 'stage 1' mock-ups were made of plywood and foam and used to enable all the pieces of the airliner to be visualised in 3D. The benefit of producing a mock-up extended to all those involved in the development process. It enabled them to appreciate what the end-product was going to look like, and helped create a sense of team effort. Others not directly involved could also benefit, including potential users and customers, and subcontractors and suppliers. These days the use of mock-ups has been replaced in many industries by computer-generated 3D models. Using CAD techniques, these produce a 3D representation of the product that can be used not only to portray the external shape of the product, but

111

also to show its internal workings. This can be particularly useful with complex products, where potential problems such incompatibility, clashes or lack of space, can be highlighted by 'walking through' the model long before the real thing is built.

Prototypes in contrast usually have little to do with form and instead are all about function. Hence the term 'functional prototypes' in Figure 6.2. They, as their name implies, function. Unlike the final product, a prototype is a version of the product constructed as a one-off. The fact that prototypes are manufactured as one-offs is important. They are normally not constructed using the manufacturing process that will eventually be used to make the final product. Instead, prototypes are typically constructed on a 'jobbing' basis using general-purpose equipment rather than specialist purpose-built equipment. They are often made with different materials from those that will go into the final product. This is normally because the materials used for prototypes are easier to work with and more flexible. James Dyson explains how he made prototypes:

> 66 And all the while I was making cyclones. Acrylic cyclones, rolled brass cyclones, machined aluminium cyclones (which looked like prosthetic limbs for the Tin Man in the Wizard of Oz – whose life was changed by a cyclone). For three years I did this alone. 99 Dyson (1997: p. 122)

Functional prototypes usually form the basis of the testing and modifications that lie at the heart of development. However, they do have other uses. A comprehensive explanation of the function of prototypes would include:

- Testing and evaluation
- Integration
- Learning
- Risk reduction

Prototypes used for testing and evaluation will be robust and capable of operation. This is a requirement because they are typically used to evaluate performance. As a result, appearances are less important and operating characteristics more important. Products like cars will typically be tested in a variety of different environments, ranging from arctic cold to desert heat, in order to ensure that the product will go on functioning in the way intended in a variety of different conditions. When performance is poor, modifications will be made and a further set of tests instituted. Innovations that fail often prove not to have been tested effectively.

Prototypes are also used to facilitate the integration of components and sub-systems. This is particularly the case with complex products which have a large number of components and sub-systems that have to interact and work together. While a sub-system may function very effectively in isolation, when put together with other sub-systems problems may occur. Airliners provide a good example. They employ a variety of sub-systems including: avionics systems, hydraulic systems, life-support systems and the like. Only when these sub-systems come together in a prototype will engineers be able to identify potential problems of interaction such as vibration. Prototypes used to explore the interaction between the product and its environment, would normally be classified as organisation/system prototypes. A pilot run with a new piece of factory equipment for example might be designed to identify potential operational problems such as excessive noise or safety problems for operators.

Learning is a crucial feature of innovation, and a function of prototypes is to facilitate learning. The process of testing enables those who have developed a new technology to learn about its properties, through the acquisition of formal technical knowledge. Where innovation is concerned, knowledge is cumulative and hours spent testing prototypes can help developers to learn about the properties of a new technology. However, learning in terms of the acquisition of tacit or informal knowledge can be just as important. User-test prototypes, in the form of working prototypes, are often used to enable firms to learn about users and user behaviour.

Finally, prototypes have a part to play in risk reduction. Tests carried out with prototypes can help to identify potential risks. It is technological rather than market risks that will be identified in this way. With products that generate significant safety issues if they do not function correctly, as with many mechanical products, this can be extremely important. Of course, having identified these sorts of risks, firms may occasionally choose, for whatever reason, to ignore them.

MINI-CASE: HYFIL CARBON FIBRE FAN BLADES

The large jet engines developed to power modern wide-bodied jets such as Boeing's 747 'jumbo jet', feature a very large fan at the front of the engine. Aero engine manufacturer, Rolls-Royce, developed a novel fan for its RB211 engine. It comprised fan blades made not of metal, but of carbon fibre. Known as Hyfil, the attraction of using what was then an entirely new technology, was that carbon fibre was much lighter than the titanium normally used. The resulting weight-saving made for an appreciably lighter engine, which in turn was much more fuel efficient. However, Rolls-Royce's expectations of carbon fibre were dashed when a prototype engine fitted with a carbon fibre fan was subjected to the 'dead chicken' impact tests. This test was designed to simulate a 'bird strike' where a bird is sucked into the engine. The test involved firing 2-lb chickens (purchased frozen from a local supermarket and duly defrosted) into the fan blades of a prototype engine operating under full power. The impact caused a blade to snap resulting in a failure to contain the broken blade within the engine casing. The prototype engine's carbon fibre blades, though immensely strong, were insufficiently strong at the edges, which meant there was a risk that in the event of a bird strike when the engine was in service, parts of the engine could have pierced the fuselage or the wings causing the plane to crash. As a result, carbon fibre blades had to be abandoned and replaced with conventional titanium ones. Rolls-Royce continued to develop carbon fibre eventually using the material for engine nacelle doors.

Source: Spinardi (2002)

By the time the development process is complete, fully functioning prototypes should be in operation. They will not necessarily look like the final product but they will have the final product's operating characteristics. By this point there should be a reasonable degree of certainly that the product will work in the way intended. However, at this point the organisation faces some crucial decisions. These centre on two particular aspects:

- How should the intellectual property embodied in the prototype be protected?
- What is the most effective way to manufacture the final product – using the organisation's own resources or getting some one else to make it?

These two decisions are closely entwined. Without effective protection for intellectual property, handing over manufacture may result in someone else gaining the intellectual property.

However with intellectual property secure, the organisation can continue to complete the innovation process.

Design

Design and designers usually come into the innovation process at various points. But for purposes of exposition in the generic model of the innovation process, design is shown as following development, and forming the first step of the commercialisation part of the innovation process. In fact, some design would almost certainly have occurred earlier. The concept sketches associated with 2D models shown in Figure 6.2 for instance, would require a designer. However, the design that forms part of the commercialisation process is likely to be at least slightly different since it is likely to be more concerned with detail. Having got the idea to work in prototype form, the designer now has to give it the attributes and features required by the consumer. Mere functionality is now no longer enough.

Detailed design is required to determine the attributes and features of the final product that will go on sale in the marketplace. This is likely to involve specifying:

- The precise shape of the product
- The tolerances to which it will be manufactured
- The materials to be used in manufacture
- The process by which the product will be manufactured

Detailed design is a process that will generate a design specification that not only includes drawings (normally computer-generated) specifying exactly the form of the product, but also gives details of the geometry, materials and tolerances of all the components that make up the final product. In more complex products there will also be system-level design taking place. This is required in order to show the systems architecture of the product, particularly the way in which the different sub-systems interact in order to achieve a fully-functioning product.

In producing a detailed design, the designer has to factor a number of constraints and come up with a design that will both appeal to the customer and enable the firm to make money. From the marketing perspective it is pretty obvious that the design has to appeal to consumers, but manufacturing will equally be concerned that the product can actually be made and preferably without too much difficulty. This is not just about technical feasibility. The design has to be capable of being produced with the resources that are likely to be available. Similarly, those in finance will be concerned that the design does not prove too expensive to produce.

These sorts of constraints are all likely to form part of the design brief. Most designers or design teams will have a design brief to work to that incorporates the requirements of the various stakeholders responsible for the product. Accordingly, the design brief will outline what is required of the design and note a number of constraints. These constraints will typically come from other functional areas of the business such as manufacturing and finance. The design brief will also include constraints derived from the work that has been done with the prototypes.

The designer's task is to take the design brief and translate it into a design that meets the requirements of the team responsible for the product. This means that it has to operate effectively, while appealing to consumers and being capable of being manufactured at a cost that will both enable the consumer to afford it, and generate a return for the firm.

In reality there may be no single designer. Instead there may well be a design team that brings together different types of designer including: technical designers able to design systems; industrial designers able to ensure functionality; and more traditional designers able to create a form that will have appeal for consumers. Often some or all of these design skills will not be available in-house and will need to be bought in, probably from a design house that has expertise and experience in the field.

Production Engineering

Essentially production engineering is concerned with making the product. Whereas prototypes are usually made on a one-off basis, the final product is likely to be manufactured in substantial quantities and this calls for quite different processes and skills.

The initial decisions surrounding production engineering concern who is to undertake manufacture. Will the product be made in-house or will it be outsourced to subcontractors? Developments in IT combined with improvements in communication mean that it is now possible to design a product on one side of the world and produce it on the other. In a whole range of industries it is now commonplace for manufacturing to be contracted out.

Assuming manufacture is going to take place in-house, a range of issues have to be tackled. Many of these issues surround the production process. Every production process involves a manufacturing system as portrayed in Figure 6.3. Every manufacturing system consists of inputs and outputs. The principal outputs are finished goods, while inputs can be divided into those that are fixed and those that are variable (i.e. they vary in relation to output). The fixed items in Figure 6.3 are equipment and tooling. Decisions surrounding equipment and tooling give rise to a series of process-choice decisions that are basically about the configuration or architecture of the manufacturing system. The type of configuration used will depend on the anticipated volume of output and is a choice between four types of process:

- Jobbing
- Batch
- Line
- Continuous Process

A jobbing system involves the use of general-purpose equipment to make one product at a time usually to a unique design. It is suitable for highly specialised products such as jewellery or process equipment. Batch systems involve general-purpose equipment combined with special-purpose tooling. Products are made in lots or batches of perhaps

Figure 6.3 A Manufacturing System

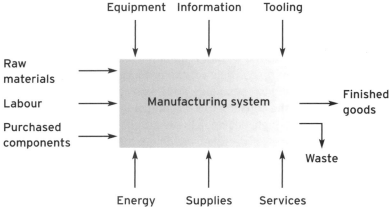

Source: Ulrich and Eppinger (2003) 'Product Design and Development', McGraw-Hill, New York

10 or 20 at a time. Tooling in the form of jigs is important because it ensures production consistency within each batch. Items like wallpaper and food are typically made in batches and the batch number will often be printed on the product to aid identification. Line systems are usually associated with mass production. Items are produced in very large numbers using dedicated equipment especially designed for the purpose. Cars and consumer goods are typically manufactured using line production. Finally, continuous-process systems are found in process industries such as oil refining and chemicals. This type of system uses a continuous flow of work through the system, so much so that the systems are in continuous operation day and night.

As well as selecting the most appropriate production system, production engineering also involves a raft of decisions surrounding the way in which the product is made. Many of these decisions will in turn be linked to the type of manufacturing system being used. There are very often big differences between a functional prototype used for development and the final product, and this reflects the fact that at the production engineering stage the product is often revised to make it easier and cheaper to manufacture. Such work typically centres on a close examination of assembly operations. With careful preparation and effective design it is usually possible to simplify the design, thereby eliminating a number of assembly operations. Ulrich and Eppinger (2003) provide a number of examples of design changes of this type:

- Reducing the parts count
- Using standardised components
- Using self-aligning parts
- Using assembly operations that require a single, linear motion

Design changes such as these reflect the volume of production that is anticipated. With the prospect of volume production it is worthwhile redesigning and revising the product in order to make assembly quicker, easier and cheaper.

Trott (2002: p. 150) gives the example of a toolbox. Manufactured as a prototype, it is produced on a one-off jobbing basis initially in which the various components are held together with industrial fasteners. With a satisfactory prototype, the production-engineering function then proceeds, with the help of the designers, to simplify the design to make it suitable for manufacturing in large batches. This entails eliminating the need for fasteners and investing in specialist machinery that will shape the individual components. The result is a big drop in the number of parts required and an assembly process that involves all the parts being held in place via 'push and snap' operations undertaken by assembly workers. Hence one gets a product that has the external characteristics and functionality of the original design and yet is capable of being assembled more quickly and more cheaply. Not only does this indicate the sort of activities carried out at the production-engineering stage, it also illustrates an important principle, highlighted by Trott (2002), namely that, as the volume of production increases so the most appropriate method of manufacture also changes.

Pilot Testing

Having ensured that the product can be made in a way that will ensure it appeals to consumers, while at the same time making money for the company, further testing has to be carried out to ensure that it can go into the marketplace. Testing at this stage will have less to do with developing the product and more to do with ensuring it is will be safe in the hands of consumers. Consequently, much of the testing at this stage will involve interaction with consumers. Much of the testing may be a statutory requirement or necessary for the product to gain type approval or certification before it can be used to provide public services.

Sabbagh (1996) describes at some length the tests that the Boeing 777 airliner had to go through before it could be handed over to the airlines that were going to put it into scheduled service. His description of the Refused Take Off (RTO) brake tests conducted at Edwards Airforce Base in California provides a graphic account of what this kind of testing is all about. The RTO test is designed to ensure that a fully loaded plane can cope with the most extreme aborted take-off. In Sabbagh's (1996: p. 336) words the test was: '...another illustration of how the plane-makers explored the far corners of the envelope in their attempts to make the plane safe'.

The test involved simulating a full load by using fuel as ballast and accelerating down the runway on full power and then putting the brakes full on. Under these conditions the brakes get so hot that they catch fire. The test was designed to ensure that a fully laden plane could stop safely if a take-off was aborted just at the point at which it was about to become airborne.

While this is perhaps one of the more dramatic examples of this kind of test, it illustrates the sort of activities that are typically undertaken at the pilot testing stage to ensure that the product is functioning safely.

Full-Scale Manufacture

Before full-scale manufacture can actually take place the equipment that forms part of the manufacturing system has to be commissioned. This is designed to ensure that not only are the individual items of equipment that make up the manufacturing system functioning as they should, but that they are also interacting effectively. With a complex manufacturing system and a sophisticated control system this can be a demanding task. Pieces of equipment will often work perfectly in isolation, but put together they become quite ineffective. Consequently, the commissioning process is intended to prove the system and ensure it is functioning as planned.

Even with a fully effective manufacturing system, those who are going to operate it have to be recruited and trained. This is something that has changed over the past twenty years as companies in the West have learnt from Japanese companies the importance of careful planning and preparation, as far as the labour force is concerned. Not only is it important to select people with appropriate aptitude and skills, they have to be trained to use the equipment.

Finally manufacturing can begin. Even this is not likely to be full-scale manufacture to begin with. Typically, firms will deliberately plan their production so that initially they are producing at perhaps 20 per cent or 30 per cent capacity. This allows those operating the system to move up the 'learning curve' as they become more familiar with the system. The learning curve concerns the way in which, particularly in batch operations, it will often take less time to manufacture the hundredth item than the first. This can be an important feature of manufacturing in some industries. Aerospace is a good example. Firms like Airbus and Boeing find that, even with sophisticated production systems, it will typically take less time to produce an airliner as output expands. This reflects the fact that this sort of learning relies heavily on tacit knowledge and is a cumulative process. Consequently, as output expands, learning increases and it can take less time to manufacture the product.

There are other reasons why firms will typically 'ramp up' (Ulrich and Eppinger, 2003) production gradually from a relatively low base. Products produced during the ramp up can be evaluated to spot potential flaws. They can also be supplied to preferential customers who will not only evaluate the product but also provide valuable marketing data in terms of their perception of the product. In addition ramping up gradually can also help with stock building. It can be disastrous to raise customer expectations by promoting a new product, only to deny customers access to it because the product has not yet reached retail outlets. Gradually building up production can allow time for distributors to build up stocks prior to the product being formerly launched.

Then, eventually full-scale production can get under way.

Market Launch

The market-launch phase brings another round of potential problems. The activities involved are likely to be quite different from those encountered earlier, however. Since the market-launch phase has a lot to do with marketing it will not be covered in detail here.

The market-launch phase essentially requires the co-ordination of a whole range of different activities. Some of the activities include:

- Ensuring that retail outlets have appropriate stocks
- Booking advertising space
- Designing and producing advertisements
- Booking exhibition space

- Ensuring that literature about the product has been designed, written and printed
- Informing the press and ensuring that they have had time to familiarise themselves with the product

The list is at best indicative of the sorts of activities that have to be undertaken. They all form part of the process of introducing the product to the public. While it may require a different set of skills, getting this phase of the innovation process right is just as important as all the others.

This reinforces one of the central features of the innovation process and one that researchers have increasingly come to recognise. It is very easy to see innovation in heroic terms. The pursuit of new ways of doing things, the struggle to get something to work, these are usually the things that people think of when they think of innovation. Yet there is actually a lot more to innovation. It is a lengthy process, even when activities are carried out concurrently. It is also a process that includes many different activities, and all of these activities are important. The feature of the innovation process that is increasingly recognised is that it is important for all of them to be carried out effectively.

MODELS OF THE INNOVATION PROCESS

While the generic model of the innovation process enables the various activities associated with innovation to be identified, it does not reflect the range of different approaches to innovation that are available. In particular, it fails to take account of some of the newer models of the innovation process that have been introduced in recent years.

Rothwell (1994) identifies no less than five models of the innovation process. Significantly, he suggests that these models form part of a continuum that has seen new models of innovation introduced over the last half century. All five of these models are presented here.

Technology Push

The technology-push model (Figure 6.4) is very much the traditional perspective on the process of innovation. It is effectively the research-led version of the generic model presented earlier, since one of the features of this model is that it is driven by developments in science and technology. It assumes that more technology, brought about by additional expenditure on R & D, will lead inexorably to more innovation. The process is entirely linear and sequential, each stage following on from the completion of the previous one. The model virtually ignores the marketplace, which is portrayed as being passive and simply taking what technology has to offer. The model is naïve as far as the process itself is concerned. We are told very little about the nature of the process. Having said that,

Figure 6.4 Technology-Push Process

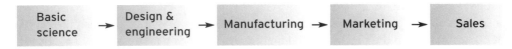

Source: Republished with permission, Emerald Group Publishing Limited

there are industries where the innovation process does take place in very much this way – for example, the pharmaceutical industry.

Demand Pull

Recognising the passive role given to marketing, theorists in the late 1960s and early 1970s came up with a new perspective on the process of innovation. In the demand-pull model, the role of the market is central. According to Rothwell (1994), the move to a more market-centred type of innovation process reflected the maturing of many technology-based industries and a growing realisation that consumer requirements were becoming more sophisticated.

In the demand-pull model (Figure 6.5), the market forms the source of ideas for new innovations. Knowledge of consumer requirements is seen as driving research and development rather than the other way around. This is a variant on the generic model, if one sees consumer needs as the source of new ideas that lead to innovation.

While there can be little doubt that the recognition of the importance of consumer needs provided a useful re-orientation of innovation, nonetheless this approach too has its weaknesses. As Christensen (1997) has shown, there is a danger that firms, especially those in mature industries, may become locked into 'technological incrementalism' (Rothwell, 1994: p. 9), where they devote their energies to providing innovations that offer very modest improvements in product performance to meet the apparent needs of their customers, while ignoring emerging new technologies that will one day lead to radical innovations. Ultimately this kind of technological incrementalism can lead to firms losing the capacity to innovate, as they become wedded to an old technology that in time is superseded.

Figure 6.5 Demand-Pull Process

Source: Republished with permission, Emerald Group Publishing Limited

MINI-CASE: BARBED WIRE

Kevin Costner got it all wrong. In the Western 'Open Range' Costner, like other Hollywood actors before him, pits himself and Robert Duvall as prairie cattlemen or 'free rangers' fighting a powerful local ranch owner. Only this time the ranch owner has the local town, including the sheriff, in his pocket. In reality ranchers did not need to rely on intimidation, they had barbed wire, and it was barbed wire rather than the six-gun that helped win the West.

As cattle ranches spread across the treeless Great Plains in the 1870s, the ranchers found their activities constrained not by prairie cattlemen like Costner and Duvall, but by a severe shortage of timber for fencing. As Basalla (1988: p. 51) notes, 'Between 1870 and 1880 newspapers in the region devoted more space to fencing matters than to political, military, or economics issues.'

Such was the scale of the problem that westward expansion across the vast open spaces of the Great Plains was constrained by the high cost of fencing (Hayter, 1939) which was itself a function of the treeless landscape. As Howells (2005) points out, there was a pressing need for a 'wooden fence substitute'. One solution was the planting of hedges made of Osage orange (Basalla, 1988). Unlike many forms of hedging, the long thorns growing at right angles to the stem on the Osage bush meant it was sturdy enough to restrain cattle. Native to Texas and Arkansas, it was planted in the Great Plains region. Unfortunately it proved too slow-growing to be entirely effective. Smooth-wire fencing provided a cheap substitute, but it was ineffective in restraining roaming livestock, which loosened the fencing poles and broke the wire by constantly rubbing against it (Hayter, 1939).

It was against this background of well-defined and articulated consumer need holding back economic development, that in the 1870s inventors in Illinois on the eastern edge of the Great Plains began to experiment with various forms of barb that mimicked the Osage bush, in order to create a new type of fencing. J.F. Gidden (Hayter, 1939) perfected and patented a form of wire that became the pattern followed by most other wire producers. Very quickly barbed wire solved the fencing problem and, with the bottleneck removed, economic development could continue westward, to the detriment of prairie cattlemen.

Sources: Basalla (1988); Hayter (1939); Howells (2005)

Coupling

For many industries both technology-push and demand-pull innovation processes are flawed. They both rely on innovation being a linear, sequential process. Unfortunately, processes like this are said by some to encourage what is often described as 'over the wall' behaviour (Trott, 2002: p. 215), where the departments responsible for each stage carry out their task in isolation, providing little in the way of guidance and help to the next department. This insular approach not only slows the process, it also means that learning is not passed on, with the result that the final outcome is not as effective as it might be. It was to overcome precisely these kind of problems that the coupling model (Figure 6.6) evolved.

Rothwell (1994) suggests that the coupling model was the result of a growing recognition derived from an increasing body of empirical studies of the innovation process that showed the technology-push and demand-pull models of innovation as standing at the opposite ends of a continuum. They were effectively more extreme forms of a more generalised process. In this process, sometimes technology was dominant while at other times the market was the more powerful influence.

Hence, the coupling model is very similar to earlier models of the innovation process. The process is still essentially linear and sequential. However, this time both technology and the market are influential. Technology enhances the state of knowledge within the broader scientific and technological community, while the market works to express wider consumer needs and expectations. New ideas are the product of both. However, a crucial difference between this model and the earlier ones is the presence of 'feedback loops'.

Figure 6.6 Coupling Model Process

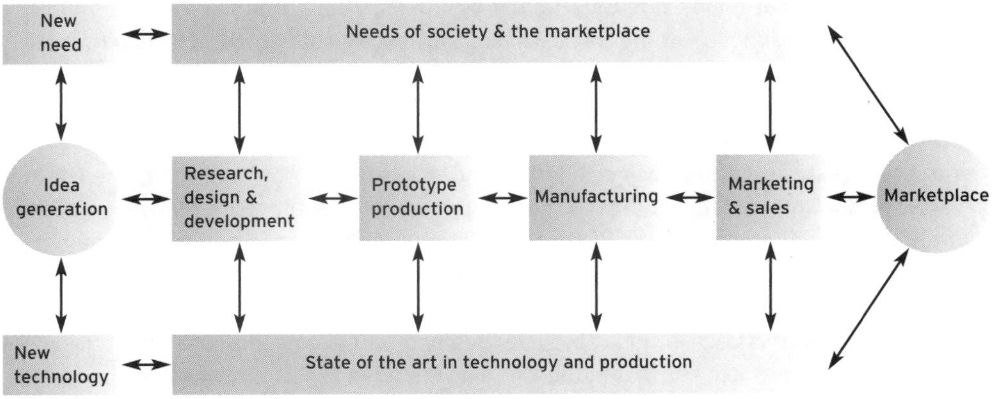

Source: Republished with permission, Emerald Group Publishing Limited

The lines of communication between the various functions carry a two-way traffic. No longer can functions operate on an 'over the wall' basis, forgetting about the process once their immediate tasks have been completed. In the coupling model one has a series of distinct functions or stages, but they are interacting and interdependent.

Integrated

The 1980s were characterised by powerful forces for change. In many fields the old order and the old certainties that had prevailed in the years since the Second World War gave way to new and much more intense competitive pressures. Developments in technology, both in the computing and the communications fields, led to the introduction of IT-based manufacturing systems that shortened product life cycles. In parallel with changes in manufacturing technology came new ideas about manufacturing management. Many of these new ideas, such as just-in-time production and set-up reduction, came from Japan. Among the most powerful ideas were notions of concurrent or parallel development. Applied to new product development, this implies an end to strictly linear and sequential processes. Japanese companies rely on project teams that integrate the various functions. Under such arrangements the functions are brought into the new product development process from the start, and joint group meetings ensure that issues such as manufacturability are considered early in the process rather than near the end. Team-based new product development therefore represents a much more integrated process (Figure 6.7).

Network

Finally, the 1990s have seen the advent of what Rothwell (1994) describes as a 'fifth-generation' innovation process. Termed the network process (Figure 6.8), this reflects the way in which some organisations increasingly rely not on their own internal resources for innovation, but instead draw on external resources through alliances, agreements and contracts with third-party organisations. The use of networks reflects continuing developments in computing and

Figure 6.7 Integrated Model Process

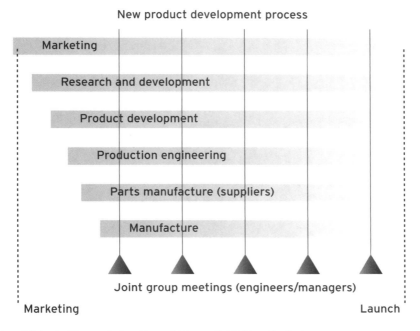

Source: Republished with permission, Emerald Group Publishing Limited

communications which have facilitated information transfer and the popularity of outsourcing whereby vertical disintegration leads organisations to forsake certain activities, preferring instead to buy them in as and when needed.

Network innovation is by no means universally practised. Only in certain industry sectors has it proved popular. Good examples are pharmaceuticals, aerospace and computing. In pharmaceuticals, developments in biotechnology have fostered the growth of small specialist biotechnology companies on which large pharmaceutical companies increasingly rely as a source of innovation. In computing, the growth of specialist companies making computer peripherals has helped to make Silicon Valley in California viable. Finally, in aerospace, the prohibitive cost of developing a new airliner or a new engine has led many aerospace giants to put together joint ventures and partnerships that bring together a number of suppliers to engage in new product development.

While developments in IT and communications have helped to make the sort of 'collaborative innovation' implied by the network model viable (i.e. through the electronic transfer of CAD-generated design data to the manufacturing function), the use of this approach is not entirely the result of facilitating factors. Rising consumer expectations and increasing emphasis on choice and variety have also helped to fuel an increasing emphasis on innovation and new-product development. Thus, companies anxious to provide consumers with ever greater choice have increasingly sought to look outside their own organisation for ideas and technologies. By looking outside they have access to a greater range of opportunities.

Figure 6.8 Network Model Process

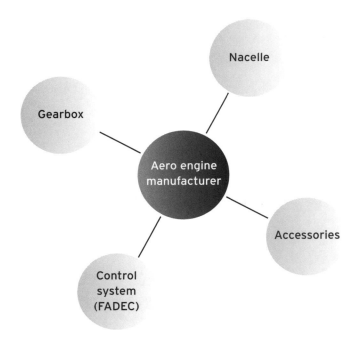

CASE STUDY: THE CHILLED MEALS REVOLUTION

What did you have for dinner last night? We are increasingly eating a range of exotic meals eaten not in restaurants or collected from the takeaway but prepared in our own homes in a matter of minutes. Chicken Tikka, Chicken Madras, Mango Chicken curry, Chicken Chow Mein, these are just a few of the wide range of ready-meals that are available in the chiller cabinets of our supermarkets these days. They offer high-quality, ready-prepared meals at reasonable prices. But it was not always so. Chilled ready-meals such as Chicken Tikka are a relatively recent innovation that only began to appear on our supermarket shelves in the early 1990s, pioneered initially by multiple retailer Marks & Spencer.

Prior to the introduction of chilled food, ready-made meals were available in our super-markets, but they came as frozen foods. Unilever's frozen food subsidiary Birds Eye introduced the first 'TV dinners' as they were known in 1969. Over the years, the freezer cabinets of Britain's supermarkets became home to a range of ready-meals. Social changes in the 1970s and 1980s, such as the increasing number of women working full time and the increasing number of single-person households meant a steady increase in the popularity of these kinds of products. Though the range of meals became steadily more sophisticated, the products themselves did not. They might look attractive, but when it came to eating them, most were nothing like the ready-meals we have today. As frozen foods, they were hampered by the fact that freezing food and then thawing it to reconstitute it had an adverse effect on both the texture and the flavour of the food that inevitably made the meals less palatable. This was a major drawback that constrained the growth of the market for ready-prepared meals sold in supermarkets. The solution was not to freeze the food but simply to chill it. Chilling involves lowering the temperature of the food to about 5 degrees centigrade but not actually freezing it. Keeping food at a low temperature helps to preserve it (for a time at least), while not actually freezing it avoids the problem of damaging the texture and flavour of the food. However as Cox *et al.* (1999) point out, chilling presents formidable logistical difficulties, since the meals are highly perishable and have a very limited shelf life with the result that the maximum period of time that can elapse between production and final consumption of such products is a few days rather than weeks or months in the case of frozen foods. Without very careful and precise co-ordination of supply and demand, the premium price associated with delivering to the consumer a superior product would be more than absorbed by high wastage rates.

By the late 1980s Marks & Spencer felt recent developments in technology, combined with their proven and long-standing skills in relational contracting in the textile and clothing sector, offered scope for offering a range of chilled products in the form of ready meals.

The technological developments that formed part of this innovation covered both consumption and production. On the consumption side, the introduction of microwave ovens in the early 1970s and their widespread use in domestic households meant that a means of quickly and easily preparing and heating chilled ready-meals was readily available. On the production side, developments in IT systems, especially in the field of communications technology and data management, helped to provide retailers with an unprecedented degree of control over their operations.

The IT developments centred on electronic point of sale (EPOS) systems based on laser-scanning technology introduced in the 1980s. These helped to transform retailers' ability to exercise detailed operational control over the goods going through their stores. EPOS systems, which scanned all the goods going through the check-outs, enabled retailers to link their inventory replenishment to consumer requirements. No longer did they have to estimate demand, since EPOS systems enabled them to link their purchases of replacement inventory directly to consumer purchases. A key feature of this was the use of electronic data interchange (EDI) systems. EDI in particular enabled retailers to manage inter-firm co-ordination, between themselves, manufacturers and distributors, in real time. Working in real time brought an unprecedented degree of precision, both to inventory management and purchasing. No longer was it necessary to estimate demand by laboriously checking the stock on the shelves to see which were empty or at least needed re-stocking. EPOS systems using bar codes on all items of stock did this automatically. The systems were also highly efficient as EDI replaced paper-based administrative systems with computer links. Effective inter-firm co-ordination required a high degree of systems compatibility in order to provide links that would facilitate data transfer between retailers, manufacturers and logistics/distribution companies. The achievement of the necessary compatibility can be directly attributed to the work of a trade association (Bamfield,

▶

1994), the Institute of Grocery Distributors (IGD). The IGD brought together retailers, manufacturers and distributors to establish a set of common standards for bar coding. Bar coding was an essentially element in ensuring rapid and easy data transfer.

To add to the wealth of data that retailers now possessed in relation to the goods going through their stores came developments in data warehousing and data mining. Retailers introduced store loyalty cards in the 1990s so that they could link the data coming from their EPOS/EDI systems to individual consumers. This in turn permitted retailers to record the activities of consumers on a regular basis. Then, using data mining they could establish consumer buying patterns and trends. Armed with this sort of data, retailers were in a position to exercise a high level of co-ordination between all the parties involved in producing and selling goods to consumers. This was particularly significant for food items where the perishable nature of the goods was an important issue.

The introduction of these various technologies created an infrastructure that provided scope for innovation in ready-meals, driven not by manufacturers but by retailers. Marks & Spencer began with a range of meat pies and quiches marketed under its *St Michael* brand name. Their strategy for chilled ready-meals was that they should be promoted as a substitute for takeaway meals or even restaurant meals, which had been increasing both in popularity and the range of products available. The key elements in promoting these products were variety, novelty and quality (Cox *et al.*, 2003). As such chilled ready-meals were marketed as high-quality, premium-priced products.

Marks and Spencer's strategy was that as quality substitutes for restaurant meals, their range of chilled ready-meals should offer an extensive and constantly changing array of new products that mirrored customer eating trends. To provide them with the necessary variety and choice, as well as new offerings, they turned to small specialist food manufacturers. These ranged from micro-kitchens employing less than five people to larger concerns such as Hazlewood Foods, though most were relatively small concerns. For retailers like Marks & Spencer the advantage of using several small manufacturers was the flexibility offered by small suppliers (Cox *et al.*, 1999). These small firms manufacture in small batches, which is highly desirable given the relatively short shelf life of the product. Being small these firms are also highly specialised. Many specialise in particular product bases (e.g. poultry, fish, etc.) and 'ethnic' recipes (e.g. Indian, Thai, Italian, etc.). Specialisation provides scope for retailers offering a very broad product range and also facilitates the rapid development of new products. Given the access to customer behaviour provided by customer loyalty schemes (like Tesco's clubcard), retailers were anxious to be able to identify new market niches and fill them with new products as quickly as possible. Of the small specialist food manufacturers used by the major retailers, S & A Foods of Derby is typical. The company was started in 1986 by Perween Warsi after she despaired of ever finding a decent samosa in her local supermarket in Derby. S & A Foods began as a micro-kitchen supplying a range of chilled Indian ready-meals sold as own label products for retailers like Marks & Spencer and has grown to the point where it has two factories manufacturing Indian meals. As a specialist food manufacturer S & A Foods does not engage in marketing, branding or distribution, focusing its efforts instead on developing new products.

Figure 6.9 The Innovation Network for Chilled Ready-Meals

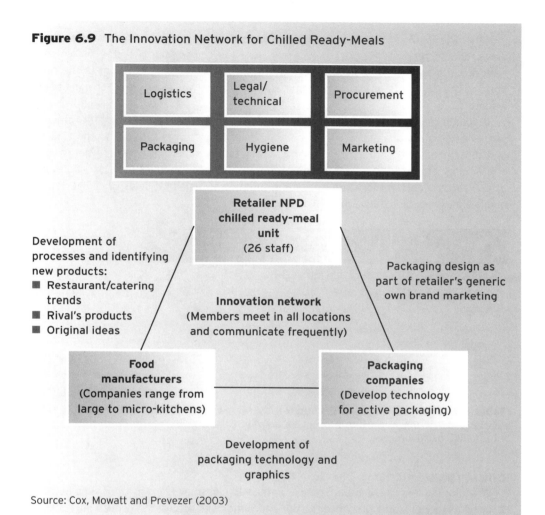

Source: Cox, Mowatt and Prevezer (2003)

In developing chilled ready-meals, firms like Marks & Spencer assembled new product-development teams that brought together employees from the specialist food manufacturers with employees from the packaging companies and their own staff drawn from their food technology and hygiene departments. The new product teams formed a network for pooling knowledge and drawing on data from a variety of sources in order to facilitate innovation. From the retailer came data on purchasing patterns and trends derived from its IT system. From the specialist food manufacturers came ideas for new dishes and guidance on which dishes would be more suitable for chilling and re-heating. From the packaging companies came guidance on packaging materials suitable for use in microwave ovens. This was particularly important as microwavable meals necessitated the development of 'active' forms of packaging designed specially for use in microwave ovens. Working on a collaborative basis the new product development teams devised a range of ready-meals suitable for chilling.

While the innovation network based on new product development teams generated new products, retailers like Marks & Spencer were able to use their IT systems to co-ordinate product and distribution to ensure that the right amount of stock got to the right store at the right time at the right temperature so that it was available when consumers wanted it.

To gauge the success of chilled ready-meals as an innovation one has only to travel a few miles on a motorway and count the number of large trucks with the words "Chilled Distribution' painted on the side. Similarly, a visit to any supermarket will reveal rows of chiller cabinets offering a wide range of Indian, Chinese, Thai, Italian and traditional British ready-meals. A more conventional evaluation reveals that sales of chilled ready-meals almost doubled between 1993 and 1999.

Year	£m	Index
1993	340	100
1994	380	112
1995	435	130
1996	475	140
1997	497	146
1998	551	162
1999	596	175

Table 6.1 UK Retail Sales of Chilled Ready-Meals 1993–1999
Source: Cox *et al*, 'Industry and Innovation' 2003. Taylor & Francis Ltd, http://tandf.co.uk/journals

QUESTIONS

1 Why did retailers like Marks & Spencer choose to use small firms as their suppliers?
2 What aspect of Marks & Spencer's prior knowledge and experience proved particularly useful in terms of the innovation process used to develop chilled ready-meals?
3 Which model of the innovation process did Marks & Spencer adopt in order to bring about the innovation of chilled ready-meals?
4 What benefits did Marks & Spencer (and the other retailers) obtain from the particular innovation process they used?
5 What alternative models of the innovation process might Marks & Spencer have used?
6 What enabling factors permitted firms like Marks & Spencer to use their chosen model of the innovation process?
7 What do you think were the critical factors in achieving successful innovation in this case?
8 Why was co-ordination vitally important and how was it achieved?
9 Using an appropriate series of market research reports such as Mintel or Key Note reports, show:
   - How the market for chilled ready-meals has grown in the last decade
   - How the shares of the chilled ready-meal market have changed over the last decade.

QUESTIONS FOR DISCUSSION

1 Why is innovation often a lengthy process?

2 Where do innovations come from?

3 What are the relative merits of technological change and the market as sources of innovation?

4 Why can it be problematic portraying innovation as a series of phases?

5 Distinguish between research and development.

6 Why is testing such an important part of the development process?

7 Which personal qualities do you think are required of those engaged in development work?

8 What do you consider to be the most important feature of the 'Coupling Model' of the innovation process and why?

9 Which do you consider provides the better explanation of innovation – technology push or demand pull?

10 Explain how the technology-push and demand-pull processes of innovation are related to different types of innovation?

11 Where do teams fit in the 'Integrated Model' of innovation?

12 Why has the network model of innovation become popular in recent years?

13 What facilitating factors have helped to make the network model of the innovation process viable in recent years?

14 Which theory of innovation best explains why firms are increasingly using the network model of the innovation process?

ASSIGNMENTS

1 Using an account of an innovation of your choice, prepare a report that describes the process innovation. The report should:

   a make clear the type of innovation

   b use one of the models of the innovation process to make clear how the innovation was conducted

   c identify and describe the various steps or stages in the innovation process.

▶

2 Explain what is meant by a prototype and use one or more accounts of an innovation to show their value to innovators.

3 Distinguish the different types of prototype that can be used as part of the development process.

RESOURCES

There are some excellent resources available that help to illustrate and explore the nature of the innovation process. These resources are not neatly labelled 'innovation'. In fact they are not necessarily about innovation, but they show, often in some detail, what is involved in developing new products. They take the reader/viewer through the various stages in the innovation process. Unfortunately this is not always immediately apparent as the steps will not necessarily be clearly differentiated. However, they are there all the same. Armed with models of the innovation process, it is possible through careful analysis to unpick the process, identify the stages and get a clear and detailed picture of the nature of the process of innovation.

Four types of resource that are particularly useful are:

Biographies

One of the best studies of the innovation process is James Dyson's autobiography *Against the Odds* (Dyson, 1997). This describes the innovation process for the dual-cyclone vacuum cleaner. Covering the 15-year period from Dyson having the idea for a cyclone vacuum cleaner to his finally getting one into production and onto the market, it explains in detail the steps involved including: how the idea arose, the building of prototypes, testing them, design of product, the acquisition of manufacturing facilities and the problems he faced in getting his new cleaner into the shops. The chapters on development, i.e. building and testing 5127 prototypes, are particularly informative.

Other biographies that can provide valuable insights into the innovation process are: Baylis (1999); Berners-Lee (2000); Kamm and Baird (2002); McElenhy (1998).

Business Histories

Business histories can also provide valuable insights into the innovation process. Of course it depends on the business! The business in question has to be one that has been active in innovation. A good example of the genre is Wood (2001). This is the story of Oxford Instruments, a firm that has been at the forefront of innovations associated with magnetism. The section dealing with the introduction of superconducting magnets in the 1960s shows how small firms innovate, particularly the ways in which they undertake the various stages in the innovation process. Given that this is a high-technology business, it provides a fascinating account of the interaction between the academic and industrial communities. It also shows how firms, especially young small firms acquire the resources to enable them to engage in research and development.

Another good example is Levy (1994) which tells the story of the personal computer maker, Apple. The section on the development of the Macintosh is particularly detailed.

CHAPTER 6: THE PROCESS OF INNOVATION

Campbell-Kelly (2003) is also useful, though it is probably more of an industrial history than a business history of a single organisation. It provides a first-class history of the computer software industry. The references alone constitute an invaluable resource for anyone interested in innovation.

While it is not quite a business history, Uttal's (1983) study (reprinted in Burgelman *et al.*, 2001) of Xerox PARC provides a fascinating insight into the work of a major research laboratory.

TV Documentaries/Videos/DVDs

Another valuable insight into the innovation process is Karl Sabbagh's book *21st Century Jet* (Sabbagh, 1996), which describes the development of the Boeing 777 airliner in the early 1990s. Based on a television programme it is also available as a video cassette or DVD. While the 777 airliner is an example of incremental rather than radical innovation, the book/video covers the whole process showing all the steps along the way. It is particularly useful in showing the use of computer-based prototyping that has replaced the use of mock-ups. It is also very good on the testing undertaken as part of the development process. A particular highlight is the RTO tests. These are the Refused Take-Off tests designed to test the aircraft's brakes fully laden and under full power.

Websites

It has already been noted that websites can be problematic, but http://www.Bricklin.com provides a wealth of historical detail, including pictures of people, places and products from all the stages of the innovation process; it even includes a downloadable version of VisiCalc, the world's first spreadsheet!

REFERENCES

Bamfield, J. (1994) 'The Adoption of Electronic Data Interchange by Retailers', *International Journal of Retail and Distribution Management*, **22** (2), pp. 3-11.

Basalla, G. (1988) *The Evolution of Technology*, Cambridge University Press, Cambridge.

Baylis, T. (1999) *Clock this: My Life as an Inventor*, Headline Publishing, London.

Berners-Lee, T. (2000) *Weaving the Web: The original design and ultimate destiny of the World Wide Web by its inventor*, Harper Collins, NY.

Brown, G.I. (1996) *The Guiness History of Inventions*, Guiness Publishing, Enfield.

Burgelman, R.A., M.A. Maidique and S.C. Wheelwright (2001) *Strategic Management of Technology and Innovation*, 3rd edn, McGraw-Hill, New York.

Campbell-Kelly, M. (2003) *From Airline Reservations to Sonic the Hedgehog: A History of the Software Industry*, MIT Press, Cambridge, MA.

Christensen, C.M. (1997) *The Innovator's Dilemma*, Harvard Business School University Press.

Cox, H., S. Mowatt and M. Prevezer (1999) 'From Frozen Fish Fingers to Chilled Chicken Tikka: Organisational Responses to Technical Change in the Late Twentieth Century', *Research Papers in International Business*, 18-99, Centre for International Business, South Bank University, London.

Cox, H., S. Mowatt and M. Prevezer (2003) 'New Product Development and Product Supply within a Network Setting: The Chilled Ready-meal Industry in the UK', *Industry and Innovation*, **10** (2) pp. 197-217.

Dodgson, M. (2000) *The Management of Technological Innovation: An International and Strategic Approach*, Oxford University Press, Oxford.

Dyson, J. and R. Uhlig (2004) *The Mammoth Book of Great Inventions*, Robinson.

Hayter, E.W. (1939) 'Barbed Wire Fencing – A Prairie Innovation', *Agricultural History*, **13**, pp. 189–207.

Hickman, R.P. and M.J. Roos (1996) 'Workmate', in E. Taylor (ed.) *Innovation, Design, Environment And Strategy*, T302 Technology, Block 2 Readings and Cases, Open University, Milton Keynes.

Howells, J. (2005) *The Management of Innovation and Technology*, Sage Publications, London.

Kamm, A. and M. Baird (2002) *John Logie Baird: a life*, National Museum of Scotland Publishing, Edinburgh.

Leonard-Barton, D. (1991) 'Inanimate Integrators: A Block of Wood Speaks', *Design Management Journal*, Summer 1991, pp. 61–67.

Levy, S. (1994) *Insanely Great: The Life and Times of Macintosh, The Computer that Changed Everything*, Penguin, London.

McElenhy, V.K. (1998) *Insisting on the Impossible: the Life of Edwin Land*, Perseus Books, Reading, MA.

Rothwell, R. (1994) 'Towards the Fifth-generation Innovation Process', *International Marketing Review*, **11** (1), pp. 7–31.

Sabbagh, K. (1996) *21st Century Jet: The Making of the Boeing 777*, Pan Books, London.

Spinardi, G. (2002) 'Industrial Exploitation of Carbon Fibre in the UK, USA and Japan', *Technology Analysis and Strategic Management*, **14** (4) pp. 381–398.

Trott, P. (2002) *Innovation Management and New Product Development,* 2nd edn, FT Prentice Hall, Harlow.

Ulrich, K.T. and S.D. Eppinger (2003) *Product Design and Development,* McGraw-Hill, New York.

Uttal, B. (1983) 'The Lab that Ran Away from Xerox', *Fortune*, 5 September 1983.

Van Dulken, S. (2000) *Inventing the 20th Century: 100 Inventions that Shaped the World*, British Library.

Wood, A. (2001) *Magnetic Venture: The Story of Oxford Instruments*, Oxford University Press, Oxford.

CHAPTER 7

INTELLECTUAL PROPERTY

OBJECTIVES

When you have completed this chapter you will be able to:

- Appreciate the rationale behind the various rights associated with intellectual property
- Identify the various types of intellectual property right (IPR)
- Distinguish the benefits conferred by intellectual property rights
- Differentiate the remedies available to those whose intellectual property rights have been infringed
- Show how intellectual property rights can be used to create value for their creator.

INTRODUCTION

This chapter is about intellectual property in general and intellectual property rights in particular. The nature of intellectual property and how it arises is considered. The different forms of intellectual property right are introduced and explained in detail. So too are the various mechanisms and procedures for registering these rights.

Having acquired an intellectual property right what do you do with it? The chapter goes on to explain the different forms of protection associated with each intellectual property right. Similarly the various institutions and individuals associated with the process of registration are analysed. There is then consideration of the forms of protection offered by intellectual property rights, and finally a short section showing how intellectual property can be used, in particular how it can be traded in order to create value for its creator.

INTELLECTUAL PROPERTY AND INTELLECTUAL PROPERTY RIGHTS

In so far as innovation involves a creative effort that gives rise to something new, it is concerned with intellectual property. However, creative effort takes many forms and is not just about new ideas for products or services, but includes designs and computer programs for instance, as well as literary and artistic works. Similarly, the concept of intellectual property extends beyond creative effort and encompasses aspects of commercial reputation. Whether one is talking about a new product, a new logo or a new publication, they all involve creative effort and as such are the result of intellectual activity on the part of individuals which requires the application of knowledge and skills leading to the creation of something new. Intellectual property may therefore be seen as the product of the application of knowledge and skills.

One of the problems with any form of intellectual activity is that the greater its potential value, the stronger the incentive for others to copy it so that they can reap some form of commercial gain. Copying has the potential to deny the creator any commercial benefit from his or her creative endeavour. This in turn acts as a major disincentive for potential creators. Why invest a lot of time, effort and resources in something, only for someone else to profit from it? With such a disincentive surrounding creative effort, society will be the poorer because those individuals with the creative intellectual powers to come up with new ideas will be less inclined to bother with innovation.

To overcome this potential loss to society, the law provides legal recognition of the ownership of the products of creative effort. In turn the proprietor can use this legal recognition to stop other people exploiting his or her property. Hence intellectual property rights create for the innovator a system by which he or she can benefit from their ingenuity. As with other forms of property, the proprietor, as the owner, may choose to sell the intellectual property rights or license them to others. However, since many intellectual property rights are monopolistic in nature, the state in many cases requires that certain rigorous tests have to be met before such rights will be granted.

In the UK a government agency, the Patent Office, is responsible for granting intellectual property rights. The main forms of intellectual property right are shown in Figure 7.1.

Figure 7.1 Forms of Intellectual Property Right

| | | Nature of the Intellectual Property Right (IPR) | |
		Creative	Reputation
Creation of IPR	Registration	Patent registered design	Trademark
	Inherent	Copyright design right	Passing off

Source: Bainbridge (1999: p. 4)

Each of these rights provides legal recognition of ownership for the creator of a new design, new product or written work. Ownership in turn gives the owner of the intellectual property right the right to stop others from exploiting his or her intellectual property, for a time at least. In this way intellectual property rights provide the creator with a system by which they can ensure that they benefit from their creative and intellectual endeavour.

PATENTS

A patent is a 20-year monopoly right granted by the state to an inventor. It is a reward for invention, designed to provide the patent holder with an exclusive right to it. Exclusivity gives the patent holder the right to prevent others from making or selling a patented product or using a patented process. It is a 'social bargain' designed to promote innovation and the spread of new ideas, and in return for exclusivity the patent holder is obliged to provide the Patent Office with full details of how the invention works (Figure 7.2).

Figure 7.2 Patents and the State

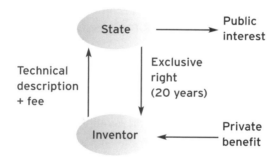

The origin of the patent system in the UK goes back to medieval times when the monarch granted individuals monopolies for a variety of purposes. The actual term 'patent' is derived from the Latin *litterae patentes* meaning an open letter intended for public display. Over time it became abbreviated from 'letters patent' to just 'patent'.

The function of patents is to stimulate and encourage innovation. Any innovator faces the problem that, if the invention is a success, it may well be quickly copied and he or she may derive little in the way of reward for their hard work and effort in developing it. If the likelihood of copying can be reduced, the chances of financial success for the innovator are greater. This is actually a matter of public policy, as the state has to weigh the benefit to the public interest of encouraging innovation against the cost (to the public) of a slower rate of diffusion (i.e. take-up) of the innovation.

In practice it has been found that a 20-year monopoly gives the innovator a sufficient incentive to invent, while ensuring that the resulting innovation does not command a premium price for too long because of the lack of competition. The state has to weigh the benefits of diffusion leading to the rapid spread of a new technology against the benefits to creativity and innovation arising from granting inventors exclusivity.

MINI-CASE: JAMES WATT

James Watt did not invent the steam engine. However, he did devise a number of improvements which dramatically improved the efficiency and performance of the steam engine. Watt was granted a patent for his improved steam engine (Von Tunzelmann, 1995). This gave Watt and his partner Matthew Boulton the exclusive right to produce and sell his improved steam engine during the opening phase of the industrial revolution in Britain. The fact that Watt held a patent did not stop others from copying Watt's design, but it enabled Boulton and Watt to take action in the courts against those engaged in copying and the courts duly ordered the copies to be destroyed and compensation paid. This enabled Boulton and Watt to establish and build up a thriving and very successful steam engine business supplying engines to mill owners anxious to build textile mills no longer dependent on water power. In the process it almost certainly slowed the rate of technological diffusion of steam power, at least until Watt's patent ran out in 1800.

In the UK the body that deals with patents on behalf of the state is the Patent Office. Before a patent is granted by the Patent Office, an inventor has to show that the invention is new. In practice this means meeting three conditions:

1 Novelty
An invention must be new. According to the Patents Act 1977 an invention may be considered new if it 'does not form part of the state of the art'. State of the art is all about whether the invention has been made public prior to the date at which the patent application is filed. Making something public is quite narrowly defined. Using an invention on a single occasion in one location would be sufficient for an invention to be part of the state of the art. In the 'windsurfer' case it was held that a 12-year-old boy who had built a sailboard and used it in public at Hayling Island in Hampshire had effectively anticipated a later patent for a sailboard (Bainbridge, 1999: p. 351). This has important implications for innovators. Demonstrating an invention, perhaps to potential investors, prior to filing a patent could easily jeopardise the eventual granting of a patent. Hence as Bainbridge (1999) advises, anyone contemplating field trials of a prototype invention ought to file a patent application before conducting any such trials, if there is a possibility of members of the public seeing the invention.

2 Inventive step
An invention must involve an 'inventive step'. Essentially this means that it must not be obvious. This raises the question of obvious to whom? The answer is that it must not be obvious to someone skilled in the art, which is to say a notional skilled worker. The notional skilled worker does not have to be an expert: rather he or she is implied to have a general knowledge of the subject. Hence, the notion of an inventive step implies that the apparatus being patented would strike someone with a reasonable general knowledge of the subject as incorporating something that constitutes a genuine invention. In the case of *Dyson Appliances Ltd* v. *Hoover Ltd* [2000] for instance it was held that, because the vacuum

cleaner industry was firmly committed to the use of bags in vacuum cleaners, a bagless cleaner was not obvious and therefore involved a genuine inventive step.

3 Industrial application

An invention has to be capable of being used in some kind of industry. This reflects both the history of patents, which were at one time referred to as 'industrial' property rather than 'intellectual' property and their practical nature. The requirement for an industrial application rules out scientific discoveries. The discovery has to be incorporated into some sort of apparatus, device or product if it is to be patentable. Alternatively, if it is produced by an industrial process it may be patentable. Thus, the discovery of the drug pencillin was not patentable as such but it could be patented when it was produced through an industrial process.

Providing it can meet these three tests then an invention is patentable. However, certain items are excluded (though this is effectively covered by the need for an industrial application). Scientific theories, mathematical models and aesthetic creations (e.g. literary or artistic works) are excluded. Computer software and business methods (at least outside the US) are among those excluded. It is possible to patent software-related products. If the software results in the introduction of a 'technological innovation' then it may be patentable. The critical point, in Europe at least, is that as long as the software brings about a 'technical effect' leading to an inventive step that goes beyond the normal physical interaction between the new program and the hardware leading to a technical improvement in the running of the computer or an attached device (such as making the computer memory usage more efficient), then it may be patentable. In the US around 15 per cent of the 170 000 patents granted each year are for software (Gapper, 2005).

MINI-CASE: BTG SUES AMAZON OVER TRACKING SOFTWARE

BTG, the British patent licensing company, is suing a group of American retailers including Amazon.com, the largest online retailer in the world, for allegedly infringing its rights over a technology to track customers' use of the Internet.

The company's lawsuit, filed in a Delaware court, claims the retailers are using a technique it has already patented to monitor when customers move from one site to another. The technology is important because retailers will normally pay a fee to other sites that direct traffic their way.

BTG is claiming an undisclosed amount of damages from the group, which also includes BarnesandNoble.com, the electronic version of America's pervasive bookshop chain. BTG buys patent rights to new technologies and licenses them to manufacturers. It also sets up its own companies to develop technology and is best known for its subsidiary Provensis, which is testing a revolutionary varicose vein treatment. The treatment ran into trouble at the end of last year when US regulators suspended its tests.

The company, originally set up by the government to protect and patent the country's inventions, has a history of taking on opponents much larger than itself to protect its wide-ranging intellectual property rights.

▶

Earlier this month BTG sued Microsoft and Apple for including patented technology in their operating systems that allows users to obtain software updates over the Internet.

In a statement about the latest lawsuit, BTG said: 'The suit asks for unspecified damages for past infringing activity and an injunction against future use of the technology.' It also said it had tried to reach an agreement but had failed. Fighting the case in court could take three years, making it possible that the parties will yet reach an agreement over the technology.

Ian Harvey, BTG's chief executive, said the patents are 'fundamental to the tracking of users for online marketing programmes', adding that the technology's commercial potential is 'significant'. BTG's shares moved ahead 4p to 140p.

Source: Griffiths (2004, p. 48)

Obtaining a patent (in the UK)

To obtain a patent, an innovator has to follow a procedure with a number of clearly defined steps:

1 Making an application to the Patent Office: the application, on forms supplied by the Patent Office has to contain:
   - A request for a patent
   - Identification of the applicant
   - A description of the invention

   The description has to be sufficiently clear and complete; otherwise the level of protection will be limited. Once the application has been received by the Patent Office it is said to be 'filed'.

2 Search and publication: once a claim for a patent has been filed and a search fee paid (within a 12-month period) a preliminary search will be undertaken by a Patent Office examiner who will go through the records of previous patents to see if the invention meets the necessary conditions and is in fact new. At this point the application is published.

3 Full ('substantive') examination: this is the final stage in the process. The applicant pays a further fee within six months of publication and detailed examination of the description then takes place to see if it meets all the relevant legal requirements of the Patents Act 1977. Attention will focus on whether documents reported at the search stage and any others which have come to light since indicate that the invention is not in fact new or is obvious. The applicant gets a report and may make amendments at this stage. Once the examiner is satisfied that it meets all the requirements the patent is issued.

What protection does a patent give?

A patent will not stop others from copying the invention. As with most forms of intellectual property, a patent is a legal right that is enforceable by legal action: that is to say, if an infringement of the patent occurs the inventor has to take legal action to secure a remedy. If the court finds that an infringement has occurred then the penalties can be heavy and might include an injunction to prevent further sales of the copy, a requirement that all copies are handed over to the patent holder, and damages. Such penalties send out a clear message deterring would-be copiers. This is the power of patents. Patents assist and aid the process of innovation. The innovator may have to resort to the courts, but if the patent has been properly drafted, it should enable the innovator to eliminate copies at least while the patent remains in force.

REGISTERED DESIGNS

There are two forms of protection available for designs: registered designs and design rights. Registered designs require registration, design rights do not. While there is considerable overlap between the two forms of protection, registered designs cover designs where outward appearance is important, while design rights cover designs that are more functional. It is important to note in both cases the protection applies to the design and not the product or article.

The purpose of a registered design is to provide protection for the look or appearance of products. Registered design tends to apply particularly to aspects of products such as shapes or surface patterns. As a form of protection a registered design is likely to be particularly appropriate for products where appearance is a key attribute, such as jewellery, glassware and furniture. However, the Designs Registry, the part of the Patent Office that deals with this particular intellectual property right, receives applications from every branch of technology including: cars, laptop computers, washing machines, tennis racquets and even such mundane items as paperclips!

What is a registered design?

In essence it works in a similar way to a patent in that it too is a monopoly right that can be bought and sold and which is used to stop copying. However, it is a monopoly right that covers the outward appearance of an article. The main features of a registered design are:

- The design must be new and materially different.
- It covers appearance resulting from the lines, contours, colours, shape, texture or materials of a product.
- Two exceptions are – must-fit and must-match.
- Two- and three-dimensional objects are covered.
- Duration – five years but it can be extended to 25 years.

As with patents, a design has to be new if it is to be registered. A design is regarded as new if it has not been made public in the UK. There is more flexibility than with patents, because a design can be shown for purposes such as marketing during the 12 months preceding registration. Just as patents involve disclosure of the details of the invention to the public, so too designs that have been accepted as registered designs are open to public inspection at the Patent Office.

The significance of appearance highlights the main distinction between this form of intellectual property right and patents. For registered designs outward appearance is important – one cannot rely on function, operation, manufacture or material of construction of an article, as in the case of a patent. What constitutes an article has recently been extended by an EC Design Directive to include, '...any industrial or handicraft item intended to be assembled into a complex product, packaging, get-up, graphic symbols, and typographical typefaces, but excluding computer programs'.

In terms of the two types of exception, the must-fit exception relates to function and means that it is not possible to gain a registered design for a design that is purely a matter of function. Therefore, one could not gain a registered design for the jaws of a spanner, because the jaws form part of the function of a spanner. Similarly the must-match exception relates to parts of a design. One cannot gain a registered design for part of a design where the part is determined by the shape of the whole. It is not possible to gain design rights for things like the front wing panel of a car, because the shape is determined by the overall shape of the car.

What does one gain from a registered design?

As with patents the main benefit is the exclusive right in the UK to make any article to which the design has been applied. The significance of this right is that the owner can then take legal action against anyone who infringes upon this exclusivity, which in turn forms a deterrent to would-be copiers.

DESIGN RIGHT

Registered design is not the only form of protection available for designs. In addition there is an automatic form of protection called 'design right'. Design right operates rather like copyright. It is a form of protection that arises automatically when a design is created. Though arising automatically, some form of tangible evidence of its creation is required.

To be eligible for design right a design must be 'original': that is to say, it must not be commonplace within its field. Similarly design right only covers the shape and configuration of an article, it does not extend to two-dimensional designs.

As an intellectual property right, design right does not provide the exclusivity that comes with a registered design. Instead it provides a right to prevent copying. Design right lasts for 10 years from the date the product was first marketed, subject to a limit of 15 years from the date the design was created.

TRADEMARKS

Introduction

Intellectual property does not only apply to the products of creative effort but also covers commercial reputations. Specifically this refers to trademarks. A trademark is a sign used to distinguish the goods or services of one trader from those of another. The law defines a trademark as,

> 66 ... any sign capable of being represented graphically which is capable of distinguishing goods or services of one undertaking from those of other undertakings. 99

Typically the term covers words, logos and pictures, although these days it has been extended to include other forms used to identify particular goods or services.

Registration of a Trademark

Trademarks have been an important feature of commercial life for a very long time. Since medieval times they have been used by traders to differentiate their goods from those of others. Trademarks have taken many different forms. In the eighteenth century, as shops became more widespread, signs were used by shopkeepers to denote the type of goods they were selling. In the nineteenth century the appearance of manufactured and standardised consumer products led to trademarks being used by manufacturers to differentiate their products. Among the first products to use trademarks in this way were everyday household items like soap, tea and chocolate.

As the means for communicating with customers have been extended and have become vastly more sophisticated (i.e. through television, animation, computer graphics, simulation and special effects), so the scope for differentiating products has expanded and trademarks have become more widely used. Developments in marketing such as branding and relationship marketing have increasingly led companies to enhance the image and reputation of their products and services, and trademarks have normally formed an important part of this process. Similarly, the increased emphasis on merchandising has also led to trademarks assuming greater importance.

Although some protection is available for trademarks without registration through common law by means of an action for 'passing off', registration of a trademark provides the most comprehensive protection for a name, brand name, logo or slogan. Registration via the Patent Office lasts for ten years and can be renewed indefinitely. It was first introduced in the UK in 1875 through the Trade Marks Registration Act of that year. The very first registered trademark was registered by the brewing concern Bass in the form of a red triangle symbol for one of its pale ales. The trademark is still in use today and Bass reckons that over the years it has had to deal with 1900 cases of infringement (Bainbridge, 1999).

In the past only words or logos or combinations of the two were registrable, but the Trade Marks Act 1994 was a landmark piece of legislation that greatly expanded the range of things that could be registered as trademarks. Among the more recent and more unusual registrations have been:

- The Coca-Cola bottle
- A Chanel perfume bottle
- Bach's 'Air on a G-string'
- The colour green
- The sound of a dog barking
- The colour yellow
- The slogan 'exceedingly good cakes'

These registrations reflect the new classes of trademark that were eligible for registration under the 1994 Act. Specifically the items that can now be registered include:

- Domain names
- Logos
- Music
- Slogans
- Colours
- Shapes

Another change introduced by the Trade Marks Act 1994 is in relation to infringement of trademarks. The act places a statutory duty upon Trading Standards officers to take action against those who trade in counterfeit goods using unauthorised trademarks. Trading Standards officers have the power to seize counterfeit goods, thereby assisting in the enforcement of trademarks.

In order for a trademark to be registered, an application has to be made to the Patent Office. This needs to satisfy a number of criteria:

- Section 1(1) of the Trade Marks Act 1994 which specifies what can be registered and now includes colours, shapes and pieces of music
- Distinctiveness – in the sense that the trademark singles out the company and its product from its competitors
- Non-deceptiveness – in the sense that it should not in any way mislead the public or lead them to believe that the product has attributes that are not in fact present
- No conflict with existing trademarks

MINI-CASE: MR MEN

More than 100 million of the Mr Men books have been sold since Roger Hargreaves published the first in 1971, making him the second-best selling author in the UK after JK Rowling, the creator of Harry Potter. The Mr Men phenomenon began when Roger Hargreaves' son Adam, then aged seven, asked what a tickle looked like. Mr Tickle was followed by a further 41 Mr Men. The books were followed by a television series narrated by Arthur Lowe which ran from 1974 to 1985. However, Roger Hargeaves, who died in 1988, perhaps by virtue of his background in advertising realised that there might be more to the Mr Men than children's books and a television series. He endeavoured to protect his intellectual property by registering the Mr Men as trademarks. This enabled him to negotiate a series of licensing agreements with manufacturers. Thus, for those with fond memories of the books it is possible to buy a wide range of Mr Men merchandise including Mr Bump plasters, Mr Perfect boxer shorts and Little Miss Naughty underwear. Trademark registration enabled Roger Hargreaves and the executors of his estate to capitalise on the intellectual property that his creative talent was able to realise. The licensing agreements are estimated to be worth £130m.

Source: BBC News, 29 October 2001

PASSING OFF

Passing off is effectively a common-law version of trademark registration. It has no requirement for registration. Instead, passing off is based on the principle that a trader must not sell goods under the pretence that they are the goods of another. To do so is to commit passing off. Essentially passing off is a form of misrepresentation. For an action for passing off to succeed, the trader whose goods have been passed off has to show not only that misrepresentation has occurred, carried out by another trader in the course of his/her trade, but that goodwill or reputation is attached to his or her goods in the first place and that his or her trade has been damaged normally through loss of sales, though it could equally be through damage to or dilution of reputation.

MINI-CASE: THE JIF® LEMON CASE

An example of a passing off action was the JIF® lemon case (*Reckitt and Coleman Products Ltd* v. *Borden Inc.* [1990]). One of Reckitt and Coleman's best-known consumer products is JIF® lemon juice, a household bakery product used in cakes and pastries. Sold in yellow, plastic, lemon-shaped containers, JIF® lemon has been a familiar sight on supermarket shelves for many years. However, in 1990 Reckitt and Coleman, the manufacturers, took action against a competitor who was also selling lemon juice in a similar, though larger, lemon-shaped container. This was prior to the Trade Marks Act 1994 which permitted the registration of colours and shapes as trademarks. Consequently, Reckitt and Coleman took out an action for 'passing off' claiming that the competitor was misrepresenting their product. To prove their case Reckitt and Coleman produced evidence from hundreds of consumers all of whom stated that they had been confused by the appearance of a very similar product. The Court found in favour of Reckitt and Coleman on the grounds that members of the public had been confused, indicating that the competitor had tried to pass off its product as genuine JIF® lemon juice. The view of the court was that the shape and colour of the product ingeniously alluded to its contents and was generally recognised by the public as JIF® lemon juice.

COPYRIGHT

Copyright is an intangible right that comes into effect through creative effort that gives rise to a wide range of creative works including literary, artistic and musical works. Other types of work are also included such as films, sound recordings broadcasts and typographical layouts, though with these, unlike the others, there is no requirement for originality. Copyright is automatic and comes into existence upon creation of the work.

Copyright confers an exclusive right to certain actions in relation to the work upon the owner. The actions concern the exploitation of the work and include selling copies, giving others permission to copy and the like. The significance of the exclusive right is that if someone who is not the copyright holder and does not have permission to copy makes copies and sells them, then the copyright owner can sue for infringement and seek redress, such as an injunction to forbid the sale of copies, together with damages.

The range of works covered by copyright is extensive. Literary works covers a great deal more than books. In the commercial field it extends to authorship of many things of a technical or commercial nature that perhaps would not at first seem to be literary works such as technical reports, equipment manuals, databases and customer lists as well as engineering and architectural drawings and plans. Of particular significance these days is that copyright extends to computer software. Given the growth of computer applications this is a rapidly expanding field. Thus, for a wide range of new products and services, such as video games for instances, intellectual property rights may be less a matter of patent protection and more a matter of copyright.

The copyright owner is normally the author who created the work, and as with many other forms of intellectual property right he or she can assign it or sell it to another. However, there are circumstances where copyright may be conferred not on the author but on others. For instance, if authorship occurs during the ordinary course of employment then copyright belongs to the employer. Similarly, a contractor who creates a work will retain copyright unless the terms of the contract specify that he or she will not retain copyright.

The exclusive right conferred by copyright extends to a wide range of activities that includes:

- Copying or reproducing
- Adapting
- Distributing
- Issuing and renting
- Public performance
- Broadcasting

However, copyright legislation does provide for certain activities of an educational or academic rather than commercial nature to be undertaken without infringement. For example, it is permitted to copy at least part of a work for the purpose of private study or research. Similarly reviews and other works of criticism can copy part of a work.

Figure 7.3 Copyright Timescales

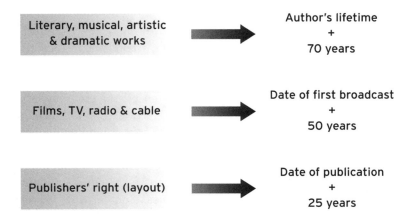

<table>
<tr><td>Literary, musical, artistic & dramatic works</td><td>→</td><td>Author's lifetime
+
70 years</td></tr>
<tr><td>Films, TV, radio & cable</td><td>→</td><td>Date of first broadcast
+
50 years</td></tr>
<tr><td>Publishers' right (layout)</td><td>→</td><td>Date of publication
+
25 years</td></tr>
</table>

The duration of copyright varies according to the nature of the work. As Figure 7.3 shows, the longest period of copyright protection relates to literary, musical, artistic and dramatic works where the protection lasts for 70 years beyond the lifetime of the author.

MINI-CASE: PETER PAN

27 December 2004 was the hundredth anniversary of the first public performance at the Duke of York's theatre in London of J.M. Barrie's children's classic, Peter Pan. The anniversary was marked by a number of special performances of Peter Pan as well as a number of radio documentaries about the author. Many of these detailed some of the famous performances of Peter Pan not just in Britain but in other countries such as the US. For the Great Ormond Street Hospital for Sick Children in London, the renewed interest in and performances of Peter Pan were of particular interest. Barrie donated the copyright of Peter Pan to the hospital, a move that was to benefit thousands of children over the years. However, for the hospital the anniversary itself was perhaps a mixed blessing since it was a reminder that the period covered by copyright is limited. J.M. Barrie died in 1937, so that the year 2007 would be the year in which Peter Pan went out of copyright thereby depriving the hospital of this source of income.

Source: Strachan (2004: pp. 20-21)

LICENSING

One of the key features of intellectual property rights (IPR) is that they provide scope for licensing, that is to say where the IPR associated with an invention has been legally established through a patent, the holder can then permit someone else to produce the invention in return for a fee. Such an arrangement is usually known as a 'licensing agreement'.

One of the attractions of licensing agreements is that the inventor does not have to complete the final stages of the innovation process, such as manufacturing and distribution. Where the inventor is an individual or a small company this can be a very important consideration. Licensing provides a means whereby small start-up businesses, lacking financial resources and complementary assets (e.g. reputation and brand name, marketing expertise, merchandising capability, product support facilities, etc.), can commercialise their technological innovation. Licensing not only means that the innovator does not have to find the capital expenditure required to build or buy the assets required, it also reduces the risk (Teece, 1986).

Both James Dyson with his dual-cyclone vacuum cleaner and Ron Hickman with his Workmate® portable workbench were individual inventors. They did not work for a company or have the backing of a company behind them. Consequently, neither planned to produce their invention themselves. Both tried to interest large, well-established consumer-product companies in their invention and persuade them to take out a licence. Unfortunately both Dyson and Hickman, despite a great deal of effort, found it extremely difficult to persuade a company to adopt their invention and agree to purchase a licence, possibly because in both cases the invention was unlike anything then on the market. What is significant is that both men felt that licensing was the most sensible course of action.

They recognised that they did not have the expertise or the resources to undertake the final commercialisation phase of the innovation process. Equally both men recognised the importance of asserting and protecting their intellectual property (i.e. their inventions) through patents. As it turned out both men did eventually find a company willing to take out a licence. James Dyson persuaded a Japanese company, Apex, to take out a licence for his dual-cyclone technology, while Ron Hickman had to start manufacturing and selling his portable workbench on a small scale before Black & Decker agreed to take out a licence.

MINI-CASE: INVENTIVE EMPLOYEES COULD BE AWARDED ROYALTIES

Employees who dream up lucrative inventions could share in their employers' profits under government plans to radically change patent law.

The move would be the biggest shake-up for a quarter of a century. The Department of Trade and Industry is considering ways for inventors to share the fruits of their own research, such as awarding them a percentage of royalties.

Employees currently have little entitlement to the benefits brought by inventions where their employer owns the patent.

The proposal comes in a consultation launched by the DTI and the Patent Office as part of moves to bring the UK into line with changes to the European Patents Convention. Patent laws in Britain have not been significantly updated since 1977.

The DTI said it had an 'open mind' on how such a scheme could work and would listen to industry's views on the subject – due by 19 February – before forming an opinion.

But drug groups said the moves would be almost impossible to implement. 'Hundreds of people contribute to a product's development. It would be extremely difficult to identify specific individuals who should be rewarded,' said one GlaxoSmithKline manager.

In theory, employees can claim compensation if they can prove the patent resulting from their invention has brought 'substantial benefit' to the company.

But Jeremy Philpott at the Patent Office said: 'That has proved to be an impossibly high barrier. I think there have been only two cases in 25 years and both failed.'

Source: Sherwood, Financial Times 14.12.02

CASE STUDY: TRAMPLED UNDERFOOT – HOW BIG BUSINESS HIJACKED THE UGG BOOT

Tony Mortel's hair is standing on end, an effect created by equal doses of gel and outrage. 'Who do they think they are?' he fumes. 'Telling us what we can and can't call our product, trying to stop us from making a living. Well, they can stick their demands where the sun doesn't shine.'

Tony comes from seven generations of boot-makers and for the last 45 years his family has been making Uggs, the once dowdy sheepskin boots now worn by the likes of Gwyneth Paltrow and Kate Moss. Their factory in Australia's Hunter Valley turns out 16 000 pairs a year. At least it used to, before a large US company across the Pacific Ocean began taking an unwelcome interest in their affairs.

Staff at Mortels Sheepskin Factory had just returned from their Christmas break when a letter arrived from the Melbourne solicitors of Deckers Outdoor Corporation, a California-based conglomerate. The letter, which was sent to 19 other Australian firms, informed them that Deckers owned all rights to the name Ugg and instructed them to stop using it or face litigation.

'I just laughed,' says Tony. 'I thought they were crazy. I threw it in the bin.' But it was no laughing matter. Soon afterwards, at the instigation of Deckers, Mortels was ejected from eBay, the Internet auction site where it had been selling Uggs to American consumers. Last Wednesday, it was ordered by Icann, the Internet regulatory body, to stop using 'Ugg' in its domain name.

The two dozen traders affected by such legal moves are reeling from shock and disbelief. For decades, they have been part of a thriving cottage industry founded on an Australian product that – according to folklore – dates back to the 1920s, when shearers used to wrap sheepskins around their feet to keep warm in the sheds.

Uggs, they argue, have always been called Uggs, originally an abbreviation of Ugly. No one bothered with trademarks, because Ugg was a generic term. Everyone knew it meant a comfortable, flat-heeled sheepskin boot, although – until the current fashion craze – few people admitted to owning a pair. Brian Iverson, owner of Blue Mountains Ugg Boots, says of Deckers' demands: 'It's like saying you can't call a car a car.'

The problem is: someone did bother with trademarks. In 1971 a local surf champion, Shane Steadman, decided to capitalise on the growing popularity of Uggs among Australian – and visiting US – surfers, who were starting to recognise the appeal of a snug boot when they emerged shivering from the ocean. He began selling Uggs and registered the name.

Steadman was not the only Australian wave-rider with a sharp eye for a business opportunity. In 1979, so the story goes, Brian Smith arrived in New York with a few pairs of Uggs in his backpack. He set up a company, Ugg Holdings Inc, registered the Ugg trademark in 25 countries and in 1995 sold out to Deckers.

For a long time not a peep was heard from the new American owners of the iconic Australian boot. The company sent out a flurry of warning letters five years ago, but did not follow them up. According to Middletons, its Melbourne lawyers, it was only when Australian manufacturers began selling Uggs on the Internet to meet soaring demand overseas that Deckers felt obliged to crack down.

Not surprisingly the Australian firms – most of them small family outfits with a handful of employees – are unimpressed with the Santa Barbara-based company's arguments. They say Brian Smith was awarded the trademarks in error and are planning to have them rescinded, at least in Australia.

Their only other choice is to give up and go under – for without the name Ugg, they say, they cannot sell their boots. 'People around the world know them as Ugg boots', says Tony Mortel. 'My family has been marketing them as Uggs for 45 years. For Deckers to say we should give it all up, without compensation, is borderline monopolisation.'

The Australian traders have united under the banner of the Ugg Boot Footwear Association and set up a fighting fund to finance the forthcoming legal battle. Those

▶

waiting in limbo include Westhaven Industries, a disabled charity that employs 65 people at its factory in Dubbo, a small town in New South Wales. Ugg boots are the charity's most profitable product and, without them, the business would not survive.

Employees include Dougie Stewart, who has been making Ugg boots at Westhaven for 30 years and travels more than 60 miles each day to work. 'He's a brilliant worker and he loves what he does', says Gordon Tindall, the charity's general manager. 'If we had to close as a consequence of this, it would be devastating for our workers. This is all they know, and they won't get a similar job elsewhere.'

Gordon insists that Ugg is 'as generic as meat pie or tomato sauce', and says he has every right to use it. 'If it waddles like a duck and quacks like a duck, then it is a duck and not a chook (chicken) in my book', he says, adapting an oft-used Australian phrase. 'It's like the Ford Motor Company claiming that they own the word "sedan".'

When Westhaven received the letter from Middletons in New York, he says, 'my first thought was "bugger, we'll have to comply". Then I thought "why should I?" Our industry has agreed that we won't be bullied by these guys. We'll carry on doing what we've always done and let Deckers take some of us on.'

At Mortels, situated on a light industrial estate outside Maitland, about 100 miles north of Sydney, the latest Uggs – in this season's colours of pale blue, pale pink, lavender and denim – are arrayed in a shop emblazoned with 'Ugg Boots And Slippers' in huge lettering. Tony Mortel has been told to remove the word 'Ugg' from the window. He has not complied.

Inside the small factory, machinists are discussing conspiracy theories about Princess Diana's death and periodically checking the temperature; if it rises above 40 degrees Centigrade, they can go home early. It hovers, irritatingly, at 39.9 degrees. Beneath the laughter and good-natured banter, there is an edge of anxiety. 'If we can't make Uggs anymore, I'll have to find another line of work', says Marewa Lamb, stitching a pair of tan boots.

She and the others operate a mini-production line. Tony (the 'clicker') cuts out the pieces of tanned and dyed skin and passes then to Marewa, known as Ma, who sews on the heel support. Next in line is Wanda Herickwitz, who attaches the inner sole, and Andrew Cook stitches the whole thing together. Angela Daley binds the boot and adds the finishing details. Damien Lambert glues on the sole.

Cheerful, down-to-earth people, they have one word to describe the notion that they should stop calling an Ugg an Ugg. 'Stupid', says Angela. 'Everyone knows them as Ugg boots. If you changed the name people wouldn't know what you were talking about.'

Their views are shared by Tony's father Frank – now 71 and retired, but furious about the turn of events. Frank emigrated from Holland in 1958, bringing a few sewing machines, and set up a tiny sheepskin factory. Descended from a long line of orthopaedic boot makers, he made his first pair of fur-lined slippers for his wife, Rita, who wanted something to keep her feet warm. He then started making the slippers and boots commercially.

'We called them Uggs from the start', he says. 'Although I recall other names such as "woolly hoppers." I'm sure this American company is just trying to frighten people off.'

If that is true, the tactics have had the desired effect. Some manufacturers have excised the offending word from their trading names or websites. Westhaven no longer uses the word Ugg in its catalogues and price lists. Others, such as Uggs-N-Rugs in

Western Australia, are standing firm, but with trepidation. Brian Iverson, whose family has made Uggs for three generations, is resisting. 'Uggs are as Australian as the Harbour Bridge', he says.

Tony Watson, a partner with Middletons, says the portrayal of Deckers as 'some big bad aggressive American company that likes squashing small businesses' is unfounded. 'We don't want litigation, but people have to understand the bigger picture', he says. 'It was Deckers, he says, that transformed Uggs into a high-fashion item, spending $7 million on marketing over the past decade and sending the boots to personalities such as Oprah Winfrey. Now others are reaping the benefits. My client has developed a marketplace and is now trying to protect it', he says. 'They are certainly not going to throw their hands up and say, "We've invested all this money, we've built up the brand and registered the trademark, now we're just going to walk away."'

Among Deckers' competitors, those who sell over the Internet are the most vulnerable. Without 'Ugg' in their domain, or trading names, they will not be located by consumers searching the web. All searches will lead to Ugg Australia, the brand name under which Deckers sells the boots around the world.

The company appears determined to protect its dominant position in the US as well as among European consumers. Yet Australian traders say they have been exporting to the US and elsewhere for decades. 'Between us, we must have spent far more than Deckers on marketing', says Tony Mortel.

The irony is that, while Deckers is trying to prevent Australian traders from calling an Australian product a name by which it has always been known in Australia, it brazenly exploits Ugg's Australian origins through its choice of brand name. Claims that it uses American (rather than Australian) sheepskins are flatly denied by Tony Watson, although he admits that, as of a few months ago, 'some' Uggs are manufactured in China, with the rest produced in Australia and New Zealand. He compares 'Ugg' with 'Biro' and 'Hoover' which, although commonly used generically, are protected by trademark.

A false comparison, says Tony Mortel. In those cases, a product was developed and marketed and a name invented and trademarked. In the case of Ugg, all the hard work was put in by others, then Deckers came along and bought the name. 'We've put our heart and soul into this product', he says. 'It's our livelihood, our heritage.'

Tony Watson does not have an answer to this point. 'We'll no doubt get to the bottom of it if the case comes to court', he says. He adds that Australian traders should accept reality and develop another brand. 'How about "Surfer" Sheepskin Boots?"' he suggests.

Tony Mortel refuses to acknowledge the possibility of defeat. 'We're going to carry on fighting', he says. 'We know we're in the right, and we know we're going to win. It's just a matter of time.'

Source: The Independent, 17.02.04 by Marks

QUESTIONS

1 Why was Tony Mortel told to remove the word Ugg from his shop window?
2 What is the intellectual property in this case?
3 Why have problems with the use of the word 'Ugg' only recently come to a head?
4 How can Deckers instruct Australian manufacturers to stop using the name Ugg?

▶

5 In what ways has the Internet affected trademarks?

6 Why is it important for Australian firms to be able to use the name 'Ugg', especially in terms of their trade names?

7 What was Brian Smith's contribution to the present problem?

8 What are Deckers cracking down on and why?

QUESTIONS FOR DISCUSSION

1 **Which of the following items would not meet the criteria required for a registered design?**
   - **A portable CD player**
   - **A rubber sealing ring for the door of a washing machine**
   - **A toilet disinfectant container**
   - **A corkscrew**

2 Why is computer software not normally patentable? What other forms of protection are available?

3 Why are trademarks an increasingly important piece of intellectual property?

4 Why do companies accused of infringing a patent often mount a defence based on a counter-petition claiming that the patent is not valid?

5 What is meant by diffusion? What impact does the patent system have on the rate at which new technological advances are diffused?

6 What is meant by 'novelty' where patents are concerned and why is it important?

7 What is the Windsurfer test?

8 Why do inventors need to take particular care before publicising their inventions?

9 **Which of the following can be registered as a trademark?**
   - **A brand name**
   - **The shape of a container**
   - **A smell**
   - **A colour**
   - **A domain name**

10 What is meant by the term 'passing off'?

11 Why has the Internet been a very significant development as far as copyright is concerned?

ASSIGNMENTS

1 Why have trademarks become an increasingly important form of intellectual property?

2 What is intellectual property and how can a firm or individual protect it?

3 What remedies are available to holders of patents who find that their patents have been infringed?

4 Why do some patent holders choose not to license their invention/innovation to others?

5 Carly Fiorina, the former head of Hewlett-Packard, when asked whether the company was living up to its creative traditions under her leadership, would point out the company filed 11 patents per day under her leadership compared with only three per day when she arrived. Comment critically on this statement.

RESOURCES

Books

There is a comparatively large legal literature on intellectual property rights (IPR) that extends to some major legal texts dealing specifically with the subject. Notable examples are Bainbridge (1999) and Bentley and Sherman (2001). Comprehensive and authoritative though these tomes are, they provide a somewhat difficult read for those lacking a formal legal training. Much more user-friendly is Black (1989) which specifically focuses on IPR and is written for readers with a business rather than a legal background.

Newspapers

As one might guess from the references at the end of this chapter, newspapers can be a surprisingly good source of information about IPR. Most of the serious newspapers report important commercial cases about IPR. Less accessible but probably more important are the sections of these papers that deal specifically with Law Reports.

Websites

There are a number of websites that are also very informative. Several large legal firms provide excellent briefings on various types of IPR. Typical examples are http://lloyd-wise.com which provides: frequently asked questions; an overview of IPR; specific details of a whole range of aspects of IPR; and timelines detailing the process of patent registration. Other similar sites are http://humphreys.co.uk and http://berwin.co.uk. As well as the websites of legal firms there are the websites of patent agents. Their professional body, the Chartered Institute of Patent Agents (CIPA) is probably the best place to start. It provides a number of 'simple insights' into patents and other forms of IPR. A particular feature is a detailed account of the patenting process. A number of public agencies maintain websites that provide details of IPR. One of the best is: http://patent.gov.uk, the website of the UK Patent Office. It provides an excellent overview of IPR as well as information on specific types of IPR such as patents. The latter gives details of patent specifications held by the British Library as well as a step-by-step guide to the patenting process. There is also the website of the British Library itself at: http://www.bl.uk. The Department of Trade and Industry provides useful background information at http://www.innovation.gov.uk. This website is dedicated to helping individuals and organisations understand innovation. It provides explanations of the various forms of IPR using user-friendly language. Other similar websites include: http://www.invent.org.uk and http://businesslink.gov.uk. The last-named is the Business Link website which as well as giving details of various aspects of innovation, includes a section devoted to 'protecting your IPR'. Among the other aspects covered are:

- The importance of securing your intellectual property
- Conducting an audit of intellectual property
- Getting legal protection for your intellectual property
- Keeping your ideas secure
- Business names and domain names
- Respecting other people's intellectual property.

The websites of companies actively engaged in a variety of aspects IPR can also be useful. Some notable ones are: http://www.arm.com, http://www.BTGPLC.com and http://www.pilkington.com.

REFERENCES

Bainbridge, D. (1999) *Intellectual Property*, 4th edn, FT/Pitman Publishing, London.

Bentley, L. and B. Sherman (2001) *Intellectual Property Law*, Oxford University Press, Oxford.

Black, T. (1989) *Intellectual Property in Industry*, Butterworths, London.

Dyson, J. (1997) *Against the Odds*, Orion Busines, London.

Eaglesham, J. (2001) 'Hoover loses Dyson appeal', *Financial Times*, 5 October 2001.

Gapper, J. (2005) 'Europe is right to take a patent risk', *Financial Times*, 23 June 2005, p. 19.

Griffiths, K. (2004) 'BTG sues Amazon over tracking software', The *Independent*, 16 September 2004, p. 48.

Henry, J. and D. Walker (1991) *Managing Innovation*, Sage Publications, London.

Marks, K. (2004) 'Trampled underfoot: How big business trampled the Uggboot', The *Independent*, Review Section, 17 February 2004, pp. 1-3.

Sherwood, B. (2002) 'Inventive employees could be awarded royalties', *Financial Times*, 4 December 2002.

Strachan, A. (2004) 'The Lost Boy', The *Independent*, 27 December 2004, pp. 20-21.

Teece, D.J. (1986) 'Profiting from technological innovation: implications for integration, collaboration, licensing and public policy', *Research Policy*, **15** (6) pp. 285-305.

Tidd, J., J. Bessant and K. Pavitt (2001) *Managing Innovation*, 2nd edn, John Wiley and Sons, Chichester.

Von Tunzelmann, G.N. (1995) *Technology and Industrial Progress: The Foundations of Economic Growth*, Edward Elgar, Cheltenham.

HOW DO YOU manage INNOVATION?

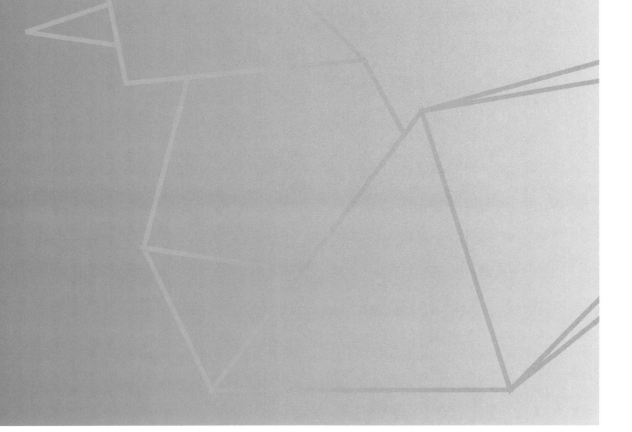

chapter 8

technology strategy

OBJECTIVES

When you have completed this chapter you will be able to:

- Distinguish between the different types of strategy
- Appreciate the part that business strategy plays in an organisation achieving its long-term objectives
- Explain the nature of technology strategy
- Recognise the various innovation strategies that are available as a means of exploiting innovations.

INTRODUCTION

Much of the attention directed towards innovation has tended to focus on the process of innovation. This is not surprising since this is where much hard work and effort takes place and where the real battle to develop something new occurs. However, having got something to work, having developed a prototype, perhaps even having got as far as patenting it, the innovator, whether as an individual or an organisation, is then faced with some important choices in terms of how best to exploit the innovation. It is these choices that form the focus of this chapter.

Because the choices have long-term consequences in terms of the life cycle of the innovation, they represent strategic decisions: that is to say, decisions that affect the long-term future of the innovation. They may even affect the long-term future of the organisation. Consequently, this chapter focuses on a variety of strategic decisions. It begins with strategic decisions that are the most wide ranging and that have the greatest impact. These form part of what is usually termed 'business strategy'. The chapter then proceeds to explore more specific strategies in particular those associated with the development of an organisation's technology. Technology strategy is dealt with in some detail first because it relates to a whole raft of issues to do with innovation. Only when we are clear about the nature of technology strategy does the chapter finally get round to focusing on innovation strategy.

Why does strategy matter when it comes to innovation? Often the resources required to bring an innovation to market require the organisation concerned to 'bet the company'. This means that the investment associated with the innovation, in terms of time, effort and money, is on such a scale that if the innovation fails, the future of the organisation is at risk.

MINI-CASE: BETTING THE COMPANY: EMI AND THE CAT SCANNER

EMI, or Electrical and Musical Industries to give the company its full title, is probably most familiar as the owner of several well-known record labels, including one that in the 1960s signed a then little-known band called the Beatles. However, in reality the electrical industries that also formed part of EMI were in their time as important as the musical ones. EMI had a strong record for innovation, having pioneered airborne radar during the Second World War and taken an active role in the development of early computers. Nottingham-born Godfrey Houndsfield, one of the company's senior research engineers on the computing side, was one of the first to develop software for pattern recognition – one of the early precursors of artificial intelligence. This work led him to explore the scope for linking pattern recognition to the processing of images including images generated by X-rays.

What Houndsfield did was to link together X-ray equipment and a computer. Conventional X-ray equipment was used to generate a succession of images, taken by moving in a 160° arc around a patient's head, which were then stored on a computer. He and his team then used the pattern-recognition software they had developed to process the data and display an integrated picture of the cross-section of the human brain. The resulting three-dimensional image of the brain was a big advance on the conventional two-dimensional X-ray image. EMI recognised that it had a significant 'invention' on its hands and steps were taken to obtain appropriate patent protection.

However, in exploiting its CAT scanner technology, EMI faced some big challenges. It had no experience of the medical equipment market. While it had manufacturing facilities, these produced defence products not medical equipment. It was also clear that developing a commercially viable CAT scanner was going to be expensive. In the event EMI decided to exploit the technology itself by investing directly in manufacturing and marketing. EMI enjoyed early success. Within four years of launching its first CAT scanner, the sales of its electronics division had quadrupled to £207 million. But EMI soon faced stiff competition from established manufacturers like General Electric and Toshiba and soon found itself losing ground, even though it had a technically superior product. It was not long before EMI had been overtaken in terms of market share. Although Godfrey Houndsfield was awarded a Nobel prize, the electronics division of EMI began making losses. With problems in its other divisions as well at this time, a financially weakened EMI was acquired by the Thorn group and within a year EMI withdrew from the medical equipment market and sold its scanner interests to General Electric of the US. For EMI exploiting a major technical advance in the form of the CAT scanner proved to be a case of betting the company on an innovation – and losing.

Source: Martin (1994)

THE NATURE OF STRATEGY

Most of the decisions that most managers make, even those at or near the top of an organisation, have nothing to do with strategy. Instead, most managers spend their time dealing with what are usually described as 'operational' decisions.

Operational decisions cover day-to-day things. Often they are very mundane. They are important nonetheless. Typically operational decisions cover things like:

- How much of a particular material should be ordered?
- Who is going to work which shifts?
- What is our production target for this plant this week?
- How big a batch of a particular product should we make?
- Should we change today's menu?

There are plenty of operational decisions surrounding innovation too. For example decisions surrounding the material to be used on a new prototype are operational, as are decisions surrounding the kind of training to be used to train staff who will deliver a novel new service. However, with all of these decisions, the consequences are basically short term. They will come into effect quickly and for the most part they can be overridden or changed fairly easily.

In contrast strategic decisions have long-term consequences and very often impact upon a lot of people. Hence strategic decisions are big ones. There is often a lot of money involved and many people are affected. Because their consequences can be so significant, strategic decisions are normally left to the most senior managers within an organisation, although particular individuals and groups lower down the organisation may be highly influential.

Strategic decisions and their associated strategies can typically be ordered in most organisations into a hierarchy (see Figure 8.1) that has business strategy at the apex, functional strategy in the middle and product/service strategy at the base.

Business strategy as its name implies is concerned with the strategy of the whole business. As a strategy, business strategy is mainly concerned with answering the question: how does the business compete?

Coming beneath the business strategy, one should find a suite of functional strategies corresponding to the various functional aspects of the business. These should complement the business strategy and give a more detailed picture of how particular parts or functions within the organisation will contribute to achieving the business strategy.

Figure 8.1 Hierarchy of Strategies

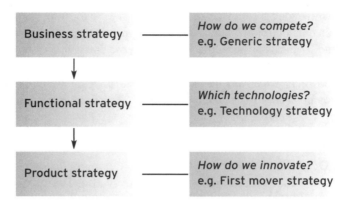

Examples would include a marketing strategy, a human resource strategy and an operations strategy. An organisation would have a marketing strategy that covers the marketing aspects of the business and an operations strategy that covers production and operations aspects. Included among the suite of functional strategies would be a technology strategy. Typically a technology strategy would show how the technology of the organisation is to be nurtured and developed. Beneath the functional strategies would be a number of product strategies, each of which would cover the long-term development of a product. For new or planned products there would be an innovation strategy that dealt with the various aspects of innovation required to achieve successful commercialisation of the product.

BUSINESS STRATEGY

Business strategy, as its name implies, is about the strategy of the business. In the case of a large organisation a 'business' may be a particular division that competes in a particular industry sector. It is this sectoral aspect that gives the business its coherence, which in turn is why it is possible for it to have a strategy. As a strategy business strategy is inevitably concerned with the long-term future of the business.

How long is the long term? It is normally a year or more, because this takes in decisions that do not fit the one-year budget cycle within which most organisations operate. The length of time required to bring decisions within what is usually regarded as the long term will vary from industry to industry and reflect the behaviour of consumers who buy the products of the industry. In an industry such as textiles which is closely linked to fashion the long term may be as short as six months. In contrast, for public utilities the long term may be ten years if that is how long it takes to design, build and commission major additions to capacity. Typically however the long term means three to five years.

Business strategy is concerned with defining where the organisation wants to be at some future point in time, such as five years hence and then selecting an appropriate means of bringing this about. In terms of where the organisation wants to be, business strategy is concerned with the means of achieving long-term objectives. Typically this involves deciding how the organisation is going to compete with its rivals. It is 'how' the organisation intends to bring about the desired future to which it aspires. Texts on strategic management lay down potential business strategies that organisations can employ. Two of the best known of the business strategies prescribed in this way are Ansoff's Product-Market matrix (Figure 8.2) and Porter's Generic strategies.

According to Ansoff's prescription, business strategies have to consider two crucial aspects: markets and products. In terms of markets the aspects to be considered are whether to seek new customers within the existing markets served by the organisation or whether to seek entirely new markets. The same logic applies to products where the choice is between seeking new products or services, or sticking with existing ones. The resulting two-dimensional matrix therefore provides four possible strategies:

Figure 8.2 The Ansoff Product-Market Matrix

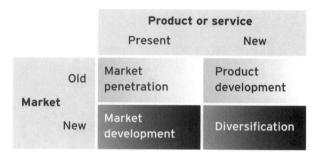

Source: Burgelman *et al.* (2001) 'Strategic Management of Technology and Innovation', 3rd edition, McGraw-Hill, New York

While the third of these strategies, product development, embraces innovation, it is important to stress that this is not what innovation strategy is concerned with. All of these are examples of possible business strategies that an organisation might employ in order to realise its long-term goals. They tell us nothing about how to conduct innovation. Clearly they do have implications for technology strategy, simply because some of these business strategies, such as product development, may well require a big investment in technology.

Porter's prescriptions are quite different. They focus not on markets and products as such but on the way in which organisations compete. Termed 'generic' strategies, to denote the fact that they can be applied in very different ways, according to Porter they comprise:

1   Cost leadership
2   Differentiation
3   Focus

In each case they prescribe 'how' the organisation should compete. In the case of cost leadership the notion is that the organisation should endeavour to be highly efficient so that its cost base is lower than its less efficient competitors. This clearly has major implications for how the organisation operates. Efficiency in terms of working practices and the eradication of waste are paramount. Differentiation is quite different. This time the emphasis is on trying to persuade consumers that the organisation's products or services are significantly different from those of its competitors. If consumers can be persuaded, this strategy offers the prospect of being able to charge customers a price premium in order to obtain the product. Focus on the other hand involves homing in on a narrow range of specific market segments rather than trying to compete on a much broader basis.

All of these are business strategies in the sense that they indicate what the business, as a coherent and viable commercial entity located within an industry sector, should do in order to get to where it wants to be.

TECHNOLOGY STRATEGY

The field of technology strategy has attracted considerable attention from researchers in recent years. Dodgson (2000: p. 134) in a recent study stresses the link between technology strategy and competitive advantage, suggesting that, 'technology strategy comprises the definition, development and use of technological competencies that constitute their competitive advantage'.

While this clearly suggests that technology strategy is of critical importance to organisations and their ability to compete effectively, another study (Clarke et al. 1995) of technology strategy in the UK notes that surveys in both Europe and the US reveal a failure on the part of top management to appreciate the significance of technology, suggesting that perhaps technology strategy is neither well understood nor well practised.

What is a technology strategy? Essentially technology strategy is concerned with decisions about the technology that an organisation uses in order to deliver products and services to customers. These decisions are likely to include:

- Which technologies should an organisation employ?
- How much money should the organisation invest in technology?
- How should the technology be developed?
- How can the technology be commercialised?

Burgelman et al. (2001) argue that technology strategy is to do with the set of technological capabilities that the firm chooses to develop. Virtually all organisations employ one or more technologies. However, only certain of these technologies will be crucial to an organisation and capable of materially influencing its competitive advantage. It is these core technologies that form the focus of technology strategy. Technology strategy is concerned with the long-term development of the core technologies that make up the technology base of the organisation. Development has to address two issues: the breadth of the technologies that are core technologies and their depth. Breadth refers to the range of technologies, which may be set narrowly where the technology base is highly specialised, or broadly if it encompasses a number of different technologies. Similarly, depth refers to the level of expertise associated with technology. This can range from a comparatively superficial level with only modest expertise where depth is limited, to extensive expertise where greater depth is present.

Where technology is a crucial feature of a product or service, as in high-technology industries such as computing, aerospace or biotechnology, technology strategy is likely to be critical to the competitiveness of the organisation.

While technology strategy is the responsibility of senior management, like other types of strategy it will be subject to external as well as internal influences. These influences will be the same as those that operate for other functional strategies, but with one exception. In the case of technology strategy, the picture is made more complex because one of the key external factors is the evolution of technology itself. Burgelman *et al.* (2001) set technology evolution alongside industry context as the main determinants of technology strategy (Figure 8.3).

Figure 8.3 Determinants of Technology Strategy

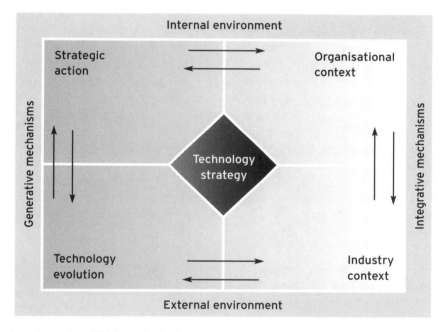

Source: Burgelman *et al.* (2001) 'Strategic Management of Technology and Innovation', 3rd edition, McGraw-Hill, New York.

In their model of technology strategy (Figure 8.4), technology evolution is one of four determinants of technology strategy. These determinants include internal ones in the form of strategic action and organisational context and external ones in the form of technology evolution and industry context. Technology evolution is subject to external forces such as the emergence of new technologies, that transcend the individual firm. Though the forces are different, the same is true of industry context. Strategic action is in the hands of management and will tend to interact with the prevailing organisational context. The significance of this perspective is that it shows technology strategy, not as something that is purely deterministic, but as the product of a number of interacting forces some of which can be controlled by the organisation and some of which cannot.

What does technology strategy consist of? Clarke *et al.* (1995) suggest that an organisation's technology strategy is concerned with three aspects of technology: acquisition, development and exploitation. To these three aspects one can add a fourth, namely – selection.

Figure 8.4 Technology Strategy

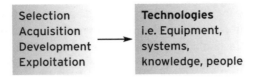

Selection

Selection is concerned with the choice of the most appropriate technology or technologies to enable the organisation to achieve its long-term goals and aspirations. Selection decisions can range from choices that affect small incremental changes to existing products, all the way to the identification and selection of emerging technologies with the propensity to transform not just an individual company but whole industries. Since investments in technology can prove very expensive and time consuming to reverse (because of the long-term consequences of such investments), choosing the most appropriate technology is a critical aspect of technology strategy.

The selection of the most appropriate technology or technologies to facilitate the business strategy of the organisation is likely to be largely a technical matter that has either to be left to technologists or at least is highly reliant on technical assessments provided by them. As Dodgson (2000) observes, technology selection is very much easier to do *ex post* compared to *ex ante*. Particularly when technologies are young and volatile, it is difficult to know how they will develop.

MINI-CASE: THE PERSONAL COMPUTER

When the first personal computers such as the Apple II appeared in the early 1980s there were those who predicted dramatic changes to the way we work, how we spend our leisure and how we organise our lives. Others on the other hand predicted that personal computers were just a passing fad and that they would never displace large mainframe computers.

Decisions concerning technology selection also come up against the problem of *sunk costs*. These are costs associated with transferring from one technology to another. Over time organisations make a big investment in a given technology. This investment extends to physical items such as capital equipment and systems, and stocks of materials, components and work-in-progress. It also extends to people, skills and know-how. The significance of these costs is that in all likelihood they cannot be transferred to the new technology and are in that sense sunk. If these costs have to be taken into account it may make decision-makers unwilling to opt for a new and untried technology.

Another point about new technologies is their source. Many studies have shown how new technologies often come from outside the existing industry or sector. The McLaren MP4 mini-case in Chapter 4 shows how, when McLaren's chief designer John Barnard wanted to innovate by employing carbon fibre technology in Formula One for the first time, he had to go outside the automotive sector, instead getting an aerospace company to build his first prototype. Last, it is worth remembering that technology selection decisions are very often strategic ones. The impact of moving to a new technology can be very far reaching, resulting not only in new products and services but new processes, new systems, new working practices, new skills and a whole lot more. When Formula One racing teams like McLaren switched to carbon fibre technology they had to acquire entirely new skills and equipment. With aluminium construction they had used techniques such as riveting and bonding to build the chassis or 'tub' that forms the basis of a modern racing car. With carbon fibre they had to acquire large autoclaves or ovens to 'bake' the chassis. Not only that, technology selection may also have profound significance for innovation, if the technology selected leads the way to new products and services. Selecting a new technology may set a company on a path to new innovations. In the light of this it is relatively easy to see how technology selection is a key element in any technology strategy.

Acquisition

Although any organisation will at any time normally have an existing technology base, comprising the skills and knowledge of its staff, machines and equipment, systems, patents and the like, nonetheless it is constantly having to replace and replenish it. Consequently, organisations are constantly making decisions about acquiring technology. Much of the time this aspect of technology strategy takes place at a relatively mundane level. Henry and Pinch (2000), for instance, in a study of Motor Sport Valley in Oxfordshire show how firms involved in the manufacture of racing cars rely heavily on recruiting staff, and the knowledge and expertise they bring with them, as means of acquiring technology. From time to time, however, every organisation is faced with major decisions surrounding technology acquisition. It may be in the form of a decision about capital investment where the organisation has to decide whether to purchase a major piece of capital equipment. Alternatively it may involve setting up and installing a new system or deciding whether to continue with a particular form of research and development. In each of these cases the organisation is likely to be faced with choices regarding the kind of technology that is involved. Should it stick with its existing technology? Should it consider a new or a different technology? It is at this point that a technology strategy comes into play. The technology strategy should guide those making these kinds of decision.

Clarke *et al.* (1995) point out that the acquisition element of technology strategy typically involves decisions concerning the most appropriate means of acquiring technology. Technology can be acquired via a variety of mechanisms. Each of these mechanisms has important organisational consequences and involves different degrees of integration: that is to say, the extent to which the technology is internalised and integrated within the acquiring organisation can vary. In organisational terms there is a spectrum between complete internalisation and complete externalisation:

- Direct investment
- Joint venture
- Subcontract R & D

Direct investment requires no third-party involvement while subcontracting entails a high degree of third-party involvement. Direct investment may take a variety of forms. It can include capital investment where the organisation buys new equipment incorporating the technology as well as the recruitment of staff with the necessary skills and expertise. It can also involve the purchase of the technology as a package through the purchase of another company that already has the technology and capability required. The use of mergers and acquisitions (M & A) in this way has the advantage that it is often easier and quicker, and this explains why many large firms in high-technology sectors possess a business development function whose specific task is to identify and evaluate new technology based firms (NTBFs) as potential acquisitions that can provide the technologies they are keen to acquire. Once acquired such firms then have to be assimilated into the organisation and this is not always easy.

A joint venture represents a more collaborative route to technology acquisition. It is likely to take the form of setting up a new company whose equity is jointly owned by the organisation with the technology and the one that wants to acquire the technology. Obviously the level of integration is lower and thus the extent to which the technology is actually absorbed into the organisation may be less.

Subcontracting involves the lowest level of integration. Technology is externalised as the acquirer effectively relies on another organisation's technology. Subcontracting has the advantage that it is a low-risk strategy, because the acquiring organisation commits little in the way of its own resources, and it is flexible since the scale of the subcontracting can be varied according to market conditions. This may explain why in recent years, with the growth of innovation through networking, it has proved increasingly popular. As the McLaren MP4 case study shows, subcontracting work to a technologically capable subcontractor as McLaren did when it got Hercules Aerospace to build a carbon fibre chassis (Henry, 1988) can be an effective way of acquiring a technology.

Development

Having acquired a given technology, by whatever means, it then has to be developed. Typically this is going to involve in-house research and development (R & D). The research element is likely to be geared to improving and enhancing the technology, while the development aspect is more about adapting the technology so that it can be incorporated into commercial products and services. Given that there is often a very large gap between a technology that is effective in the laboratory or workshop and one that is effective in the marketplace, development can be an expensive and time-consuming process. Even when the R & D function is successful in ensuring that the technology is robust enough for the rigours of the marketplace, the job is not over. Technology does not stand still. It has to be improved and enhanced so as to provide consumers with a better and more effective product when in a couple of years time the competition has caught up.

Although the development of technology is usually an activity that takes place inside an organisation, the degree of integration can vary. Car manufacturers for instance will sometimes join together to develop a new technology. Diesel engine technology has been around for a great many years having been widely used in commercial transport, marine and railway applications. However, in the 1990s an increasing number of car manufactures began to offer cars fitted with diesel engines. Given the greater fuel efficiency of diesel engines, diesel technology was attractive. However, most of the world's car manufacturers had little experience of diesel technology and so in a number of instances they formed strategic alliances, usually in the form of joint ventures, to develop new diesel engines. Alliances had the advantage that manufacturers could thereby share the high cost of developing new engines in a technology they were not familiar with. Similarly, pharmaceutical manufacturers increasingly rely on biotech companies to assist them in developing new drugs.

In the same way it is possible for firms to contract out the development of technologies they have acquired. However, it is unlikely that they will do this on a large scale because of the danger of 'hollowing out', where they find that they no longer retain control of one of their key assets, namely technology. The significance of this can be gauged by the fact that these days most firms devote considerable resources and effort to protecting technologies that they have developed, by means of patents and other forms of intellectual property right protection.

Exploitation

Having developed a technology and got it to work, exploitation is the means by which an organisation gets a financial return on the investment it has made in developing the technology. Essentially exploitation is where innovation comes in (see Figure 8.5). The exploitation and commercialisation of a technology requires innovation. However, exploitation can be carried out in various ways and there is not a single formula for innovation.

Those seeking to exploit a new technology through innovation have two important hurdles to overcome: risk and funding. Risk arises because innovation is an uncertain business. Given a high level of uncertainty with new technology, there is likely to be a considerable risk attached to innovation. The history of innovation is littered with products and services that failed to catch on. Two types of risk are associated with innovation: market risk and technological risk. Technological risk arises because of the possibility that the technology itself may fail. It simply may not work or fail to work in the way intended when the product gets into the marketplace. Market risk on the other hand is associated with consumer behaviour. Quite literally no matter what the marketing people say, consumers may not want the product or the service. The issue of funding arises because there is likely to be a substantial investment required to bring about successful exploitation and there may well be a considerable delay between the investment being made and the new product or service starting to generate revenue.

In the light of these two hurdles the would-be innovator faces some important choices. These choices are critical. Making the right choice may result in the innovation being successful and the innovator being well rewarded. Alternatively, making the wrong choice can easily lead to the innovator getting nothing. It is also the case that failure to make a decision can be just as bad. It can mean delay, resulting in others being the first to get the technology into the marketplace.

The main issue is whether to internalise the process or whether to externalise it. This is the choice that faces most innovators. Internalising means the organisation or the individual carries out the innovation. Externalising exploitation involves getting someone else to carry out the innovation. If the risks are high it may well be preferable to take the external route. Similarly, if funding is difficult, the external route has attractions. If risk and funding are not a problem then taking the internal route may be more appropriate and lead to a better return on the investment.

Figure 8.5 Technology Exploitation

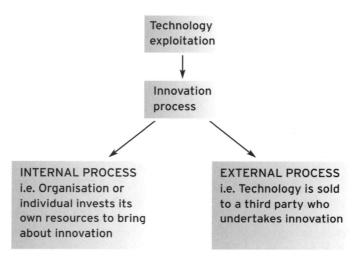

Before proceeding down either route the individual/organisation has also to address the vitally important issue of intellectual property rights (IPR). It has to protect its investment in developing technology by securing the rights to the intellectual property that the technology represents. This usually means patenting the technology. A patent does not stop someone from copying, but as Chapter 7 showed, if they do copy, a patent allows the holder to ask the courts to call upon the copier to desist. This can be a very powerful deterrent to would-be copiers and imitators. However, it is essential that the patent describes the technology accurately and effectively. Otherwise those who copy it may simply claim that they have not infringed the patent.

Internal Exploitation

Direct Investment
Internal exploitation through direct investment in the commercial application of the new technology to products and services is the most obvious way to exploit a new technology. Having acquired and developed the technology, the natural instinct is to exploit it, by manufacturing and marketing the product or service and selling it to consumers in order to generate an income stream.

Nonetheless, exploitation through direct investment on the part of the individual or organisation that developed it is never easy, and will almost certainly prove to be more complicated and more time-consuming, as well as more expensive, than expected. To exploit a technology internally requires not only manufacturing facilities (i.e. factory premises, equipment, stocks of materials and skilled labour), but experience and expertise of manufacturing, as well as access to appropriate marketing and distribution facilities. Where consumer goods are concerned the latter may be extremely expensive to acquire. Consequently, many firms, particularly if they are small and lacking the necessary facilities or access to resources, will seek an alternative means of exploiting technology.

The generic problems of finance and risk have already been outlined earlier in the chapter. However there are further obstacles that those contemplating internal exploitation via direct investment have to face, namely:

- Resources
- Timescale

Resources lie at the heart of direct investment. The main resource is clearly finance, because most resources have to be paid for. Money has to be found to lease premises, buy equipment, pay wages and buy stocks of materials. However, it is not purely a matter of money. Those who exploit technology frequently find they can acquire resources without having to pay for them, initially at least. As Chapter 10 outlines, financial 'bootstrapping' is a recognised route by which innovators ease the financial burden, especially where the incorporation of new technologies into new products is concerned. The example of Steve Jobs and Steve Wozniak, the creators of the most successful of the first generation of personal computers, is well known. Their first makeshift production line for the Apple II personal computer was in the garage (Linzmayer, 2004) of Jobs' parents' home. Nor was this particularly unusual. Garages and garden sheds have formed the initial manufacturing base of many innovations. The case of Martin Wood, the co-founder of Oxford Instruments, the company that pioneered the manufacture of superconducting magnets, shows that it is not only premises that can be borrowed. When Martin Wood began manufacturing his first superconducting magnets, he managed to borrow a variety of items of equipment, from his employer – Oxford University – and most of his design calculations were undertaken on the university's computer (Wood, 2001).

Even if resources can be acquired cheaply, there is still the issue of timescale. Exploiting a new technology takes time, which means that companies often have to cope with months if not years when the cashflow is all one way – out of the business. Alistair Pilkington's development of the 'float glass' process for making plate glass by drawing it across a bed of molten tin provides a salutary lesson in the problems of timescale. His employers had to wait seven years before this new technology resulted in a commercially viable process. Only when the new product or service goes on sale will cash start to flow into the business. Until that time arrives the negative cashflow has to be financed.

Derivative Strategy

A derivative strategy is something of a hybrid. Essentially it involves applying a new technology to an existing product. Clearly it is not a strategy that can be used by new firms, as there has to be an existing product and it has to have a reasonably strong position in the marketplace. However, for established products with a strong reputation it can be an attractive strategy. Keeble (1997) notes that the exploitation of new technologies is not confined to new-product development. It can also occur through what Rothwell and

Gardiner (1989a) describe as 're-innovation'. Rothwell and Gardiner note that in many industry sectors there are relatively few completely new products entering the market. Instead, one often finds a large number of what Rothwell and Gardiner (1989a) describe as 'post launch improvements'. In some cases these improvements extend to a complete re-design of the product. Rothwell and Gardiner (1989b) give the example of the development of a heat gun for paint stripping by the consumer products manufacturer Black & Decker. In this instance the product was new and Black & Decker were uncertain of whether it would be accepted by consumers. Given the uncertainty, they utilised the outer casing, motor, fan and switch from an existing product – an electric drill, thereby substantially lowering the development cost and the scale of the investment in innovation.

This was an example of a derivative strategy being used, though in this case the technology was not particularly new. Smith and Rogers (2004) show how in the aerospace industry manufacturers often use derivative strategies by adding a new technology to an existing product, citing the case of Rolls-Royce's Tay engine. This was a derivative of the company's well-established Spey engine which had been in service for some 20 years. By adding a new fan utilising the advanced, wide-chord fan technology of the company's RB211-535 engine, Rolls-Royce was able to develop an engine that was substantially more fuel efficient, quieter, lighter and more reliable. The use of derivative strategy in this way greatly reduced the development cost and the time taken to get the new engine into service.

External Exploitation

External exploitation essentially involves selling the technology to a third party so that they can exploit it. The individual/organisation that originated the technology will often want to retain an interest in the technology. To do this they will retain the patent rights while granting someone else the right to use the technology, providing certain conditions are met. Exploiting the technology in this way can come about for a number of reasons (Ford and Ryan, 1981):

- Lack of resources
  The individual or organisation that developed the technology (i.e. the patent holder) may lack not just the necessary finance to exploit the technology, but also the facilities or the staff. This may well occur with individuals or SMEs. Where finance is concerned it is not necessarily a case of having the necessary finance; access to finance may be just as important.

- Lack of knowledge
  The individual or organisation that developed the technology may not have sufficient knowledge of manufacturing, marketing, or distribution channels. This can often occur where scientists and technologists have developed a new technology, but lack the commercial background to exploit it. Given that innovation is often initiated by outsiders, it is perhaps not surprising that lack of knowledge can be a problem.

- A poor fit with the company's strategy
  The technology may be one that has applications in markets that are too small, too remote, or too specialised to be of value for the organisation that has developed the technology. Under these circumstances exploitation of the technology will not fit comfortably within the strategy of the organisation.

■ Lack of reach

If the technology has applications that span markets across the world, then it may be that the company does not have the global reach to market technology applications in all these markets. Under such circumstances it may prefer to sell the technology, but on terms that confine it to very specific markets.

If a firm decides to sell the technology that it has developed rather than exploiting the technology itself then it has a number of routes through which it can make the sale. Two of the commonest routes for selling a new technology are:

■ Licensing
■ Spin-offs

Licensing

As we saw in the previous chapter, licensing is open to organisations that can exercise control over their intellectual property rights. With patent protection in place one firm can grant another a licence to manufacture products using its technology. Under the terms of a licence the patent holder normally retains intellectual property rights over the technology but allows the licensee to use the technology in the products or services it develops, in return for a royalty fee. However, it is normal for the licensing agreement to provide for a royalty payment that will be a percentage of the purchase price of the product. Typically this is ranges between 3 per cent and 10 per cent. Licensing agreements will usually include a minimum level of royalty that is not a function of sales, so that the inventor is at least guaranteed a minimum return. But the exact financial arrangements will vary according to circumstances. If the firm selling the technology thinks there is a high level of uncertainty surrounding the products or services the licensee is planning, it may seek a significant initial payment for the licence and a smaller royalty fee.

Licensees will typically be organisations that possess assets the owner of the technology does not have, such as:

■ *Knowledge*

This might be market knowledge, resulting from a substantial market presence, or the result of experience working in a particular market. Alternatively it might be production knowledge derived from years of working in the trade. Whatever the field, it is likely to be 'tacit' knowledge based on experience rather formal knowledge resulting from qualifications. It is precisely because tacit knowledge can take a long time to acquire and is not easily codified or assimilated through conventional learning, that it may be better to grant a license to someone who does have this kind of knowledge.

■ *Access to Finance*

This might be cash, but is probably likely to mean loan capital or equity capital. Certainly it needs to be some form of long-term capital since this is what the innovation is likely to require. Sometimes conventional sources of finance will not be appropriate especially if there is a high degree of risk. In these circumstances it may particularly be 'patient capital' that is required: that is to say, the investor providing it needs to be willing to wait a long time for a return on their investment as innovations frequently take a long time to generate a return. Of course, it is not so much a case of having such capital as having access to individuals or organisations who themselves have access to it.

- *Motivation*
  Finally those seeking to exploit a technology have to have motivation. They particularly have to have the motivation to carry out innovation themselves. Innovation is a long and difficult process and requires the necessary motivation to see it all the way through. This is particularly important when one considers that innovation requires considerable commercial acumen. Inventors for their part frequently like inventing and may not be much interested in what are often commercial decisions. Under circumstances such as these the exploitation of technology through innovation is probably best left to some one else, i.e. a licensee.

According to Cesaroni (2003) the case for licensing, as opposed to in-house development, as a means of exploiting a proprietary technology, rests on three factors:

- Complementary assets in production and marketing
- Transactions costs associated with acquiring complementary assets
- Competition in the final product market

Complementary assets are the assets required to support the production and sale of products incorporating the technology and might include manufacturing expertise, marketing expertise, product support or training. If a company does not have these complementary assets, as was the case with Ron Hickman and the Workmate®, licensing is more appropriate than in-house development. Transaction costs are the costs of transactions/exchanges associated with in-house development (i.e. the purchase of complementary assets) or licensing the technology. If the transactions costs of licensing a technology are lower than the cost involved in purchasing the required complementary assets, then licensing is the more logical strategy. Finally, licensing the technology may be appropriate depending on the extent of competition in the final product market.

Pilkingtons, the glass manufacturers who developed the revolutionary 'float glass' process for manufacturing plate glass, for instance, relied heavily on licensing. Their thinking was that licensing the technology would both provide the company with an income and prevent other companies from developing an alternative process. The strategy proved highly effective. The first foreign licence was issued to the Pittsburgh Plate Glass Company in 1962 and by the 1990s the float glass process had been licensed to 35 companies in 29 countries. This was in addition to the 14 plants operated by the company itself (Henry and Walker, 1991).

In recent years there has been renewed interest in licensing as a strategy for the exploitation of technology. One study (Kollmer and Dowling, 2004) noted that licensing was no longer confined to small companies lacking the resources to exploit a technology fully, with many large well-established concerns using it to exploit their more peripheral technology assets while focusing their internal resources on core activities.

MINI-CASE: ARM HOLDINGS

Acorn Computer is a British computer firm that was among the first to develop a commercial Reduced Instruction Set Computer (RISC) processor or chip. The RISC chip is a central processing unit (CPU) that exchanges versatility for processor speed. Essentially the CPU executes a reduced number (i.e. set) of commonly used instructions very fast, thereby enhancing the overall speed of the processor (Khazam and Mowery, 1994). Hitherto CPUs employed a Complete Instruction Set Computer (CISC) chip that got the hardware of the CPU to do as much as possible per instruction (Afuah, 2003). RISC technology operates on a quite different basis with simple instructions that get the CPU to do less per instruction.

Unlike other chip manufacturers such as Intel and Motorola, ARM Holdings chose to exploit the new technology in a very particular way. By licensing, rather than manufacturing and selling RISC chip technology, the company established a new business model that redefined the way in which microprocessors were designed, built and sold. Licensing meant that ARM Holdings could focus on design work as a core activity, leaving others to undertake manufacturing. It also enabled ARM Holdings to quickly establish a market presence that in turn enabled the company to exercise a very powerful influence over the sorts of microprocessor used in a variety of consumer products including: automotive, entertainment, imaging, security and wireless applications. Among the everyday items using ARM Holdings RISC technology are mobile phones, digital cameras, DVD players, smart cards, set-top boxes, SIM cards, scanners and desktop printers. Some 80 per cent of the mobile phones shipped worldwide utilise ARM technology. All this from a company that makes nothing, preferring instead to license its technology.

Among the companies who are licensees of ARM technology are such household names as Motorola, Philips, Sharp, Sony and Texas Instruments, as well as a large number of specialist manufacturers of computer peripherals and similar devices. ARM Holdings now employ more than 700 people in plants in four countries, with design centres in Blackburn, Cambridge and Sheffield in the UK, Sophia Antipolis in France, Walnut Creek, California and Austin, Texas.

Spin-offs

Though they too represent an external form of technology exploitation, spin-offs require the organisation that wants to exploit the technology to take a more active role. A spin-off is where one organisation quite literally creates another organisation in order to exploit the technology. It is likely to be an attractive option where the technology is not closely related to the core technology of the firm that possesses the technology. It also requires individuals able to effectively manage the new enterprise. However, where the technology is not core, it can be a very attractive option because it avoids unnecessary distractions. The technology is exploited, but by somebody else.

In order to spin off the technology through the sale of a subsidiary company, it is necessary to 'package' the technology alongside the staff who have developed it and the associated corporate resources (e.g. equipment, facilities, etc.) and sell it off. The normal way of doing this is to locate the technology and the relevant human and other resources in a separate company and then sell off the company. This is what is meant by a 'spin-off' where the parent company divests itself of the technology by selling off the subsidiary company where it is based. There are a variety of ways in which it can be sold off including:

- A company flotation via an initial public offering (IPO)
- A management buy-out (MBO) where the company is sold to its managers
- Sale to a venture capital (VC) organisation who will invest in the company with a view to selling it off at some time in the future
- Sale to another company

Spin-offs have the attraction that they can generate a substantial lump sum, rather than the future income stream associated with licensing. If the parent company is anxious to re-invest the proceeds in other ventures (e.g. core business, new ventures, etc.) then clearly a spin-off has attractions.

PRODUCT/INNOVATION STRATEGY

Since innovation is all about the commercialisation of new inventions and new technologies, there is a very close link and a lot of overlap between the exploitation of technology and innovation. However, there are important issues surrounding the way in which an innovation is carried out, in particular regarding the timing of the entry of an innovation into the marketplace.

An innovation strategy is a strategy for carrying out innovation. Major issues surround *who* does the innovating and *when*.

The most distinct and recognisable innovation strategies are:

- First mover
- Follower/imitator

First Mover Strategy

The first mover strategy, as its name implies, is about being first to market a new product or service. The logic of this strategy is that by being first to introduce an innovation to the market an organisation will gain a dominant and enduring market position (Kerin *et al.*, 1992). In fact, there are other benefits to be conferred by being first. Gaining a premium price for the product because consumers are willing to pay more for novelty, or 'scooping the pool' as it is sometimes described, is one such benefit. Others include setting consumer expectations and building a strong market share.

There is no shortage of examples of organisations that have successfully employed a first mover strategy for innovation. Sony's Walkman and the Polaroid instant camera are two good examples. Sony revolutionised the audio equipment market when it brought out the Walkman music player. Though it has been copied by many other manufacturers, being the first to market such a product not only helped confirm Sony's reputation for innovation but also established the company as a major player in the audio equipment field. It also helped Sony to sell 20 million units in a little over five years (Martin, 1994).

In fact, whether or not a first mover strategy is successful is likely to depend on a number of factors. Teece (1986) suggests three such factors are likely to be critical:

appropriability, complementary assets and product standards. Appropriability refers to the extent to which the innovator is able to appropriate the benefits of innovation for himself/herself and stop others from copying. This is likely to be a function of the innovator's control of his/her intellectual property rights (IPR) through patents and the like. If the innovator is able to exercise a high degree of control over his/her IPR then the appropriability regime will be strong and a first mover strategy stands a good chance of success, because it will be difficult for other firms to enter the market with copycat products. The case of *Dyson* v. *Hoover* shows how a strong appropriability regime can lead to a successful first mover strategy. Complementary assets, as we saw earlier in the chapter, are the other assets, in addition to the innovation itself, that the innovator has to put in place to market his/her product. If these things are important in the market then the chances of a first mover strategy succeeding are diminished, simply because such facilities can easily be put in place by would-be imitators. Not only that, firms which are good at providing such facilities may find they can compete with copycat products that provide a lower level of performance, as was the case with EMI and the CAT scanner. Product standards refer to the extent to which the creation of a common standard facilitates interchangeability or training. An example of this is the home video cassette recorder (VCR) market where Sony's Betamax system which was introduced in 1975 (Cusumano *et al.*, 1992) was a commercial failure while JVC's VHS system introduced a year later succeeded. The success of VHS can be explained in terms of the success of JVC in persuading other companies to use the VHS standard, most notably through a strategic alliance with Matsushita which possessed massive VCR manufacturing capacity, but also through collaborative agreements with leading European manufacturers.

Follower/Imitator Strategy

If the first mover strategy is all about being first to market, the follower/imitator strategy is about taking a 'wait-and-see' approach. While innovation may seem to be primarily concerned with being first to market, in fact a surprising number of successful innovations have resulted from a wait-and-see approach.

This is no mere accident. There are circumstances when the follower/imitator strategy has attractions. These circumstances are:

- Free rider effects
- Imitation costs
- Scope economies
- Learning effects

The free rider effect occurs where the pioneer firm has to bear costs that followers do not have to bear. These might include the cost of putting an infrastructure in place, gaining regulatory approval or locating buyers. For the pioneer firm or firms the snag is that, if others can use or easily copy these facilities once they have been put in place, followers have an advantage. Imitation costs are costs associated with the development of a copycat product. If they are low, followers again have an advantage. Scope economies arise where several versions of a product share a common ancestry. If follower firms can produce new products that share elements in common with existing products then their costs will be lower. The very act of sharing helps to reduce costs, and firms that can develop new products in this way are likely to enjoy lower costs. Finally, learning effects occur where pioneer firms encounter problems and follower firms are able to learn from these mistakes and produce a better product. An instance of this would be Boeing's 707 airliner. It was not the first jet airliner. The De Havilland Comet entered service six years before Boeing's rival offering. However, the Comet suffered a number of early crashes due to metal fatigue. Boeing's engineers were able to learn from these mistakes and produce a more robust and reliable aircraft with the result that the 707 outsold the Comet more than tenfold and became the airline industry standard for commercial jet airliners.

If these sorts of factors are in evidence, then the follower/imitator strategy may well prove the most appropriate. Despite the apparent attractions of being first, when it comes to innovation, pioneering may well have its limitations. What matters is being able to judge when these conditions are likely to apply.

CASE STUDY: DOLBY LABORATORIES

On the back of most DVDs you will see a little box that that says 'Dolby Digital'. This is the trademark of Dolby Laboratories. It also appears when the credits roll at the end of a film, after the names of the actors and technicians, and on audio equipment and audio cassettes. But who are Dolby Laboratories and what do they do?

Dolby Laboratories developed a range of techniques for eliminating the extraneous noise that appears when making sound recordings. So effective are the noise-reduction techniques that they developed and so widely have they been used that today most of us have little idea that noise is a major problem when making sound recordings.

Dolby Laboratories was founded by Ray Dolby. Dolby himself is an American who initially worked for Ampex Corporation, one of the earliest manufacturers of magnetic tape recorders. He graduated from Stanford University before doing a PhD at Cambridge University in the UK. A keen amateur recording engineer, he began experimenting with ways of reducing the noise inherent in the process of recording audio signals on to magnetic tape. He established Dolby Laboratories in England in the late 1960s where he developed the A-type noise reduction system (A is for audio). This system was designed for a wide variety of audio noise-reduction applications, especially those where the noise arises when music is being taped at a recording studio.

At the time recording studios used analogue technology when recording. Dolby devised a technique that compressed the signals generated by noise during the recording process. His innovation was to break the audio spectrum down into different sections. This made it possible to eliminate the noise when the signal was de-compressed during the playback process.

Novel though Dolby's solution was, he faced a sceptical recording industry, where sound engineers had encountered a variety of noise-reduction systems but none had proved capable of reducing noise without degrading the sound in some way.

He originally saw his company as a research laboratory that would provide specialist recording equipment for the recording industry. Having invested $25 000 of his own savings and loans from friends, Dolby's fledgling company only had the financial resources to meet the demands of selling to this small, specialist niche market. He lacked the resources to take on consumer markets. On the other hand, the smallness of the professional recording equipment market (i.e. recording studios) meant he could exploit the technology without a massive capital investment and without attracting a lot of competition. Accordingly, Dolby Laboratories began by producing noise-reduction modules that could be incorporated into the complex recording equipment used to produce master tapes at recording studios. This quickly proved popular, particularly with studios recording classical music where sound quality was at a premium. Word spread within the comparatively small professional recording community that Dolby's noise-reduction system really worked, and the record industry began to make increasing use of the Dolby A noise-reduction technology as part of the recording process. From classical recordings it spread to other types of music recording, especially as more use was made of multi-track recording. Early records made using the Dolby A type noise-reduction system were often identified as such and the improvement in sound quality was commented upon favourably by music critics and reviewers in magazines and newspapers. Gradually the record-buying public became aware that the Dolby Laboratories name was associated with high-quality sound recordings.

Having established a reputation for his technology, Dolby was in a position of strength in terms of marketing, when he came to consider entering the consumer market for audio equipment. For this market Dolby developed a simplified version called Dolby B-type noise reduction. However, the consumer market was dominated by large companies like Sony, Phillips and JVC. Dolby realised that it was unlikely that the strategy he had used in the professional recording equipment market would work in the consumer audio products market. Quite apart from anything else, the massive scale of the market would have called for massive investment. Instead Dolby decided to license his noise-reduction technology to all manufacturers. Mindful of the reputation he had already built up, the terms of the licence agreement required manufacturers not only to pay a royalty fee for each item sold, but also to display the Dolby Laboratories trademark on their products. Although manufacturers of audio equipment paid a royalty fee on all tape players using the Dolby noise-reduction system, pre-recorded tapes were royalty free in order to help establish Dolby B as the standard, audio cassette, noise-reduction system. Thus it was that audio tapes (that preceded CDs) nearly always had a Dolby Laboratories trademark on them to signify that the recording could be played on equipment using the Dolby B noise-reduction system.

Licensing not only generated a substantial income, it allowed Dolby Laboratories to focus on delivering and extending its core knowledge and expertise – noise-reduction technology. Hence, when analogue sound recording technology was superseded by digital technology, Dolby Laboratories were at the forefront of the technology. Dolby Laboratories developed a further range of noise-reduction techniques. These much more sophisticated techniques can be applied in a variety of contexts including film recordings. In exploiting this technology, Dolby Laboratories again chose to go down the licensing route. Hence why the Dolby Laboratories trademark – Dolby Digital – appears on the back of our CDs.

Source: Ford and Ryan (1981); Hardcastle (1983); Wadsley (2005)

QUESTIONS

1 What is the nature of the intellectual property created by Dolby?
2 Why was it feasible for Dolby Laboratories to enter the professional recording equipment market as a manufacturer, but not the consumer audio equipment market, even though they both used the same technology?
3 Why was Dolby able to license his noise-reduction technology?
4 What are the merits of licensing as a means of exploiting a technology?
5 Why would licensing probably not have been a feasible strategy in the professional recording equipment market?
6 What choices was Dolby faced with in exploiting his noise-reduction technology?
7 How did technology evolution affect Dolby's technology strategy?
8 Which of Porter's generic strategies has Dolby pursued in terms of business strategy?
9 How did the changes in the industry context affect Dolby's technology strategy?
10 How did Dolby Laboratories develop its noise-reduction technology?

QUESTIONS FOR DISCUSSION

1 **What is the difference between strategic and operational decisions?**

2 What is meant by the term 'functional strategy'?

3 **What is the long term where business decisions are concerned?**

4 How is it possible for the long term to vary?

5 **What is business strategy mainly concerned with?**

6 What were the various business strategies put forward by Igor Ansoff?

7 **What are Porter's generic strategies?**

8 What is technology strategy?

9 **How do businesses acquire technology?**

10 What are sunk costs and how do they affect technology strategy?

11 **What are the different ways in which firms acquire new technologies?**

12 When is licensing likely to be an attractive technology strategy?

ASSIGNMENTS

1 Prepare a presentation outlining the case for a technology strategy based on either the internal or the external exploitation of a new technology. Assume that the presentation is being made either to the board of directors of the company that has developed the technology or members of the financial community who are planning to invest in the company.

2 Prepare a report that compares and contrasts the relative merits of internal and external approaches to the exploitation of technology.

3 Choose an innovation, and write a report showing how the innovation has contributed to the 'business strategy' of the organisation that developed it. You may well have to make some assumptions about the organisation's business strategy. Thus, if the company has a mature product portfolio that has changed little over the last ten years but has recently engaged in innovation, one might infer that the organisation is following what in terms of the Ansoff matrix is usually described as a 'product development ' strategy. Alternatively, if the company has embarked on innovation in a field quite unrelated to its current product portfolio, you might infer that its business strategy is one of 'diversification'.

4 As an employee of a company that has developed and patented a new technology, prepare a report for the company's senior management explaining how licensing might be used to exploit this technology. Pay particular attention to the circumstances where licensing may be appropriate and the benefits and pitfalls associated with it.

5 Select a well-known trademark. Prepare a report showing how and why owners of trademarks use them to promote their products and explain what potential the trademark offers when it comes to licensing.

RESOURCES

The field of strategy is extremely well catered for when it comes to texts. Texts such as Johnson and Scholes (1999) provide a comprehensive guide to most aspects of strategy. However, most of such texts place comparatively little emphasis on either technology or innovation. One of the exceptions is Grant (1998). This text not only considers the place of technology in strategy, it also covers technology strategy.

Technology strategy is a more specialised field and this is reflected in a very limited literature. Dodgson (2000) focuses on technology and strategy and has a substantial section devoted explicitly to technology strategy. Similar coverage is provided by Burgelman *et al.* (2001) and Afuah (2003). Though now a little dated, Dodgson (1989) provides a useful collection of readings on technology strategy.

Specific forms of technology strategy are comparatively well catered for. The literature on licensing for instance is extensive and includes key contributions from Grindley and Teece (1999) and Kollmer and Dowling (2004), as is the literature on joint ventures where important contributions include Dussuage and Garrette (1999) and Lorange and Roos (1993)

REFERENCES

Afuah, A. (2003) *Innovation Management: Strategy, Implementation and Profits*, 2nd edn, Oxford University Press, Oxford.

Burgelman, R.A., M.A. Maidique and S.C. Wheelwright (2001) *Strategic Management of Technology and Innovation*, 3rd edn, McGraw-Hill, New York.

Cesaroni, F. (2003) 'Technology Strategies in the Knowledge Economy: The Licensing Activity of Himont', *International Journal of Innovation Management*, **7** (2), pp. 223-245.

Clarke, K., D. Ford, M. Saren and R. Thomas (1995) 'Technology Strategy in UK Firms', *Technology Analysis and Strategic Management*, **7** (2), pp. 169-190.

Cooper, A. (1999) 'Material Advantage', *Motor Sport*, **LXXV**, 3, pp. 32-37.

Cusumano, M.A., Y. Mylonadis and R.S. Rosenbloom (1992) 'Strategic Maneuvering and Mass Market Dynamics: The Triumph of VHS over Beta', *Business History Review*, **66**, pp. 51-94.

Dodgson, M. (1989) *Technology Strategy and the Firm: Management and Public Policy*, Longman, Harlow.

Dodgson, M. (2000) *The Management of Technological Innovation: An International and Strategic Approach*, Oxford University Press, Oxford, p. 134.

Dussuage, P. and B. Garrette (1999) *Cooperative Strategy: Competing Successfully through Strategic Alliances*, J. Wiley and Sons, Chichester.

Ford, D. and C. Ryan (1981) 'Taking technology to market', *Harvard Business Review*, March-April, pp. 117-126.

Grant, R. (1998) *Contemporary Strategy Analysis: Concepts, Techniques, Applications*, 3rd edn, Blackwell, Oxford.

Grindley, P.C. and D.J. Teece (1997) 'Managing Intellectual Capital: Licensing and Cross-Licensing in Semiconductors and Electronics', *California Management Review*, **39** (2), pp. 8-41.

Hardcastle, I. (1983) 'Commercializing the Dolby System', *Les Nouvelles*, December 1983.

Henry, A. (1988) *Grand Prix Car Design and Technology in the 1980s*, Hazleton Press, Richmond.

Henry, N. and S. Pinch (2000) 'Spatialising knowledge: placing the knowledge community of Motor Sport Valley', *Geoforum*, **31**, pp. 191-208.

Henry, J. and D. Walker (1991) *Managing Innovation*, Sage Publications, London.

Johnson, G. and K. Scholes (1999) *Exploring Corporate Strategy: Text and Cases*, 4th edn, Prentice Hall, Hemel Hempstead.

Keeble, D. (1997) 'Small firms, innovation and regional development', *Regional Studies*, **31**, pp. 281-293.

Kerin, R.A., P.R. Varadarajan and R.A. Peterson (1992) 'First Mover Advantage: A Synthesis, Conceptual Framework and Research Propositions', *Journal of Marketing*, **56**, pp. 33-52.

Khazam, J. and D. Mowery (1994) 'The commercialisation of RISC: Strategies for the creation of dominant designs', *Research Policy*, **23**, pp. 89-102.

Kollmer, H. and M. Dowling (2004) 'Licensing as a commercialisation strategy for new technology-based firms', *Research Policy*, **33,** pp. 1141-1151.

Linzmayer, O.W. (2004) *Apple Confidential 2.0: The Definitive History of the World's Most Colorful Company*, No Starch Press, San Francisco, CA.

Lorange, P. and J. Roos (1993) *Strategic Alliances: Formation, Implementation and Evolution*, Blackwell, Oxford.

Martin, M.J.C. (1994) *Managing Innovation and Entrepreneurship in Technology Based Firms*, John Wiley and Sons, New York.

Rothwell, R. and P. Gardiner (1989a) 'The strategic management of re-innovation', *R & D Management*, **19** (2), pp. 147-160.

Rothwell, R. and P. Gardiner (1989b) 'Design management strategies', in M. Dodgson (ed.), *Technology Strategy and the Firm: Management and Public Policy*, Longman, Harlow.

Smith, D.J. and M.F. Rogers (2004) Technology Strategy and Innovation: The Use of Derivative Strategies in the Aerospace Industry, *Technology Analysis and Strategic Management*, **16** (4), pp. 509-527.

Teece, D. (1986) 'Profiting from technological innovation: Implications for integration, collaboration, licensing and public policy', *Research Policy*, **15**, pp. 285-305.

Wadsley, P. (2005) Personal interview with the author, 25 June 2005.

Wood, A. (2001) *Magnetic Venture: The Oxford Instruments Story*, Oxford University Press, Oxford.

CHAPTER 9

technical entrepreneurs

OBJECTIVES

When you have completed this chapter you will be able to:

- Explain the nature of entrepreneurship
- Distinguish different categories of entrepreneur
- Identify the key characteristics of technical entrepreneurs
- Categorise technical entrepreneurs
- Analyse the factors that lead to the growth and development of technical entrepreneurs.

INTRODUCTION

The phenomenon of Silicon Valley in California has helped over the last half century to counteract what was at one time the received wisdom: namely, that large firms were the source of most innovations. Silicon Valley has served as an exemplar to promote the notion that small firms have a part to play in innovation. The small firms that formed the basis of Silicon Valley were primarily technology based. Other locations around the world have their own versions of Silicon Valley but on a more modest scale. Singapore for instance has 'Winchester city' (Brown, 1998) while Taiwan has its own technology city, Hsinchu (Castells and Hall, 1994). The UK has no equivalent of Silicon Valley, but there are concentrations or clusters of small technology-based firms. Notable high-technology clusters in the UK include Cambridge (Castells and Hall, 1994) and Motor Sport Valley (Henry and Pinch,

2000) in Oxfordshire – both relatively recent creations and both populated by small technology-based firms. The growth of small, highly innovative, technology-based firms throughout the world from the beginning of the last quarter of the twentieth century onwards has resulted from the activities of what have come to be known as 'technical entrepreneurs'. Technical entrepreneurs are individuals who establish technology-based businesses that typically are active in innovation. They stand in marked contrast to the large, vertically integrated firms that were the leaders in innovation in the 'modern' period of the mid-twentieth century. In the twenty-first century, in many industry sectors, technical entrepreneurs play a leading role in innovation, making them very much a product of the fifth Kondratiev.

INNOVATION AND ENTREPRENEURSHIP

There are numerous examples of individuals who are high-technology entrepreneurs. Some of the better-known ones, the businesses they founded and their innovations include:

Bill Shockley	Fairchild Semiconductor	Transistor
Steve Jobs & Steve Wozniak	Apple Computer	Personal computer
Joseph Bamford	JCB	Hydraulic excavator
George de Mestral	Velcro SA	Fastener
Owen Maclaren	Maclaren Ltd	Child's pushchair
Alex Moulton	Moulton Cycles	Small-wheel bicycle
Dan Bricklin	Software Arts	Spreadsheet
Bill Gore	W.L. Gore & Associates	Gore-Tex fabric
James Dyson	Dyson Appliances	Dual-cyclone vacuum
Jeff Bezos	Amazon.com	Online book shop

These examples have a number of features in common. First, all of them had an idea or created an invention and then proceeded to innovate by developing their creation into a commercially viable product that went on to achieve commercial success. In some cases the commercial success was such that the innovation brought about major changes in the industry and the market. In Christensen's (1997) terminology they were 'disruptive' innovations.

Second, the innovations were not necessarily based on scientific breakthroughs. However, technology does lie at the heart of the innovation whether it is the adaptation of a new technology to an existing application, or the use of a new material or the use of technology for a new purpose, or the creation of a new technology for a new application. Hence the technical element is strong in all of these cases.

Third, what is striking about all of these cases is that the innovator had to create a business in order to innovate. None of the products is from an established large-scale manufacturing concern. Many of these individuals worked for a large organisation or tried to persuade a large organisation to take up their invention in order to bring about innovation. Unable to persuade someone to back their idea/invention, they had to go it alone and create a new venture (i.e. a new business) in order to innovate. It is this that makes them entrepreneurs as well as innovators and the fact that technology was an important part of their innovation means that they are all examples of technical entrepreneurs.

Technical entrepreneurs are therefore individuals active in technologically based innovation who create a business in order to innovate. Inevitably the business starts small. It has to because the individuals concerned lack the resources to create a big one. As the innovation achieves commercial success so the business grows. It is the growth that tends to lead to the individuals concerned becoming well known. But not all businesses created by technical entrepreneurs grow. Most actually remain small. They provide a specialist product or service to a highly specialised market niche and as a result the opportunities for growth are limited. Hence technical entrepreneurs are typically associated with small businesses.

INNOVATION AND LARGE CORPORATIONS

In the twentieth century the idea persisted that large corporations were the repository of innovations. In what many term the 'modern' era of mass production, scientific management (Taylorism), economies of scale and mass markets, the large corporation was not only seen as the most efficient form of business organisation, it was also seen as essential to innovation. As noted in Chapter 5, large size was identified as necessary for a high rate of innovation (Mowery and Rosenberg, 1998). The general view was that 'big is best' on the basis that the research and development facilities required for success in innovation were expensive, so that large corporations with access to massive financial resources would be much better placed to carry out innovation. Other supporting arguments favouring size, included the ability of large corporations to undertake a portfolio approach thereby spreading the risk of innovation, and the complementary assets available to large corporations, such as marketing expertise and product support. This was backed up by the practices of large corporations. Companies such as General Electric, Xerox, ICI and others maintained large research laboratories. Many too had strong records when it came to patenting their discoveries.

At a time when the dominant technologies were 'big ones' such as aerospace, pharmaceuticals, automotive and electrical equipment, size appeared to be an essential prerequisite for innovation.

During the last quarter of the twentieth century the tide began to turn against the large corporation. Studies in both the US and the UK showed that small firms could be as successful as large firms when it came to innovation (Van Dijk *et al.*, 1997). Some large firms have a poor record on innovation. Scholars have pointed out that size has to be seen in the context of industrial evolution (i.e. the industry life cycle). Thus, at the birth of a new industry, especially one based on a transforming new technology, small firms are often to be seen at the forefront of innovation. As Christensen (1997) notes, under these conditions large firms actually find it difficult to abandon well-established routines and practices, while small firms being unencumbered in this way, can be much more flexible and adaptable. When in time industry evolution leads to the technology diffusing and a dominant design emerging, the emphasis typically shifts to cost reduction and associated process innovations, and large firms having the advantage in terms of innovation.

SMALL FIRMS

The re-appraisal of the role of small firms in connection with innovation coincided with renewed interest in small firms in general. This was particularly marked in the UK and was the result of a number of factors:

- Bolton Report (1971)
- Thatcherism (1980s)
- Structural changes in the UK economy
- New institutional arrangements
- Technology and small firms

In the 1960s and 1970s the prevailing wisdom in the UK was that small firms were no longer particularly relevant to economic progress (Beaver, 2002). While the Bolton Report noted that the small-firm sector was in decline, it nevertheless drew attention to the role of small firms and highlighted their contribution to the economy. The report particularly highlighted the importance of new small firms as a seedbed for successful high-growth firms of the future. The Bolton Report was enormously influential. It marked the beginning of a major shift in policy. From the 1970s onwards small firms were no longer seen as something of an anachronism, but rather as an important and dynamic element within the economy.

If the 1970s marked a turning point for small-firm policy, the 1980s and the political changes which were such a feature of the decade in the UK, marked a clear revival in the fortunes of small firms. In political terms the decade was dominated by Thatcherism, and it was a powerful force that brought about a significant restructuring of the UK economy. This in turn led to a revival in the small-firm sector.

| Firm size | 1979 | | 1989 | | 1999 | |
	Quantity (000)	Share of Employment (%)	Quantity (000)	Share of Employment (%)	Quantity (000)	Share of Employment (%)
0-9	1597	19.2	2802	28.6	3490	30.2
10-19	109	7.6	92	6.0	109	7.0
20-49	46	6.9	57	7.6	47	6.7
50-99	16	5.3	18	5.8	15	4.6
100-199	15	10.2	9	7.2	8	5.2
200-499	5	8.1	6	10.6	5	6.8
500+	4	42.8	3	34.2	4	39.4
Total	1791	100*	2988	100*	3677	100*

* may not add up due to rounding

Table 9.1 Firm Size and Share of UK Employment, 1979-99
Source: Deakins and Freel (2003) 'Intrepreneurship and Small Firms', 3rd edition, McGraw-Hill, Europe

As Table 9.1 shows, during the course of the 1980s the decline in the number of small firms was sharply reversed and between 1979 and 1989 small firms employing less than 50 people increased their share of employment from 33.7 per cent to 42.2 per cent, a quite remarkable turnaround and one that probably owed something to the political agenda of the time.

Alongside the reported increase in the number of small firms and the increase in their share of total employment, a number of research studies in the 1980s also pointed to small firms playing a more important role in innovation than had hitherto been thought to be the case. Pavitt *et al.* (1987), for example, found small firms' (i.e. firms with less than 200 employees) share of innovations over the period 1945–1983 had increased by over 50 per cent, to account for over a quarter of the total number of recorded innovations in the UK.

The 1980s and the years since then have also been characterised by structural changes in the UK economy that have had an impact on the small-firm sector. These changes have included: the growth of the service sector and the decline of manufacturing; increased use of outsourcing and contracting out; and 'creative destruction' as old industries like textiles and coal mining declined and were replaced by new knowledge-based and technology-based industries. Structural changes such as these have tended to foster a decline in large firms and an increase in small ones. They form part of a move away from large centralised organisations relying on vertical integration and hierarchy which are usually good at delivering efficiency gains but poor at innovation.

The new 'institutional' arrangements are usually associated with the renewed interest that emerged in the 1990s in external, rather than internal, in-house, sources of knowledge and expertise. In the context of small firms and innovation this focused particularly on networks as a source of external knowledge and expertise, where the institutional arrangement might well take the form of a strategic alliance such as a joint venture or collaborative agreement. As Freel (2005: p. 123) noted,

> by participating in innovation networks small firms are able to gain access to sophisticated technology and technological expertise, whose direct employment is precluded by internal resource constraints.

The origins of this change are diverse. Piore and Sabel's (1984) work on flexible specialisation was among the first to alert researchers and policy-makers to the scope for networks as a means by which small firms could keep abreast of technological developments. Freel (2000) links the change to a renewed interest in Marshallian externalities during the 1990s. Similarly, Nelson's (1993) work on systems of innovation helped to portray innovation less as the product of isolated individuals and organisations and more as one of interconnected systems that serve to make knowledge and expertise widely available. Whatever the sources, ideas about the value of networks and networking were readily taken up by policy-makers. The UK government's White Paper on competitiveness (DTI, 1998: p. 29) noted,

> Firms may have to collaborate and network more, particularly on new technologies, and strengthen their links with the science and engineering base, sharing equipment and often people.

Indeed Freel (2000) remarked how UK competitiveness throughout the 1990s was premised on the basis of enhancing inter-organisation linkages. As policy-makers extolled the virtues of networks, so an increasing body of academic work during the 1990s provided empirical evidence to show the value of networks for small firms (Rothwell, 1991; Rothwell and Dodgson, 1991).

Finally small firms have come increasingly to be associated with high technology. Leading the way on this has been Silicon Valley in California. For some 40 years now it has been a symbol of advanced technology, and a crucial part in the development of that technology have been a myriad of small firms, some of which have in time gone on to be quite large firms. With semiconductors in the 1960s (Rothwell, 1986) and personal computing in the 1980s (Saxenian, 1983), Silicon Valley demonstrated the power of small firms in terms of innovation. In the development of the semiconductor industry in California, small firms were crucial in the early stages and it was a very similar story in the development of personal computing, where firms like Apple and Microsoft played a crucial role while established industry giants simply looked on. In both industries small firms demonstrated an outstanding capacity for innovation.

Nor has this been a purely American phenomenon. Work by Jones-Evans and Westhead (1996) showed how in the late 1980s the total stock of small high-technology firms in the UK had increased sharply and such firms had increased their share of high-technology employment, at a time when large high-technology firms were losing employment. The reason for this growth was primarily an increase in the number of high-technology services firms. Much of the growth was through an increase in the number of computer services firms, reflecting advances in computing and the spread of computing applications to more specialist applications.

Similarly Aston and Williams (1996) showed how in the world of Formula 1 motor racing it was small firms that pioneered the application of new technologies associated with mid-engine layouts, monocoque chassis construction, 'ground effects' and semi-automatic gearboxes. Significantly Aston and Williams (1996) also point out that these small firms operate through a series of networks.

MINI-CASE: JCB

Today JCB is a well-known brand, so much so that the letters 'JCB', the initials of Joseph Cyril Bamford, the brand's creator, have become a word in their own right, defined in the Oxford English Dictionary as 'a type of mechanical excavator with a shovel at the front and a digging arm at the rear'. The first mechanical excavator was the 'steam shovel' developed by William Otis in 1837. By the 1920s steam shovels were increasingly diesel-powered but they were still large and heavy with a cable mechanism whose action was to push a front-mounted bucket or shovel through the earth, scooping it up in the process.

After the Second World War a new technology, hydraulic actuation, began to be emerge. The hydraulic excavator was pioneered in Britain by J.C. Bamford. Bamford trained as an engineer with the machine tool manufacturer Alfred Herbert in the 1930s. After wartime service in the RAF and a short spell with the family firm, Bamford set up

▶

his own business in 1945 working from a lock-up garage in Uttoxeter rented for £1.50 a week. He began producing agricultural trailers and in 1948 became the first European manufacturer to apply hydraulics to farm trailers and loaders. In 1953 came his biggest innovation, the 'hydraulic backhoe excavator'. This was effectively a tractor with a detachable digging tool. Unlike the existing cable-actuated mechanical excavators it had a rear-mounted shovel that was hydraulically actuated. Facing the tractor, the shovel was operated by extending the digging arm and drawing it back towards the tractor.

These early hydraulic backhoe excavators were less powerful than conventional cable-actuated mechanical excavators. Their shovel capacity was limited to a quarter of a cubic yard owing to the limited power of existing hydraulic pumps and seals. Cable-actuated mechanical excavators by contrast had a capacity of three cubic yards. Being relatively small, the reach of early hydraulic backhoe excavators was limited to about two metres, and because they were tractor-mounted, they could only rotate through 180° against 360° rotation on cable-actuated excavators.

Because its capacity was limited, the hydraulic backhoe excavator held little appeal for construction companies, especially those working on major construction projects. Instead Bamford had to seek new markets. Being small, Bamford's hydraulic excavator was mobile (cable-actuated excavators required a special transporter), and easily manoeuvred. It proved well suited to digging narrow trenches for foundations of domestic houses and for laying water and power lines. The boom in house-building that took place in the 1950s as part of postwar reconstruction meant there was a demand for a small, mobile excavator. Their mobility and flexibility was underlined by the fact that they were actually attachments fitted to the back of a tractor. In time Bamford's hydraulic backhoe excavator evolved into a purpose-built excavator as he moved on to sell first Ford tractors with a hydraulic backhoe excavator attachment permanently attached and then by the late 1950s a complete excavator to his own design, the bright yellow 'JCB'. It was this product that in the 1960s underpinned a dramatic rise in the company's sales and profits.

As the excavator evolved so did the hydraulics. Advances in materials meant they became steadily larger and more powerful, and as they did so the greater reliability of hydraulics technology began to tell. Firms like Bucyrus only survived by specialising in the production of very large cable-actuated excavators used for opencast and strip mining applications. An indication of the industry transformation is that JCB is now the fifth largest construction equipment manufacturer in the world, ranking alongside Caterpillar and Komatsu. By 1999 it had seven factories making 22 000 machines a year, including one in Savannah, Georgia in the US, 3 400 employees and sales of $1.22 billion. But much of its success it owes to pioneering the innovation of the hydraulic backhoe excavator.

Source: Christensen (1997); Hancock (1995)

ENTREPRENEURSHIP

The increasing emphasis that has been given to small firms and innovation in recent years helps to explain why innovation is increasingly being linked to entrepreneurship. In particular one type of entrepreneur – the technical entrepreneur – occupies a central role where new technology-based firms (NTBFs) are concerned.

chapter 9: technical entrepreneurs

In order to understand the role and function of the technical entrepreneur it is necessary first to look at entrepreneurship as a whole. Entrepreneurs are not managers nor are they inventors. True entrepreneurs need to know about management and they very often have to manage, but the terms 'entrepreneur' and 'manager' are not synonymous. There is much more to being an entrepreneur than simply managing an enterprise. To try and understand just what it is that makes the entrepreneur different, it can be helpful to look at some of the different perspectives on entrepreneurship. These have emerged over many years and taken together they provide valuable insights into the role of the entrepreneur and the nature of entrepreneurship. The perspectives can be broadly categorised into:

- Economic
- Psychological
- Behavioural/processual

Economics

Cantillon was one of the first scholars in the emerging discipline of economics to consider entrepreneurship in his 'Essay on the Nature of Commerce' published in 1755 (Glaister, 1988). Cantillon portrayed the entrepreneur as a pivotal figure within markets. Cantillon viewed the entrepreneur as essentially an organiser of production (Deakins and Freel, 2003): that is, someone who brings together factors of production such as land, labour and capital so that goods and services can be brought to market. In the late nineteenth century Marshall took a similar line arguing that in a world of small firms, the entrepreneur had a vital role to play as an organiser of production. In the twentieth century Knight (1921) took the analysis further, arguing that the entrepreneur's function was essentially one of risk taking. Since factor inputs are purchased at known prices and sold at prices that are as yet unknown, the entrepreneur has to bear risk. A key feature of Knight's analysis was distinguishing between risk which is insurable and uncertainty which is not. The combination of time lags in production, unforeseen changes in household circumstances and the absence of forward markets in the product exposes the entrepreneur to uncertainty. The entrepreneur's function is to bear uncertainty and as Knight (1921: p. 310) noted,

> 66 The only 'risk' which leads to (entrepreneurial) profit is a unique uncertainty resulting from an exercise of ultimate responsibility which in its very nature cannot be insured nor capitalised nor salaried. 99

A marked change of emphasis comes in the work of Schumpeter, who stressed the link between entrepreneurship and innovation. According to Schumpeter the entrepreneur is one of the prime movers in economic development and his/her function is to innovate. Schumpeter (1936: p. 88) was quite clear that innovation was not the same thing as invention,

> 66 Economic leadership in particular must be distinguished from 'invention'. As long as they are not carried into practice, inventions are economically irrelevant. And to carry any improvement into effect is a task entirely different from the inventing of it, and a task, moreover, requiring entirely different kinds of aptitudes. 99

Schumpeter pointed out that the knowledge underlying innovation may not be newly discovered. It may perfectly well be existing knowledge that has not previously been used in products or services. For Schumpeter what matters is that the entrepreneur gives rise to some form of new combination. In a world that is generally resistant to change, the entrepreneur is a disruptive influence. By initiating new products and new processes (which may or may not include new knowledge/discoveries) the entrepreneur initiates change. In this way the entrepreneur creates opportunities. This new combination results in technological advance and it is this that makes Schumpeter's entrepreneur a key figure in economic transformation.

In the 1940s A.H. Cole portrayed the entrepreneur in a very different light. Reflecting the development of managerial ideas he linked the entrepreneur to decision making, including decisions made by teams. Finally, in the 1970s Kirzner (1973) portrayed the entrepreneur as someone who identifies and exploits opportunities for profit, resulting from an imperfect distribution of knowledge. Rather like a middleman Kirzner's entrepreneur finds opportunities that arise because knowledge is incomplete.

Psychological

In the 1960s a rather different perspective on entrepreneurship emerged. This switched the agenda way from the function of the entrepreneur within an economic system to look instead at the individual. This body of work suggested that certain individuals had a particular aptitude for entrepreneurship by virtue of certain distinctive personality factors or personality 'traits'. These traits include:

- A need for achievement (McClelland, 1961)
- A high internal locus of control
- A willingness to take risks
- A need for autonomy and independence

While many successful entrepreneurs, especially those who come across as 'heroic' figures, do possess many of these characteristics, the place of personality within entrepreneurship has been the subject of extensive critical comment. Specific traits have not proved particularly good at predicting behaviour. Other criticisms levelled at this type of approach include: personality factors can and do change over time; the search for a single factor provides a very limited perspective on the nature of entrepreneurship; it ignores learning and it takes no account of the fact that personality factors can and do change over time. Perhaps the most serious flaw is that it ignores the influence of the environment or context in which the entrepreneur operates. Consequently, work on personality traits has to an extent been overtaken by research focusing on entrepreneurial behaviour, which places considerable emphasis on context.

Behavioural/Processual

These criticisms have led some researchers to look at the broader context within which entrepreneurship takes place. In particular efforts have been directed at exploring business behaviour and business context. These have looked at factors such as: ethnicity, gender, occupational background, culture and family structure related to entrepreneurial activity. This has led some to look at the different stages in the process of business development as instances of different contexts. Chell *et al.* (1997: p. 5) for instance point out that the 'stages of development' associated with the process of business growth provide a way of 'conceptualising the context in which owner-managers carry out their business affairs and enables a deeper understanding of what is driving their behaviour'. People develop 'repertoires' of behaviour in response to the situations and circumstances they find themselves in. This has led to the identification of a number of typologies designed to reflect differing contexts of entrepreneurship. These typologies have included: entrepreneur, quasi-entrepreneur, administrator and caretaker (Chell *et al.*, 1991); craft owner, promoter and professional manager (Beaver, 2002) and artisan, entrepreneur and manager (Blundel and Smith, 2001). One category of entrepreneur that has not figured prominently in these typologies and yet which resonates strongly with the notion that there are different types of entrepreneur each of which inhabits a different context is: the technical entrepreneur.

THE TECHNICAL ENTREPRENEUR

The term 'technical entrepreneur' is synonymous with small, high-technology firms or, as they are sometimes known, new technology-based firms (NTBFs). These are small businesses that are dependent on a high level of technological knowledge and expertise. The term is often used to describe a technology-based 'spin-off' business, formed by scientists and engineers leaving their current employment in a university, research-based institution or an industrial company and going it alone by setting up their own independent company. This sort of enterprise is typically to be found in industries such as electronics, computer services or biotechnology. A study by Jones-Evans and Westhead (1996) for instance found that in the UK there were large numbers of small, high-technology businesses in the computer services sector comprising specialist software companies and others providing a range of other similar services. Other sectors where there were significant numbers of this type of enterprise included medical equipment, electrical equipment, precision instruments, and pharmaceuticals (i.e. biotechnology). In these sectors there is scope for individuals who have acquired technological knowledge and expertise through their work in a large organisation to set up on their own. As an independent firm they can then specialise in the supply of specialist components or specialist services. As a number of researchers have indicated, being able to do this has been facilitated in recent years by the growth of networks as a way of linking together both large and small organisations and groups of small organisations. Clearly, becoming an independent firm involves a considerable degree of uncertainty, which is why these ventures are described as 'entrepreneurial'. Nevertheless, they involve a particular form of entrepreneurship, one that relies heavily on technology and the creation of new products and services. As a result this is the form of entrepreneurship that is probably most closely connected to innovation.

Definitions of technical entrepreneurs tend to emphasise the technical at the expense of the entrepreneurial. Thus Cooper (1971) quoted in Jones-Evans (1995: p. 29) describes this type of business as,

> 66 a company which emphasises research and development or which places major emphasis on exploiting new technical knowledge. It is often founded by scientists and engineers, and usually includes a substantial percentage of professional technically trained personnel. 99

While this definition rightly highlights the importance of technology for this sort of enterprise and the connection to innovation, it tends to ignore entrepreneurship. Hence perhaps a better definition comes from Jones-Evans (1997) when he describes technical entrepreneurs as,

> 66 ...small technology-based firms (that) display a distinct form of entrepreneurship, mainly because of the dependence of the venture on the owner-manager's high degree of a technological expertise, translated into new technologies, products and processes. 99

This definition does rather better at including some of the key elements of the technical entrepreneur. As studies of locations populated by large numbers of technical entrepreneurs have shown, these businesses are not merely technical. Henry and Pinch's (2000) study of Motor Sport Valley in Oxfordshire, for instance, highlighted the high rates of entry and exit into the industry as well as the dynamism of these businesses in terms of their ability to respond rapidly to change. Similarly, this type of business is likely to be closely involved in the application of technology to innovations, be they new products, new processes or new services. This was something highlighted in a major study by Pavitt *et al.* (1987) which noted that small firms had significantly increased their share of recorded innovations in the postwar period.

Consequently perhaps the most effective definition of the technical entrepreneur comes from Autio (1995) who suggested that technical entrepreneurs possess four features:

1   The founders of the company have been affiliated with the source of the technology before establishing the company.
2   The business idea of the company is essentially based on exploiting advanced technological knowledge developed or acquired in a source of technology.
3   The company is independent.
4   The company is entrepreneurial, that is, it is controlled and managed by an entrepreneur or a group of entrepreneurs.

While this definition incorporates the essential attributes of a technical entrepreneur, there are other features that are sometimes associated with this kind of business. For example, technical entrepreneurs typically make extensive use of networks comprising contact with universities, large companies and research institutes, as a source of both knowledge inputs and market opportunities. The mechanisms that help to maintain these contacts can take a variety of different forms. Autio (1995) notes that strategic alliances and R & D partnerships are often used as well as other forms of collaborative agreement. Devices such as these help to provide an interface that can be used to convey information and knowledge through the network. Similarly, the knowledge, particularly the technical knowledge, that plays such an important part in these businesses tends to be tacit rather than explicit. It is very often knowledge associated with skills and capabilities which resides in individuals. One of the great strengths of technical entrepreneurs is that they are very effective both in accessing this kind of knowledge and in applying it to practical problem-solving situations where it can contribute to the development of new products, processes and services.

Occupational Background

Early work on technical entrepreneurs tended to portray the technical entrepreneur as an academic in a university or a researcher in a non-profit-making laboratory who had set up on his/her own, giving rise to what some have described as the 'scientist entrepreneur'. This model implied an entrepreneur as someone with little or no business experience and business knowledge. Later work (Cooper, 1971) extended the analysis to include entrepreneurs who had worked for large industrial concerns and therefore did have a business background. This has generally been the view of technical entrepreneurs that has prevailed. Only recently has the analysis been extended further. Work by Jones-Evans suggests that in fact technical entrepreneurs come from a more diverse range of backgrounds than hitherto noted.

Jones-Evans' (1995) typology (Figure 9.1) identifies four categories of technical entrepreneur.

Figure 9.1 Typology of Technical Entrepreneurs

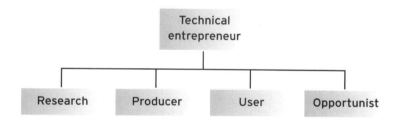

Research Technical Entrepreneur

The research category of technical entrepreneur includes individuals whose previous employment experience was in universities or public research laboratories. In Jones-Evans (1995) about one third of the technical entrepreneurs interviewed had this background. All had a scientific background that included engineers and scientists. Many continued to retain links with their former employer. While this category was clearly the equivalent of the scientist entrepreneur or academic entrepreneur of earlier studies, there were differences. Jones-Evans noted that not all had spent the whole of their previous career in an academic environment. Some had a small amount of commercial experience, and this meant that they did have knowledge of business. Even some of those who had worked exclusively in research environments had some knowledge of business, typically extending to a modest knowledge of marketing, finance and interpersonal skills.

MINI-CASE: BERGHAUS

Berghaus is a very recognisable brand name when it comes to outdoor wear. One does not have to be an active mountaineer to encounter Berghaus products – these days a visit to the local supermarket on a wet day will probably be quite sufficient. Yet Berghaus is much more than a well-known brand name. It is widely recognised as supplier of high-performance clothing. This reputation comes from its pioneering work developing outdoor wear made from Gore-Tex – an innovation that has helped to transform the outdoor clothing market in the last thirty years by making breathable waterproof clothing a reality for the first time. Set up in the late 1960s to run a shop specialising in outdoor wear, Berghaus began as a partnership between Peter Lockley who had worked in marketing and sales for the chocolate manufacturer Rowntrees, and Gordon Davidson who worked as a lecturer in mechanical engineering at Newcastle Polytechnic (now Northumbria University). After an initial start in retailing, Berghaus moved into manufacturing in the early 1970s making first rucksacks and then outdoor clothing. In 1976 Berghaus was instrumental in the development of outdoor clothing made from Gore-Tex. Although W.L. Gore and Associates developed Gore-Tex fabric, it was Lockley and Davidson at Berghaus that developed the manufacturing techniques that enabled high-performance clothing to be produced using Gore-Tex, giving the consumer for the first time garments that were both waterproof and breathable.

Source: Parsons and Rose (2003)

Producer Technical Entrepreneur
This category is the equivalent of the industrial technical entrepreneur identified by earlier researchers. In essence it includes entrepreneurs whose background is industrial. Typically these are people who have been involved in the development or production of commercial products, services or processes, usually in large industrial organisations. In Jones-Evans' (1995) study these were typically individuals with an engineering background and included: works managers, development managers, and technical managers, virtually all of whom were trained engineers, as well as draughtsmen, designers and project managers. A very small number had a purely scientific background and had worked as research scientists. All had a strong technical background even if their most recent position was a managerial one. As a result most combined a strong mix of both technical and managerial experience.

What is striking about this type of technical entrepreneur is how technical and managerial aspects are brought together in a single individual. Significantly this group constituted more than half the sample in Jones-Evans' survey, suggesting that although the notion of an academic entrepreneur is one that is most readily identified with technical entrepreneurs perhaps this is a misperception, certainly as far as the UK is concerned.

User Technical Entrepreneur
Where Jones-Evans breaks new ground is in identifying two further categories of technical entrepreneur. The first of these he calls the user technical entrepreneur. In this context the term 'user' is used to denote the fact that the entrepreneur had a

background in marketing/sales or product support. Such a role would therefore be likely to bring him or her into direct contact with consumers. In fact, half of the user entrepreneurs in Jones-Evans (1995) sample had a background in marketing/sales.

The strong connection to marketing/sales helps to give credence to the notion of innovation as a demand-pull process in which consumer needs provide the stimulus to innovation. Individuals who work in marketing/sales are likely to be very familiar with consumer needs and consumer requirements. This knowledge can act as both a spur to innovation and a trigger to the decision to set up an independent business.

Obviously the user technical entrepreneur stands in sharp contrast to the two previous categories of technical entrepreneur. Whereas they have what might be described as a 'supply side' perspective, the user technical entrepreneur in contrast has clear 'demand side' perspective. This perspective can be particularly valuable when one is dealing with the diffusion of a generic technology. Under these circumstances, new applications are likely to take the form of specialist applications developed for very specific market niches. In the development of such market niches, detailed knowledge of the consumer is likely to be at a premium. Such knowledge is likely to be informal and tacit and the sort of knowledge that those in marketing/sales will possess. Furthermore, if the product in question has a significant technology input then marketing staff may well have a good knowledge of the generic technology.

Opportunist Technical Entrepreneur

The fourth and final category of entrepreneur in Jones-Evans' (1995) typology is what he describes as the 'opportunist' technical entrepreneur. These are individuals who identify technology-based opportunities but who have little or no technical education and whose previous occupational experience is with non-technical organisations. Just how far removed some of these individuals can be from a technical background is illustrated by Jones-Evans' (1995) study which included a teacher, a naval officer, an insurance clerk, a civil servant, a personal assistant and an office manager.

Given that one is dealing with high-technology businesses, the lack of any technical background in these cases is surprising and is in sharp contrast with most of the previous work on technical entrepreneurs. However, it does serve to highlight the 'opportunity' filling nature of the entrepreneur's role. A significant amount of research into entrepreneurship has focused on the opportunity- or gap-filling nature of the entrepreneur's role. Jones-Evans' work lends support to this, while at the same time suggesting that where generic technologies are concerned and where there is scope for incremental innovations then opportunity recognition may actually be more important than technical knowledge, which may be purchased in the market either directly or through people who hold such knowledge.

Markets and Technology

Although it has attracted a lot of attention, probably because many technical entrepreneurs emerge through some form of spin-off from a parent company, occupational background is only one way of classifying technical entrepreneurs. Autio (1995) presents quite a different typology that relies instead on analysis of the context in which the entrepreneurship occurs. Autio identifies technology and markets as crucial aspects of the entrepreneurial context.

Autio (1995) argues that technology is critical because it can vary considerably in terms of the degree of technological intensity associated with the venture. Where the venture is involved in the creation of new knowledge, the degree of technological intensity will be much higher than if it is concerned with the production of new artefacts, using existing knowledge, but perhaps in a new and as yet untried application. Autio labels the degree of technological intensity as 'technological novelty', which ranges from established to breakthrough. The latter implies the creation of new knowledge while the former implies the creation of a new artefact. Autio (1995) observes that knowledge-creation activities are much more likely to be associated with science-based firms, while artefact-creation activities he suggests are likely to be associated with engineering firms involved in manufacturing, perhaps as specialist suppliers of components.

Markets are the other crucial dimension because, for an innovation to be a success, markets have to be found or created in order to exploit commercially the outcome of technology-development processes. Autio (1995) distinguishes between penetrating established markets and the creation of new markets. Where the penetration of existing markets is concerned, technical entrepreneurs face a choice between providing what Autio calls 'complementary' technologies or 'revolutionary' technologies. Complementary technologies, as their name implies, will complement existing products and services while revolutionary technologies will be very different from what is currently available. Where the technical entrepreneur chooses to create markets this can be an extremely uncertain process especially if the technology involved is revolutionary.

Figure 9.2 shows Autio's two dimensions brought together into a two-dimensional matrix. It gives rise to a typology with four types of technical entrepreneur: application innovator, technology innovator, market innovator and paradigm innovator.

Figure 9.2 Technology/Market Taxonomy

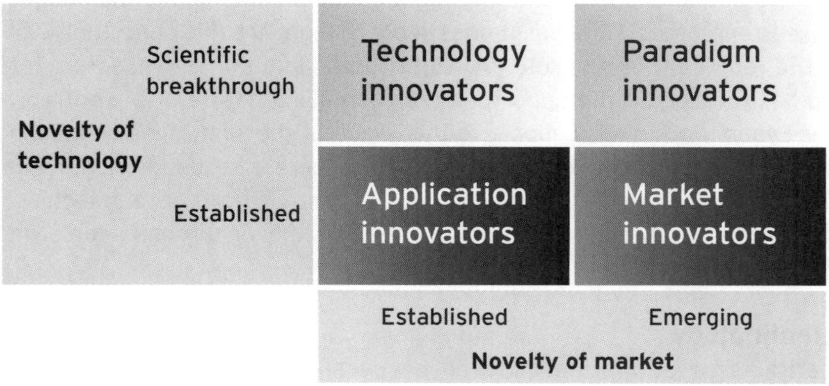

Source: Autio. Entrepreneurship and Regional Development, 1995 Taylor & Francis Ltd. http://www.tandf.co.uk/journals

Application Innovator

This type of technical entrepreneur aims to utilise established technology in established markets to produce what are essentially complementary products. This will typically involve the diffusion of an existing generic technology into new, probably much more specialised, market niches. Autio (1995) gives the example of software houses that tend to provide specialist solutions using established software principles with established software languages.

Market Innovator

Market innovators aim to develop new markets with existing technologies. The technology is likely to be generic and does not have to possess any radical features. The innovation comes typically from combining or configuring the technologies in a different way. Autio (1995) gives the example of the first Apple computer. This was the Apple II which used proprietary computer components that could be sourced 'off the shelf' but configured in a new way to create the first effective personal computer.

Technology Innovators

Technology innovators rely on new technologies. These technologies may well be associated with scientific breakthroughs. These technologies go into new products that are sold in established markets. They are, to use the terminology mentioned earlier, complementary products, as they enter the market alongside existing products. As Christensen (1997) has noted, however, they often threaten incumbent firms because they deliver sharply improved performance or meet customer requirements more effectively. Autio gives the example of biotechnology companies offering biodegradable materials, though the case of JCB (see earlier mini-case) would be just as applicable since this was an instance of a new technology – hydraulic actuation – being applied in an existing market, namely the market for mechanical excavators.

Paradigm Innovators

This is the most extreme form of technical entrepreneur. In this instance the entrepreneur delivers a new product concept based on a new technology into a new and as yet undeveloped market. This may well prove to be very disruptive. Autio gives the example of Intel, a new company set up as a spin-off from Fairchild Semiconductor, that developed and manufactured the first microprocessor which in time turned the computer industry upside down as personal computers displaced much larger mainframe machines.

Towards a synthesis

Recent work on technical entrepreneurs by Tidd *et al.* (2001) integrates earlier 'trait' theories of entrepreneurship espoused by researchers such as McClelland (1961) and Roberts (1991), with work on occupational backgrounds along the lines of Jones-Evans (1995) and research by Autio (1995) that looked at the impact of markets and technology, to provide a synthesis that draws together a variety of different strands of entrepreneurial research. The resulting model (Figure 9.3) provides a composite picture of the factors affecting technical entrepreneurship. These factors are grouped into three sets of factors labelled as:

- Antecedent factors
- Parental experience
- Environmental factors

Antecedent factors are features of the entrepreneur's personal life and cover aspects such as personality (i.e. personality 'traits' highlighted by psychological theories of entrepreneurship), home context and general background. Parental experience refers to the entrepreneur's experience derived from the 'host' organisation(s) where they were employed prior to setting up on their own, specifically the nature of the host organisation and the level of institutional support that it provides. Environmental factors refer to the commercial and technological environments in which the new venture is located, especially the degree of novelty associated with the market and the technology.

Figure 9.3 provides a more detailed breakdown of the factors that influence technical entrepreneurs. The value of this type of model is the degree of flexibility that it offers. Much of the early work on entrepreneurship was criticised because the pursuit of a rather narrowly defined set of personality traits did not fit easily with the diversity of entrepreneurs. There clearly is nothing like a standard case, and this model caters for a high degree of diversity.

The model shown in Figure 9.3 also serves to highlight some additional features of technical entrepreneurs.

Figure 9.3 Factors Affecting New Venture Formation by a Technical Entrepreneur

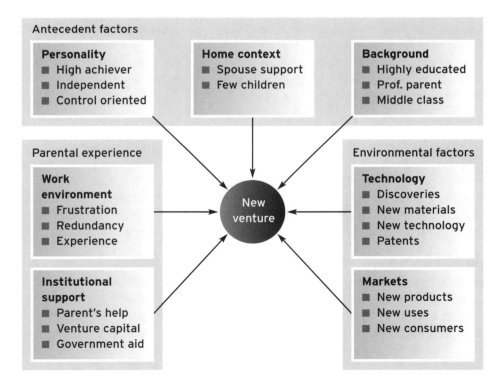

Source: Adapted from Tidd *et al.* (2001)

Technical entrepreneurs tend to be very industry specific, being found in large numbers in some industries and sectors but being almost completely absent in others. This may well be a function of low barriers to entry, which facilitate both ease of entry and exit. Thus, in sectors such as biotechnology and computing, technical entrepreneurs are to be found in abundance because entry is relatively easy.

Ease of entry into an industry may also be a function of the scope for technical entrepreneurs being 'spawned' from larger organisations. Often referred to as 'spin-offs', spawning refers to the process whereby employees of a large organisation leave to start up their own business. Often those who opt to go it alone in this way have accumulated a substantial amount of knowledge and experience while working for the large organisation. Rather than explicit knowledge that is structured and codified, the knowledge involved will be tacit, residing in the individual and often difficult to articulate. Sometimes referred to as 'know-how' (Newell *et al.*, 2002), this is the sort of knowledge that is gradually accumulated from working in a specialised field. As well as including technical knowledge it can also extend to commercial knowledge, such as a network of contacts. Where this sort of knowledge is widely used then entry into the industry may be relatively easy.

In recent years the increasing popularity of vertical disintegration and outsourcing as corporate strategies have served to increase the scope for spin-offs. If the large organisation wishes to divest itself of certain activities and then buy back the service, this creates valuable opportunities for the more entrepreneurially minded. These days many parent organisations (i.e. those from whom technical entrepreneurs are spun off) are often willing, given appropriate circumstances, to provide assistance to those who choose to leave and start up their own enterprise. Parents can provide a range of services extending to legal advice, loans, sale of equipment at attractive rates, premises and orders for goods and services. Of course there are times when employees depart in this way, but on far from amicable terms.

MINI-CASE: ILMOR ENGINEERING LTD

Ilmor make engines for the McLaren West Mercedes previously driven in Formula One by Britain's David Coulthard. They have also enjoyed success in supplying engines for several American racing teams. Ilmor was founded by Mario Ilien and Paul Morgan in 1983. Both were former employees of the racing engine manufacturer, Cosworth, based in Northampton. Their decision to branch out on their own was helped by the relatively low barriers to entry in the motor sport industry. Ilien and Morgan approached team owner Roger Penske who negotiated a financial package from General Motors of the US. When they left Cosworth, the pair were anxious to retain many of their former contacts in 'Motor Sport Valley' in Oxfordshire/Northamptonshire and therefore chose to locate their new business in Brixworth not far from Northampton. Their decision to set up on their own mirrored the formation of Cosworth, which was founded in 1958 by two Lotus employees, Mike Costin and Keith Duckworth. They began by tuning and modifying existing mass-produced engines and eventually, with backing from Ford, built a Formula One engine, the Cosworth DFV, which won 155 grand prix between 1967 and 1983.

Source: Robson (1999)

Another factor that can be significant is the institutional framework. In the biotechnology field, for instance, Jones (2000) notes that in recent years the increased emphasis on outsourcing on the part of large pharmaceutical companies has made them keen to buy in knowledge, rather than invest directly in R & D, thereby creating opportunities for small biotechnology companies.

A final characteristic of technical entrepreneurs is that they are often found clustered together. Given that technical entrepreneurs were first identified in Silicon Valley, probably the largest cluster or concentration of high-technology companies in the world, this is really not surprising. In Silicon Valley the clustering is commonly associated with a desire to for proximity to the parent organisation. Although developments in communications technology have made it easier for organisations to be located at a distance even when working together, where networking is concerned proximity still has important advantages, especially for small firms working in rapidly changing technologies. Henry and Pinch (2000: p. 206) in their work on Motor Sport Valley in Oxfordshire/Northamptonshire show how technical entrepreneurs in this sector choose to be located close to each other because of 'their vital need to dip into the deep stream of knowledge that keeps being generated and circulated throughout the region largely on an untraded basis.'

What is true of advanced automotive engineering is equally true of science-based sectors. Outstanding examples of universities acting as a hub that attracts a cluster of technical entreprenurs include Cambridge in the UK, Stamford University in California and Massachusetts Institute of Technology in the US.

CASE STUDY: ONECLICK TECHNOLOGIES LTD

There is something reassuring about the little standby light on a TV or a personal computer. It is so small and so faint that the energy being used by the device must surely be minimal. Unfortunately that is not necessarily the case. Electrical devices left on standby mode may only consume a few watts of energy themselves, but when you add the associated peripheral devices like printers and speakers for PCs or the tuners, turntables, cassette decks and speakers that form part of a hi-fi system, the watts add up and when the system is left on standby 24 hours a day, the energy involved is considerable.

It was knowledge of this that inspired Peter Robertson to develop the 'OneClick Intelligent Mains Panel', a six-socket extension block that can be used to power a hi-fi or PC and up to five peripheral devices which, as it senses when the main device is on standby, switches off all the peripherals, thereby ensuring that energy consumption drops to less than five watts an hour. The beauty of this device, which looks pretty much like an ordinary six-socket extension block, is that it is automatic. Quite literally one click is all that is required to put the principal device on standby and shut down all the peripherals.

Peter Robertson had the idea for an extension block of this type, when he worked for Yamaha selling hi-fi products. With the advent in the late 1980s of sophisticated audio-visual products incorporating Dolby 'surround sound', domestic consumers were adding CD players and widescreen TVs to their hi-fi systems. As Peter himself says, 'I was tired of switching off various things'. He felt there was a need for an extension block into which all these devices could be plugged, and with which, with the inclusion of some sort of switching mechanism, it would be possible to switch off the peripherals along with the hi-fi. Quite literally 'one click' would do the whole thing.

Peter had no formal training in electronics. Born and brought up in Nottingham where his father worked as a bus driver, he left school at 16 and had a variety of jobs. He worked first in shops and then catering, eventually becoming the catering manager of a motorway service area. In time he decided to move back to Nottingham where he got a job as an administrator, which he himself admits he 'found immensely boring'. As a result he soon switched to working as a coach driver for a couple of years. Peter then went into sales. To begin with he sold advertising space, then office equipment, and then hi-fi equipment. The last of these jobs was in retailing, where Peter worked as an area manager for the hi-fi manufacturer, Yamaha.

It was while working for Yamaha in the late 1980s that Peter became convinced that the proliferation of peripheral devices for hi-fi systems meant there was scope for some form of switchable extension block. Having had the idea he decided to try building a prototype. He bought a standard six-socket extension block and a heavy industrial relay. Then, using the kitchen table as his workshop, he set about building a prototype. By connecting the relay to the switched outlet fitted at the rear of virtually all hi-fi amplifiers, he was able to get the extension block to sense when the device was on standby and cause the relay to shut down the remaining five sockets. When the amplifier was switched on, it took the sensing from the back of the amplifier and fired the relay, which then duly switched on the current going to the peripheral devices. As Peter acknowledged 'I knew there was an application there.' He had proved the concept. However, the relay made the extension block heavy and unwieldy and far too expensive to mass produce. With a well-paid job in retailing Peter decided that this was not the time to become a fully fledged innovator.

He stayed with Yamaha for six years. As Peter said of his time at Yamaha, 'I learnt a lot about how multiples work, buying and selling and that sort of stuff and I learnt a bit about electronics.' But when Yamaha decided to re-structure their activities Peter opted for redundancy. There then followed a stint at EchoStar, a US-based manufacturer of satellite equipment. From there Peter spent a couple of years working as a consultant for the Franchise Centre in Manchester where he wrote franchising manuals and advised would-be franchisers and franchisees before going back into sales and marketing, working for a Danish company. When the company decided to relocate to South Wales, Peter again opted for redundancy. At this point he decided to take a career break, opting for a year out in which he studied for an MBA at Nottingham Trent University. It was while he was at university that Peter revisited his prototype intelligent mains panel. For part of his course he had to put together a business plan for the introduction of a new product and his idea for an intelligent mains panel, in the form of a switchable extension block, seemed an ideal new product. For the business plan he had to investigate costings, product viability and the market for the product. By now relay technology had moved on apace. Developments in electronics had caused relays to shrink dramatically in size, to the point where it was feasible for one to be fitted inside the extension block. The

▶

technology also meant that an electronic version was feasible which would suit any application and not just hi-fi, but Peter felt that at this stage it was important to test the market first. He opted to develop a passive version using proprietory technology, and found an extension block that could be easily adapted so that a proprietory relay could be fitted inside the space taken by one of the sockets. The intelligent mains panel had become a commercial proposition.

All the main components were proprietary products. The extension block was bought in while the relays were supplied by an electrical distributor. With a fly lead to the back of an amplifier and a similar lead to the USB socket of a computer, the intelligent mains panel could be used for either hi-fi or PC applications. By the time he came to finish the course, Peter was manufacturing 20 extension blocks a week. However, he quickly found he could not cope with making them this way, and began trying to find someone who could manufacture them for him. It did not take long but came about quite by accident. Unhappy that he was paying full price, Peter asked the distributor from whom he was buying relays for a trade discount. They refused, saying the quantities he wanted did not warrant a trade discount. However, they did suggest a firm in Hinckley that might be more interested in the sort of quantities he needed. Not only that, they mentioned that the firm might be able to do some manufacturing. And so it turned out. The firm in Hinckley could not only supply relays, but were willing to manufacture the whole extension block, and Peter placed an order for 3000 sets.

With proper manufacturing facilities in place, he now turned his attention to marketing. With help from friends, he soon had a small website in place. Using his extensive knowledge of retailing, Peter began to target hi-fi shops. Yamaha his former employer agreed to place an order for 300 sets. To increase his exposure in the marketplace he decided to exhibit at a national electronics exhibition in Birmingham. However, he baulked at trying to find the £5000 cost of hiring an exhibition stand and the additional cost of then fitting it out and staffing it. Aware that exhibition organisers often end up with unfilled stands, he approached the organisers about negotiating a discounted rate, stressing that he was a student. They agreed to rent him a stand in one of the less attractive locations for £1000 all in. To staff the stand he roped in some of his fellow students and to provide some publicity he purchased a supply of T-shirts and knickers with his 'Oneclick' logo printed on them. The knickers proved particularly successful. With his stand in a somewhat remote location, he needed something to get potential customers talking and they certainly provided a talking point. Orders came flooding in. They covered the cost of attending the exhibition in the first day. Several of the large electrical retailers ordered 1000 or more.

Having successfully developed and launched his innovation, Peter began to think about expanding the market. Initially, sales were confined to the specialist hi-fi market for which it had been developed and where it had obvious applications. In the hi-fi market the attraction of the intelligent mains panel was that it helped the hi-fi enthusiast control a wide range of peripherals. There were other potential markets, however, and one was the environmental-friendly, energy-conscious market. So he approached the Energy Saving Trust, hoping that an endorsement from them would give his product an opening in this market.

They pointed out that in its present form the intelligent mains panel had some weaknesses: some consumers would not know where the USB port was; similarly the use of the switched outlet on amplifiers limited its application in the consumer electronics market. Quite literally the intelligent mains panel was not intelligent enough! This prompted Peter to commit to the development of the more advanced electronic version. Whereas the current product was a passive version that responded to external signals, he now set his sights on developing an active 'current sensing version' able to actively managed the power supply.

For active power management some form of auto-calibration was needed with the capability to read a device's power consumption. Peter undertook a detailed examination of computer power-supply systems. He found that the power being used could vary enormously. Meanwhile, a patent search revealed that a similar device had been patented in Germany but had not proved successful as it was relatively primitive, working with fixed values for the power being used. Peter's solution to the problem was to use some form of programmable integrated chip linked to a software program that monitored the amount of power being used and then used a formula to decide whether to switch the slave sockets on or off. Having produced a solution was one thing; getting it to work was quite another. Development was not without its difficulties. On one occasion the only prototype blew up the night before a vital demonstration to the Energy Saving Trust. Over the course of several months more than 30 electronic prototypes of different designs either blew up or melted before Peter was finally able to come up with one that was stable and functioned effectively. At this point, having taken appropriate legal advice, Peter filed a patent for his intelligent switching device.

Since the development of an active version of his intelligent mains panel meant a major redesign of the product, Peter took the opportunity to explore alternative manufacturing facilities. He negotiated for production to be subcontracted to a British firm which arranged for manufacturing to be undertaken in China, thereby reducing the manufacturing cost by about one third, together with a commensurate improvement in product quality. At the same time Peter set about marketing the new active version of his intelligent mains panel. An order from Powergen for 20 000 items quickly showed the potential of the improved product. The original passive version had only achieved sales of 5000 during the period of slightly less than a year that it was on the market. Significant orders from major electrical retailers such as Maplin, Lakelands, PC World and even B & Q soon followed. When the active version went on the market in October 2003 it was soon selling at an average rate of 3000 items a month, evidence if such were needed that the intelligent mains panel was a successful innovation. Not that Peter Robertson felt innovation had to stop there. He was soon planning an improved version that would combine his patented intelligent switching system with additional features such as surge protection, full range calibration and split phone and modem outputs. He was even beginning to explore a version for the US market that would work with power supplies ranging from 100 to 250 volts, making the intelligent mains panel a product capable of taking on global markets.

Source: Personal interview with Peter Robertson, 2 June 2005

▶

QUESTIONS

1 To what extent was Peter Robertson helped by developments in technology?
2 What kind of technical entrepreneur would you classify Peter Robertson as and why?
3 Outline the aspects of his prior working experience that you feel helped Peter Robertson with his innovation.
4 Which aspects do you think were most helpful?
5 Using the technology/market taxonomy derived in this chapter classify this innovation.
6 Give examples of institutional support that contributed to the success of the venture.
7 Outline the market developments that you consider contributed the most to success.
8 What forms of intellectual property were associated with this case?
9 Why do you think Peter Robertson registered the patent in his own name using a separate company to manufacture and market the intelligent mains panel?
10 What does the case tell us about the value of personal networks for innovation?

QUESTIONS FOR DISCUSSION

1 **What is the role of the entrepreneur?**

2 What is the difference between a manager and an entrepreneur?

3 **Using an example of your choice identify what you consider to be the main characteristics of an entrepreneur?**

4 Identify the different types of entrepreneur making clear the ways in which they differ.

5 **What is a technical entrepreneur?**

6 Using one of the mini-cases build up a profile of a technical entrepreneur

7 **Why have policy-makers become increasingly interested in technical entrepreneurs in recent years?**

8 Why have small firms increasingly been recognised as an important source of innovation?

9 **What is the connection between Silicon Valley and technical entrepreneurs?**

10 According to Schumpeter, what is the role of the entrepreneur in the innovation process?

11 **How, according to Kirzner, does the activity of the entrepreneur lead to market equilibrium?**

ASSIGNMENTS

1 Take one example of a location within the UK where technical entrepreneurs are present (e.g. Cambridge or Motor Sport Valley in Oxfordshire). Outline why in your view the location has tended to attract technical entrepreneurs.

2 Outline the different functions that economists have suggested for entrepreneurs.

3 Why are entrepreneurs and small firms not necessarily the same thing?

4 Select an example of a business founded by a technical entrepreneur. Develop a profile of the founder (or founders if there is a team involved) and indicate which factors you think were most influential in the creation of the business.

5 In what ways do technical entrepreneurs differ from entrepreneurs in general?

6 Select a technical entrepreneur of your choice and analyse the part played by institutional support (i.e. from government, parent organisation, public agency, etc.) in the creation of a successful new business venture.

7 How are small high-technology firms able to compete with multinationals?

8 'Innovation is integral to the entrepreneurial process.' Discuss.

RESOURCES

Roberts (1991) was one of the first to identify 'technical entrepreneurs' as an important element in the field of entrepreneurship. His study of technical entrepreneurs remains a classic. However, over the last decade and a half a number of other important studies (e.g. Jones-Evans, 1997) have emerged that help to provide a broader overview. With this in mind a diverse range of sources help to illustrate the phenomenon

Business Biographies

There can little doubt that business biographies provide a fascinating and detailed insight into the careers of well-known and successful technical entrepreneurs. They possess the great advantage that they are generally cheap and very accessible. Most are available as inexpensive popular paperbacks. Classic examples include biographies of James Dyson (Dyson, 1997), Trevor Baylis (Baylis, 1999), Alec Issigonis (Nahum, 2004), Frank Whittle (Golley, 1996), Akio Morita (Nathan, 1999) and Tim Berners-Lee (Berners-Lee, 2000)

Business Histories

Business histories can also provide a substantial amount of detail about technical entrepreneurs, though they approach the subject matter from the perspective of the venture itself rather than the individual. Audrey Wood's (2001) *Magnetic Venture: The Story of Oxford Instruments* is a classic. It provides sufficient detail to analyse a range of factors that help to account for the success of this venture, ranging from personality to traits through to the technological context of the time. Other examples include a study of Hewlett-Packard (Packard, 1996) and Apple Computer (Levy, 1996).

Industry Studies

There are plenty of studies of high-technology industries. One of the best examples is Lee *et al.*'s (2000) book about Silicon Valley entitled: *The Silicon Valley Edge: A Habitat for Innovation and Entrepreneurship*. One particularly illuminating aspect of the book is the chapter on Fairchild Semiconductor which shows how this innovative company spun off a whole series of technical entrepreneurs. Sometimes industries that are not so high tech can also be useful. The cycle industry is not obviously high tech but Rosen (2002) does cover both technology and entrepreneurs extensively.

Videos/DVDs/Other Audio Visual Materials

Given that technical entrepreneurs are usually well-known figures, they are from time to time the subject of TV documentaries and the like. As a result the AV collections of libraries often contain videos recordings of these programmes which can be an invaluable source of data about these people.

REFERENCES

Aston, B. and M. Williams (1996) *Playing to Win: The Success of UK Motorsport Engineering*, Institute for Public Policy Research, London.

Autio, E. (1995) 'Four types of innovators: a conceptual and empirical study of new, technology-based companies as innovators', *Entrepreneurship and Regional Development*, **7**, pp. 233–248.

Baylis, T. (1999) *Clock this: My Life as an Inventor*, Headline, London.

Beaver, G. (2002) *Small Business, Entrepreneurship and Enterprise Development*, FT Prentice Hall, Harlow.

Berners-Lee, T. (2000) *Weaving the web: The original design and ultimate destiny of the World Wide Web*, Harper Business, NY.

Blundel, R. and D.J. Smith (2001) *Business Networks: SMEs and inter-firm collaboration: A review of the research literature with implications for policy*, Small Business Service, Sheffield.

Brown, R. (1998) 'Electronics Foreign Direct Investment in Singapore: A Study of Local Linkages in "Winchester City",' *European Business Review*, **98**, pp. 196–210.

Castells, M. and P. Hall (1994) *Technopoles of the World: The Making of 21st Century Industrial Complexes*, Routledge, NY.

Chell, E., J.M. Haworth and S.A. Brearley (1991) *The Entrepreneurial Personality: Concepts, Cases and Categories*, Routledge, London.

Chell, E., N. Hedberg-Jalonen and A. Miettinen (1997) 'Are Types of Business Owner-Manager Universal? A Cross Country Study of the UK, New Zealand and Finland', in R. Donckels and A. Miettinen (eds) *Entrepreneurship and SME Research: On its Way to the Next Millenium*, Ashgate, Aldershot.

Christensen, C.M. (1997) *The Innovator's Dilemma: When New Technologies Cause Great Firms to Fail*, Harvard Business School Press, Boston, MA.

Cooper, A.C. (1971) *The Founding of Technologically-based Firms*, Center for Venture Management, Milwaukee, WI.

Department of Trade and Industry (DTI) (1998) *Our Competitive Future: Building the Knowledge Economy*, HMSO, London.

Deakins, D. and M. Freel (2003) *Entrepreneurship and Small Firms*, 3rd edn, McGraw-Hill, Maidenhead.

Dodgson, M. (2000) *The Strategic Management of Technological Innovation*, Oxford University Press, Oxford.

Dyson, J. (1997) *Against the Odds*, Orion Business, London.

Freel, M.S. (2000) 'External linkages and product innovation in small manufacturing firms', *Entrepreneurship and Regional Development*, **12**, pp. 245-266.

Freel, M.S. (2005) 'Patterns of innovation and skills in small firms', *Technovation*, **25**, pp. 123-134

Glaister, K.W. (1988) 'The Entrepreneur: Enigma of Economic Theory', *Economics*, Spring 1988, pp. 2-6.

Golley, J. (1996) *The Genesis of the Jet*, Airlife Publishing, Marlborough.

Hancock, M. (1995) *JCB: The First Fifty Years*, Special Event Books.

Henry, N. and S. Pinch (2000) 'Spatialising knowledge: placing the knowledge community of Motor Sport Valley', *Geoforum*, **31**, pp. 191-208.

Howells, J. (2005) *The Management of Innovation and Technology: The Shaping of Technology and Institutions of the Market Economy*, Sage Publications, London.

Jones, O. (2000) 'Innovation Management as a Post-Modern Phenomenon: The Outsourcing of Pharmaceutical R & D', *British Journal of Management*, **11**, pp. 341-356.

Jones-Evans, D. (1995) 'A typology of technology-based entrepreneurs: A model based on previous occupational background', *International Journal of Entrepreneurial Research and Behaviour*, **1** (1), pp. 26-47.

Jones-Evans, D. (1997) 'Technical entrepreneurship, experience and the management of small technology-based firms – exploratory evidence from the UK', *Entrepreneurship and Regional Development*, **9,** pp. 65-90.

Jones-Evans, D. and P. Westhead (1996) 'The high technology small firm sector in the UK', *International Journal of Entrepreneurial Research and Behaviour,* **2** (1), pp. 15-35.

Kirzner, I. (1973) *Perception, Opportunity and Profit: Studies in the Theory of Entrepreneurship*, University of Chicago Press, Chicago.

Knight, F.H. (1921) *Risk, Uncertainty and Profit*, University of Chicago Press, Chicago.

Lécuyer, C. (2000) 'Fairchild Semiconductor and its Influence', in C.M. Lee, W.F. Miller, M.G. Hancock and H.S. Rowen (eds) *The Silicon Valley Edge*, Stanford University Press, Stanford, CA.

Lee, C.M., W.F. Miller, M.G. Hancock and H.S. Rowen (eds) (2000) *The Silicon Valley Edge*, Stanford University Press, Stanford, CA.

Levy, S. (1996) *Insanely Great,* Penguin Books, Harmondsworth.

McClelland, S. (1961) *The Achieving Society*, Van Nostrand, Princeton, NJ.

Mowery, D.C. and N. Rosenberg (1998) *Paths of Innovation: Technological Change in 20th century America*, Cambridge University Press, Cambridge

Nahum, A, (2004) *Issigonis and the Mini*, Icon Books, Cambridge.

Nathan, J. (1999) *Sony: The Private Life*, Houghton Mifflin, Boston, MA.

Nelson, R.R. (1993) *National Innovation Systems*, Oxford University Press, Oxford.

Newell, S., M. Robertson, H. Scarborough and J. Swan (2002) *Managing Knowledge Work*, Palgrave, Houndmills, UK.

Packard, D. (1996) *The HP Way: How Bill Hewlett and I Built Our Company*, Harper Collins, NY.

Parsons, M. and M.B. Rose (2003) *Invisible on Everest: Innovation and the Gear Makers*, Northern Liberties Press, Philadelphia, PA.

Pavitt, K., M. Robson and J. Townsend (1987) 'The size distribution of innovating firms in the UK: 1945–1983', *Journal of Industrial Economics*, March 1983, pp. 43–54.

Piore, M.J. and C.F. Sabel (1984) *The Second Industrial Divide: Possibilities for Prosperity*, Basic Books, NY.

Roberts, E.B. (1991) *Entrepreneurs in High Technology: Lessons from MIT and Beyond*, Oxford University Press, Oxford.

Robson, G. (1999) *Cosworth: The Search for Power*, Haynes Publishing, Sparkford.

Rosen, P. (2002) *Framing Production: Technology, Culture and Change in the British Bicycle Industry*, MIT Press, Cambridge, MA.

Rothwell, R. (1986) 'The role of small firms in the emergence of new technologies', in C. Freeman (ed.) *Design, Innovation and Long Cycles in Economic Development*, Frances Pinter, London, pp. 231–248.

Rothwell, R. (1991) 'External networking and innovation in small and medium-sized manufacturing firms', *Technovation*, **11**, pp. 93–112.

Rothwell, R. and M. Dodgson (1991) 'External linkages and innovation in small and medium-sized enterprises', *R & D Management*, **21**, pp. 125–137.

Rothwell, R. and W. Zegveld (1985) *Reindustralisation and Technology*, Longman, Harlow.

Saxenian, A.L. (1983) 'The Genesis of Silicon Valley', *Built Environment*, **9** (1) pp. 7–17.

Schumpeter, J.A. (1936) *The Theory of Economic Development*, Harvard University Press, Boston, MA.

Schumpeter, J. (1950) *Capitalism, Socialism and Democracy*, 3rd edn, Harper Row, New York.

Tidd, J., J. Bessant and K. Pavitt (2001) *Managing Innovation: Integrating Technological, Market and Organizational Change*, 2nd edn, John Wiley and Sons, Chichester.

Van Dijk, B., R. Den Hertog, B. Menkveld and R. Thurik (1997) 'Some new evidence on the determinants of large and small firm innovation', *Small Business Economics*, **9**, pp. 335–343.

Wood, A. (2001) *Magnetic Venture: The Story of Oxford Instruments*, Oxford: Oxford University Press.

CHapter 10

funding innovation

OBJECTIVES

When you have completed this chapter you will be able to:

- Appreciate and understand the funding problems associated with innovation
- Analyse the means employed by innovators to reduce and minimise capital requirements
- Distinguish the different sources of capital available for innovation
- Differentiate the various forms of capital available for innovation
- Evaluate the circumstances when particular forms of capital are most likely to be appropriate
- Identify some of the agencies available to assist in handling funding innovation.

INTRODUCTION

From a financial perspective innovation is a problem. Innovations, even small ones, require an outlay of funds for development which has to take place prior to generating a corresponding intake of funds. The result is a significant negative cashflow which may persist for a considerable time, before funds come in and eventually cashflow turns positive. If the innovation is associated with the setting up of a new venture, the situation may be made all the more difficult.

This chapter explores the various ways in which organisations tackle this problem. The nature of the problem is examined. The various different types of funding that are available and the sources from which they are derived are also identified and analysed. Although the focus tends to be on the case of the new venture and how it can raise the necessary funding, nonetheless the overview of the capital markets that is provided is relevant to innovations developed by established ventures.

INNOVATION CASHFLOW

The Cashflow Gap

We have clearly seen that invention and innovation are not the same thing. While individuals may be able to fund the process of innovation from their own resources, this will almost certainly not be the case with innovation. Few innovators are likely to have the resources to fund the whole of the innovation process. It is no coincidence that highly successful innovators (who also happen to be technical entrepreneurs) such as James Dyson and Ron Hickman initially tried to license their technology to large companies with substantial financial resources, rather than fund manufacturing and distribution themselves. Since innovation is about exploiting ideas and inventions to turn them into commercial products, the exploitation part of the process has to be funded and it is very expensive.

Figure 10.1 Innovation Cashflow

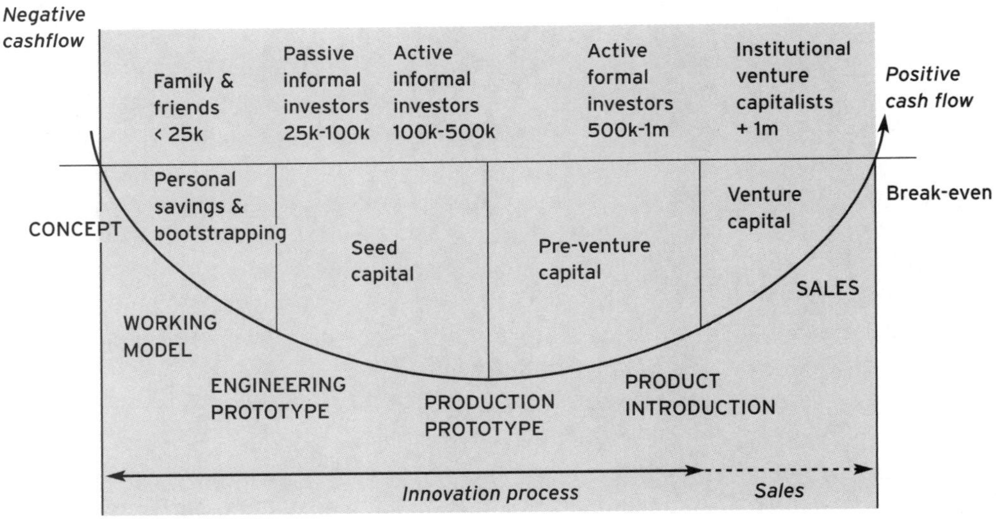

Source: Adapted from Rorke *et al.* (1991: p. 19)

Figure 10.1 shows how the various phases of the innovation process (e.g. construction of a working model, prototype, etc.) generate a negative cashflow. Unlike a business deciding to start production of a well-established product, the various phases of R & D have to be undertaken prior to the start of sales. This results in a substantial negative outflow of cash without a corresponding inflow from sales. If the innovation is radical and involves an entirely new technology, the duration of this period of negative cashflow may take months or even years before the product reaches the market and sales begin to generate an inflow of cash. Even then, as Figure 10.1 shows, it may take time before break-even is achieved and the actual cashflow turns positive.

When Pilkingtons, the Lancashire-based glassmaker, developed the float glass process for the manufacture of plate glass in the 1950s, it took seven years of research and development and cost millions of pounds. The company not only built several small-scale pilot plants, it even built a full-scale plant based on the new process and ran it for a year before it would produce glass of commercial quality. Only then could sales commence and the company begin to see a cash inflow. Similarly James Dyson took three years (Dyson, 1997) and built more than 5000 prototypes before he had perfected the dual-cyclone technology for his bagless vacuum cleaner. During this time he not only had to bear the cost of his own time (i.e. he had no income) but he had to pay for materials, power, lighting, legal fees and the cost of patenting his invention. Even when he had got his technology to work successfully and secured a patent he still had no inflow of cash as he was unable to persuade any of the existing vacuum manufacturers to take out a licence to manufacture a bagless vacuum cleaner. It was to be a further three years before a Japanese company finally agreed to take out a licence and Dyson at last had a modest inflow of cash in the form of royalty payments.

Given that there may be a substantial period when the innovator has only cash going out and nothing coming in, it is perhaps not surprising that as Figure 10.1 shows there are a variety of different types of capital that can be employed according to how near the innovation is to market.

Most innovators initially at least will make use of personal savings, usually supplemented by financial inputs from family and friends. As Figure 10.1 shows these funds are likely to be supplanted by financial bootstrapping – quite literally finding ways of acquiring resources either without having to pay for them or paying a lot less for them. But this is only a small part of the story. Again, as Figure 10.1 shows, activities such as the development of production prototypes, setting up manufacturing facilities and launching a product all have to be undertaken, and the scale of these activities is likely to be much greater. Consequently, innovators require so-called 'seed capital' to fund further development (see Figure 10.2). At this point they may well turn to 'informal investors' willing to contribute either loan or equity capital. Informal investors, or business angels as they are known, are high net-worth individuals willing to put some of their financial resources into new ventures. They do this not to obtain an income from dividends, but in the hope of capital gain at some later stage. Eventually, however, even this source may not be sufficient and, with the innovation nearing the point where it is ready for the market, venture capitalists may be approached. They provide larger sums than informal investors but operate on the same basis – that is to say, they invest in the expectation of capital gain. Finally, depending on the size of the venture, the innovator may turn to the equity market and seek additional capital through an initial public offering (IPO) perhaps on a junior market such as the Alternative Investment Market (AIM).

All of these sources of capital contribute in different ways to helping the innovator bridge the 'cashflow gap' that occurs while an innovation is under development prior to

Figure 10.2 Sources of Funding

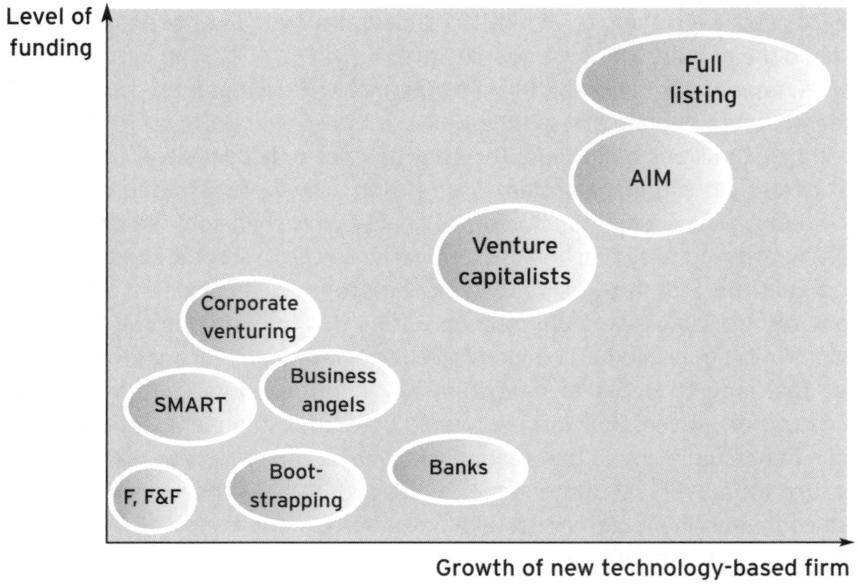

Growth of new technology-based firm

its being launched on the market and generating sales. This chapter explores the different ways in which organisations and individuals fund this period of negative cashflow.

FINANCIAL BOOTSTRAPPING

It has already been noted that business biographies provide a valuable insight into the innovation process. Such biographies usually note how difficult the process is and the obstacles and hurdles with which the innovator has to cope. At the same time business biographies also provide an invaluable insight into the ways in which innovators fund the innovation process. What is particularly revealing is the extent to which many innovators fund at least the early stages of the process, by not paying for it. Strapped for cash they look to other ways of acquiring the necessary resources: in short, they rely on bootstrapping.

Harrison *et al.* (2004: p. 308) define bootstrapping as involving, 'imaginative and parsimonious strategies for marshalling and gaining control of resources'.

They go on to suggest that such strategies typically take two forms:

1 Using creative ways of acquiring finance without recourse to banks or raising equity from traditional sources
2 Minimising or eliminating the need for finance by securing resources at little or no cost

The first of these overlaps with founder, family and friends, and is accessed through the founder's social capital network. Nonetheless, it is worth recounting some examples of this form of bootstrapping. Many innovators have relied on modest personal savings. James Dyson (Dyson, 1997) utilised this form of bootstrapping in the early stages of his work on the dual-cyclone vacuum cleaner, as did Jeff Bezos when developing his online bookshop Amazon.com.

The other category in Harrison *et al*.'s typology is probably the one that comes closest to the essence of bootstrapping, namely acquiring resources at little or no cost: in short, managing to acquire resources for free or at least for much less than one would normally expect to pay. This can take a variety of forms. One of the commonest is using domestic facilities. Examples proliferate. It is well known that Apple Computer manufactured its first personal computers in the garage of Steve Jobs' parents' home at 11161 Crist Drive, Los Altos in California (Linzmayer, 2004), but there is an abundance of other examples of successful businesses that started from a garage. Hewlett-Packard began in a garage and so did Amazon.com. Founded by Jeff Bezos, the latter started life in the garage of a house he was renting in Bellevue, Seattle (Cassidy, 2002). Other domestic facilities used by innovators have included: the garden shed which gave birth to Oxford Instruments; the home workshop which Trevor Baylis used to develop his clockwork radio; and the bedroom which was used by Dan Bricklin for his early work developing VisiCalc, the world's first spreadsheet. Whatever the location, the point is that the use of domestic premises is convenient, very flexible and above all saves on rent.

The use of facilities borrowed from or provided by the parent organisation is another means by which resources can often be obtained for nothing. When Martin Wood and his embryonic company, Oxford Instruments, was developing his first magnets back in the 1960s, he was able to borrow a winding machine from the Clarendon Laboratory at Oxford University where he worked (Wood, 2001). Again when he produced Europe's first superconducting magnet he was able to use the services of Oxford University's computer in order to carry out a series of complex calculations.

The use of second-hand equipment is another way in which innovators endeavour to keep their costs down, thereby reducing the need for finance. Notable examples of innovators who have followed this route have included Joseph Bamford of JCB fame and Martin Wood of Oxford Instruments. In Bamford's case the first piece of equipment he acquired was a second-hand welder. This proved crucial in his early experiments with hydraulic loaders. In Martin Wood's case it was second-hand machine tools bought at an auction of government-surplus equipment that enabled him to embark on the manufacture of superconducting magnets.

Another popular way of acquiring resources is to get people to work for you in their spare time preferably while paying them little or in some cases paying them nothing. This was what Colin Chapman, the innovative designer and engineer, did when he was developing the first of his Lotus cars. Crombac (2001: p. 53), Chapman's biographer, notes how, 'Chapman had attracted a motley crew of part time workers, all unpaid, but rewarded with a chance to drive the car.'

Among those who worked on this basis was Mike Costin who later worked for Lotus full time before eventually leaving to found the engine maker, Cosworth.

FOUNDER, FAMILY AND FRIENDS

As with any small firm, the initial funding is likely to come from a combination of:

- Founder
- Family
- Friends

Inevitably there is some overlap between this source of funding and the previous one. The founder will typically use his or her savings, or a redundancy package, or mortgage the family home. Some may be fortunate and have a legacy they can use, the less fortunate may sell a prized but valuable possession. Some may continue in employment. If friends and family are the source it is again likely to be past savings that are drawn on. The funding at this stage is likely to be used to fund the start-up stages particularly, as Figure 10.1 shows, initial development work such as building the first models or prototypes as part of proof-of-concept work. It is perhaps not unsurprising that innovators should draw on their families for funding. They are after all a logical source of funding. This was certainly the case with Martin Wood, the founder of Oxford Instruments and the first person in Europe to build a superconducting magnet, who gained financial support from his family in the early stages of his venture. Similarly, when his own savings ran out, Jeff Bezos, the creator of the world's first and biggest online bookshop, Amazon.com, persuaded his parents to invest $250 000 (Cassidy, 2002) in the venture. The ability to use friends is likely to be a function of an innovator's 'social capital': that is to say, his or her personal network of friends and acquaintances, derived from recreational interests, previous work experience and even school or college. James Dyson for instance was able to get funding from his friend and former employer Jeremy Fry when he set up his first company to develop his idea for a bagless vacuum cleaner (Dyson, 1997).

MINI-CASE: CYBERSENSE BIOSYSTEMS

Cybersense Biosystems was established in 2001 by Dr Tim Hart. With a PhD in Soil Botany and Microbiology and a number of years' experience of research in a university research laboratory, Hart's idea was to develop biosensors that could be used to measure land contamination. Specifically the aim was to develop sensors using bioluminescent bacteria that would permit speedy and effective identification of toxicity in contaminated land. The idea of using bioluminescent bacteria was derived from Tim's earlier research. Bioluminescence is the same principle by which glow-worms glow in the dark. The technology would potentially allow landowners, developers, environmental consultants and industrial concerns to quickly assess the toxicity of contaminated land and link the toxicity to potential pollutant triggers, thereby facilitating action to prevent further pollution.

The company's formation was linked to the award of a DTI Smart award that enabled Tim to design and build a first prototype of his toxicity measuring instrument. This was completed the following year when the company not only won a DTI Bio-wise grant but raised a substantial amount of additional capital through a syndicate of private investors. The first fully working ROTAS instrument was completed the following year and went on the market early in 2004.

Source: Talk by Tim Hart, 14 October 2004

GOVERNMENT FUNDING

SMART awards

The Small Firms Merit Award for Research and Technology (SMART) was introduced by the Department of Trade and Industry (DTI) in 1986 as a competitive scheme designed to provide support for innovation by small firms. The rationale behind the scheme is that small firms face particular problems in raising finance for R & D associated with innovation. As Caird (1994: p. 58) notes, the SMART scheme,

> ...recognises the failure of market forces to support high technology innovation as a result of the potential failure risks associated with innovation. It aims to encourage the formation of science and technology firms and help them to grow to a point where they are likely to attract financial support.

The objectives of the SMART scheme as specified by the DTI (1989) are:

- To bring forward highly innovative but commercially viable projects, now dormant because existing sources of finance do not wish to support them
- To encourage the formation of small firms which will develop and market new ideas in selected areas of new technology
- To help these small firms to mature sufficiently for private sources of funds to take a practical interest

Previously there had been other government-funded schemes designed to provide finance for innovation, but the SMART scheme was the first to be specifically designed to meet the needs of small firms. According to Moore (1993) the scheme was in fact inspired by the success of an earlier scheme in the US, the Small Business Innovation Research programme. Given that it was established by a government anxious not to engage in interventionism, the scheme was deliberately structured as a competition with relatively limited funding available. Similarly, it was a temporary programme designed to run for three years in the first instance. A measure of its success is that it is still in operation albeit in a slightly different form (the SMART scheme was re-branded as *Grants for Research and Development* in 2004).

The SMART scheme is open to individuals and small firms with up to 50 employees. It aims to develop innovative but marketable technology. According to Lawrence (1997), almost any technology with the potential to be transformed into a commercial product has a chance of winning an award. Its purpose is to provide funds for a technical and commercial feasibility study as well as taking the project on to development. Under the terms of the scheme SMART awards are available to both start-up and existing companies and come in two stages. The first stage is competitive and is designed to assist small firms and individuals with technological and commercial feasibility. Applicants for an award are required too submit a detailed proposal and in 1995 over 1100 proposals were received. Proposals are subject to detailed scrutiny by technical experts, financial advisers and commercial assessors, using a points system based on well-established criteria. The degree of innovation is an important element in the selection process, as is its technical merit, the calibre of the team putting forward the proposal, and the scope for commercial success (Lawrence, 1997). Winners receive a Stage 1 SMART award which provides 75 per cent of the first £60 000 of eligible costs in the first year up to a maximum of £45 000. This is a cash grant of which the first instalment of £15 000 is paid in advance. The Stage 2 award provides further support 'to take the idea closer to the marketplace' (Jones-Evans and Westhead, 1996: p. 30) such as assisting with the development of pre-production prototypes. The support amounts to up to 50 per cent of eligible costs up to a maximum of £150 000 (covering both stages).

Successful projects are carefully monitored to ensure that milestones and targets are met and the money is spent as planned. Lawrence (1997) notes that one of the attractions of the scheme is that the first tranche of funds amounting to £15 000 is presented as a cheque at a formal award ceremony to mark the start of the project. The attendant publicity, as Moore (1993) observes, can have powerful indirect benefits for recipients, bringing the firm to the attention of external investors and thereby increasing its chances of obtaining external funding on appropriate terms.

That the scheme has continued (albeit with some changes along the way - see below) for almost 20 years is testimony both to the fact that market failure does exist and to the success of the scheme. Not only has the scheme resulted in innovations coming to market that would otherwise not have done so, but independent evaluation of the scheme (Moore and Garnsey, 1993) has shown SMART recipients receiving injections of finance to allow further growth and unsolicited approaches from venture capitalists.

Following the success of the SMART scheme the government introduced a new programme of support for R & D called Support for Programmes Under Research (SPUR). SPUR is open to all firms employing less than 500 employees but is targeted at larger firms since projects must have eligible costs with a minimum value of £50 000. SPUR provides a fixed grant that covers 30 per cent of eligible costs (up to a maximum of £150 000) of the development of new products and processes which involve a significant technological advance for the industry.

(The SMART scheme (branded as the Grants for Research and Development scheme) is very similar, though there are now four forms of funding available: Micro Project Grant of £2500-£20 000 (up to 50 per cent of project costs) over 12 months; Research Project Grant of £20 000-£75 000 (up to 60 per cent of project costs) over 6-18 months; Development Project Grant of £20 000-£200 000 (up to 35 per cent of project costs) over 6-36 months; Exceptional Development Project Grant of up to £500 000 (up to 35 per cent of project costs).)

BANKS

❝ I went to the bank first and met a very nice chap, very friendly. It turned out he was normally doing things like lending money to people to set up newspaper shops. He did not feel really qualified to comment on my adaptive non-linear pattern recognition technology. But he did give me a good piece of advice, which I carry with me even now, which is that people will always buy confectionery. ❞ Doward (1999)

This quote from Mike Lynch of the software company Autonomy, reflects the findings of a number of studies (Bank of England, 2001) which have noted the reluctance of banks to provide finance for innovation-related, high-technology business start-ups. A study by Moore (1994) found that only 7 per cent of high-technology companies raised start-up finance from banks compared to 40 per cent of SMEs in general. This does not mean that banks do not provide finance for innovation, it simply reflects the fact that banks generally provide working capital through the provision of overdraft facilities.

However a recent report by the Bank of England (2001) highlights the fact that the UK banking sector has made efforts in recent years to improve its servicing of the innovation and high-technology market. The NatWest Innovation and Growth Unit for instance currently has some 225 Technology Business Managers operating in the field. They are assisted by a Technology Business Appraisal Service which offers a low-cost technology evaluation service via a network of independent technology marketing specialists who assess the technical feasibility, commercial viability and future potential of technology-based proposals. Similarly, HSBC has launched its own Innovation and Growth Unit, and Barclays has set up 15 Technology Centres across the UK. The creation of these specialist units indicates UK banks are taking a more significant role in funding innovation than in the past.

BUSINESS ANGELS

Business angels are high net-worth individuals seeking capital gains over the life of their investment in a company. They are informal investors in the sense that they do not belong to or form part of a recognised market. It is because there is no clearly visible market in which they operate or directories that list them (Mason and Harrison, 1996; 1997), that business angels are known as 'informal' investors. For the same reason business angels typically have a regional focus.

Recent estimates (Bank of England, 2001) suggest that there are approximately 18 000 business angels in the UK, who between them invest £500 million annually in some 3 500 companies. Although this is small compared to the £6.2 billion invested by venture capitalists, the Bank of England estimates that a high proportion of business-angel investments are in seed, start-up and early stage capital, making the 'informal' venture capital market (i.e. business angels) of equivalent importance in terms of funding innovation to the formal market, even though the latter is many times bigger (Bank of England, 2001).

A study of business angels by Coveney and Moore (1997) found that business angels are by no means a homogeneous group. Business angels vary greatly on any one of a number of different aspects of their activities including:

- Number of investments made per year
- The level of funds they have available to invest
- The size of individual investments, which varied from £10 000–£1million +
- Net worth of individual investors, which ranged from about a quarter with less than £200 000 to a similar proportion with more than £1 million
- The extent of their experience of business start-ups, which ranged from about a third with no experience to some who had been involved in several
- The reasons for investing, with only half investing primarily for financial gain

In the light of such variation, Coveney and Moore (1997) divided business angels into two broad categories – active and passive informal investors – which they then further sub-divided to create a total of six categories:

1   Active Business Angels
- Entrepreneurs
- Wealth maximising
- Income seeking
- Corporate

2   Passive Business Angels
- Latent
- Virgin

The entrepreneur business angels are generally self-made, very wealthy and interested in a broad range of business opportunities. Typically they invest in start-up situations and are motivated as much by fun and a desire to contribute to a successful venture as a desire for financial gain. For these individuals the most important criterion tends to be the personality of the innovator/founder. Surprisingly, they are often less active in the management of the venture than other types of business angel

Income-seeking business angels are generally interested in smaller investments and are looking for high rates of return. Their background is generally the least entrepreneurial and they have less net worth. Investing mainly locally and in sectors with which they are familiar, their motivation is clearly income, and for them it is a serious business rather than a matter of enjoyment and satisfaction.

Wealth-maximising business angels are, to quote Coveney and Moore (1997: p. 73), 'a contradictory bunch'. They invest mainly for financial gain and, though they normally only take minority stakes, they often get involved in company management. Richer than income-seeking business angels, they are less focused in terms of what they invest in. Their backgrounds are generally not entrepreneurial, most having acquired their wealth through inheritance.

Corporate business angels are, as their name implies, companies making investments in unquoted companies, and they are dealt with in more detail in the next section under the heading of corporate venturing.

Latent business angels are ones who have invested in the past but do not currently have any investments. Coveney and Moore (1997) found they are generally very wealthy, highly educated and older than other business angels. Their main reason for not investing is a lack of suitable investment opportunities.

Virgin business angels are those who want to invest but have not yet done so. In general their backgrounds are very varied with few having been involved in a business start-up before.

The Bank of England (2001) suggests that the presence of latent and virgin business angels is a sign of market inefficiency, reflecting information gaps and unnecessarily high search costs. These factors reflect the informal nature of this capital market. However, recent moves to create networks of business angels together with greater awareness of this source of finance (Dodgson, 2000) may have gone some way to alleviate the problem.

A key feature of the business-angel market is the extent to which it complements the formal venture capital market. Recent research by the British Venture Capital Association suggests that this is the case, with more than half of business-angel investments being for less than £50 000 (Bank of England, 2001), and less than a quarter over £100 000, compared with 86 per cent by venture capitalists. The extent to which venture capitalists concentrate on bigger deals confirms that the business-angel market is the main source of private-sector finance, after the founder's initial resources have dried up (Mason and Harrison, 2000a, 2000b). As such, business angels normally provide capital after the immediate start-up phase and before the company has reached a size where it is likely to be of interest to venture capitalists. Business-angel investment may even contribute to the due diligence required by venture capitalists. More significantly, they perform an invaluable networking function by bringing together the innovator, capital and managerial expertise.

One final word of caution regarding business angels is that in the UK they are a much rarer species than in the US. In addition, only a small proportion appear to be active in innovation through investing in high-technology companies. However the Bank of England's (2001) recent report does note that the evidence on this appears contradictory with some studies suggesting that around a third of business angel investments are in high-technology companies, while other studies put the proportion as low as 5 per cent.

MINI-CASE: AMAZON.COM

Having studied engineering and computer science at university, Jeff Bezos worked briefly for a small software house before moving to Bankers' Trust, a big Wall Street bank. Two years later he switched to another financial institution, the hedge fund D.E. Shaw. It was here, a couple of years later in 1994 that he was asked to investigate the possibility of making money from the Internet. When he came to look at the Internet he was immediately struck by the dramatic growth of traffic in the World Wide Web. Yet there were few businesses on the web, and Bezos figured that it was only a matter of time before this changed. His analysis led him to the conclusion that of all the items that could be sold over the web, books offered the best prospect.

Bezos recommended that D.E. Shaw set up an online bookstore. When Shaw rejected his idea, he left and set up on his own. At the time it was a big gamble. Bezos chose Seattle as the home for his new company. Why Seattle? It was the home of Microsoft, it was close to Roseburg, Oregon, the home of the biggest book distributor in the US and he had friends there. Bezos called his new company Amazon.com, the first use of the '.com' suffix. Bezos employed two computer scientists to develop a website and paid them out of his own pocket. When he ran out of money his parents invested $250 000. With such a risky venture raising money from outsiders was not really an option. Amazon.com began business from the garage of his rented house in Bellevue, a suburb of Seattle, in mid-1995. Amazon.com was not the first Internet bookstore, but it was the first to allow customers to search through a catalogue of a million books. At first the business grew slowly. Needing more money Bezos began to look for outside capital. With

▶

great difficulty he raised $1 million. Then Bezos got a lucky break. In May 1996, a short article on the front page of the *Wall Street Journal* reported how Bezos had quit his Wall Street job to set up an online bookshop. The day the story was published Amazon.com's business doubled and then went on doubling over a very short space of time. In no time at all he was inundated with venture capitalists offering to invest in his company.

Source: Cassidy (2002)

CORPORATE VENTURING

A recent report by the CBI (1999) describes corporate venturing as,

> ❝ a formal, direct relationship, usually between a larger and an independent smaller company, in which both contribute financial, management or technical resources, sharing rewards equally for mutual growth. ❞

Such arrangements involve the larger company taking a financial stake in the smaller one in return for a share in its development. Typically this would be connected to the development of an innovation in the form of a new product or process. This sort of arrangement most often occurs in high-technology sectors such as pharmaceuticals and software, and reflects developments in the US where the trend towards larger companies investing in smaller ones is increasingly a feature of high-technology clusters like Silicon Valley.

For the small company corporate venturing affords another means of funding innovation, although it often brings with it other benefits including access to managerial expertise and sometimes to the larger company's manufacturing and marketing resources.

The benefits however are by no means all one way. For the larger company, corporate venturing offers benefits too. It facilitates access to new ideas and skills, allows an assessment of new markets and provides the means to exploit potentially attractive returns on new technologies.

VENTURE CAPITAL

The Macmillan Report of 1931 first identified the existence of an 'equity gap', describing the inability of small firms to access long-term risk capital. At the time the gap was said to lie between founder, friends and family and the stock market. To bridge this gap the Bank of England and the UK clearing banks established 3i, the first venture capital organisation in the UK. Despite this early start, venture capital is a relatively recent phenomenon (Deakins and Freel, 2002), since it was only in the 1980s that venture capital emerged as a significant source of funding both in the UK and the US. In the US the growth of venture capital was linked closely to the success of Silicon Valley where a big increase in the number of technical entrepreneurs was facilitated by this form of funding (Howells, 2005).

Essentially venture capitalists borrow from institutional investors such as pension funds and life assurance companies and invest in companies with growth potential, with a view to making a capital gain by the time they withdraw their money some years later. As intermediaries, venture capitalists' expertise lies in evaluating the risks and growth potential of companies which they back with equity investments.

The UK venture capital industry is the largest in Europe (Bank of England, 2001). It is second only to the US and in per capita terms it is the largest in the world. Between 1984 and 1998 British Venture Capital Association members invested some £35.5 billion in more than 19 000 companies.

However, venture capital forms only a very small proportion of the external finance used by SMEs in the UK. Similarly, only a small proportion of venture capital goes into financing start-ups and early growth. This reflects what appears to be a growing preference within the venture capital industry for large deals. In fact, the average size of deal has risen from less than £1 million in 1988 to over £3 million in 1998. This rise reflects a shift away from start-up and early-stage funding in favour of later-stage deals that require larger sums (Bank of England, 2001).

A variety of factors have been put forward to explain this, including increasing reliance on pension funds as a source of finance and the greater cost of managing small-technology investments. Whatever the reasons, venture capital is of declining importance as a source of funding innovation undertaken by small companies.

INITIAL PUBLIC OFFERING (IPO): THE ALTERNATIVE INVESTMENT MARKET (AIM)

Launched in 1995, the Alternative Investment Market (AIM) is a London-based equity market aimed at meeting the needs of smaller companies, especially ones that are technology-based. As such it is designed to meet the needs of smaller companies by offering a streamlined admission process and a flexible approach to regulation (i.e. financial reporting). Specifically this means:

Accessibility

Unlike the main equity market of the London Stock Exchange, AIM does not stipulate minimum criteria for:

- Company size
- Track record
- Number of shares in public hands

Instead a nominated adviser ensures that the applicant company is suitable for entry to AIM. This type of approach is much more flexible than that taken by the main market and is therefore much more in line with the needs of small high-technology companies.

Admission Process

Again the process is a much simplified version of that required for the main market. As a result it is relatively quick, normally taking 24 weeks from the start of the process to a successful listing. The nominated adviser carries out due diligence and ensures that all the requirements are met and all appropriate information is included in the admission document.

Regulatory Regime

All AIM-listed companies are required to disclose their financial performance through the publication of interim and full-year results, as well as making appropriate disclosures regarding any issues relevant to future performance. But they are not required to publish details regarding acquisitions and disposals.

At launch there were just ten companies on AIM. By the end of 2000 this had risen to 524 with a market capitalisation of £14.5 billion. Of these, around a fifth are technology-based companies with a market capitalisation of £3.26 billion. In fact, 2000 was a record year for the admission of technology-based companies (Figure 10.3) with 62 joining the market that year.

Figure 10.3 Technology-based AIM admissions

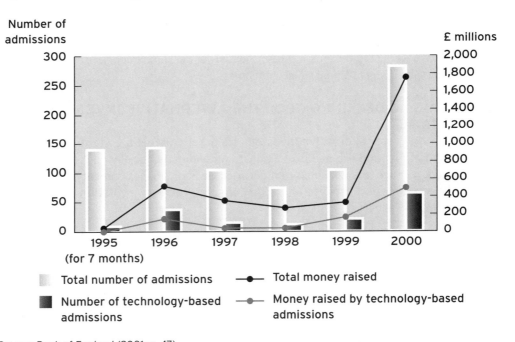

Source: Bank of England (2001: p. 47)

As with any public equity market AIM offers a number of benefits beyond raising equity capital. These include:

- An exit route for existing shareholders
- Enhanced visibility for the enterprise
- A broader shareholder base
- An objective market valuation of the enterprise

The market is aimed specifically at the unique community of innovative and entrepreneurial companies.

CASE STUDY: OXFORD INSTRUMENTS

Formation of the company

Oxford Instruments was started by Martin Wood and his wife Audrey. Martin Wood came from a solid, middle-class professional family. Required to do three years' national service before going to university, Martin Wood opted to do his national service working in the coal mines of South Wales instead of going into the army. He then gained a BSc from Imperial College, London and an MA in Engineering from Cambridge. Having finished his studies, he decided to pursue a career in industry working initially as a management trainee for the National Coal Board. However, he became frustrated with working for such a large organisation, decided to look around for something else and got a job as Senior Research Officer at the Clarendon Laboratory, part of Oxford University's physics department.

It was while he was working at the Clarendon Laboratory that he founded Oxford Instruments. In fact, the company began very much as a part-time venture, with Martin Wood continuing his work at the university, for a further ten years. To begin with the work of the new venture was all concerned with design and consultancy. This was a time when the university system in the UK was expanding and Martin Wood initially assisted and advised new departments on the equipping of laboratories. Since it did not actually make anything, Oxford Instruments did not need any premises. Most of the work came through Martin Wood's contacts at the Clarendon Laboratory.

In time Martin Wood and his wife Audrey came to realise that some of his customers, especially those in public research laboratories, such as the Royal Radar Establishment (RRE) in Malvern and the Harwell Laboratory of the UK Atomic Energy Authority (UKAEA) had a requirement for magnets that they were finding difficult to fulfil. As a result Oxford Instruments moved into the manufacture of magnets. The move into manufacturing required rather more in the way of assets than the design and consultancy business. The need for premises was met by the Woods installing a large garden shed which had once been half a postwar 'prefab' house at the bottom of their garden. For equipment the company bought an old lathe at auction and arranged to borrow the Clarendon laboratory's special machine for winding magnetic coils. Material supplies were less of a problem, because Martin Wood was dealing with the same suppliers that he dealt with regularly at the university. The company employed a retired Clarendon laboratory technician on a part-time basis and most of the rest of the jobs were undertaken by Audrey Wood.

The company delivered its first magnet, a specialist laboratory magnet, manufactured in the shed at the bottom of the garden of the Woods' North Oxford home to the Royal Radar Establishment, and other orders soon followed.

Superconductivity and magnets

Now initiated into the intricacies of manufacturing, Martin Wood continued his academic work. He and his wife attended an academic conference at Boston in the US. This was a routine event for university staff such as Wood, but the conference was to have a far from routine impact on the new company. At the conference several research groups reported research into new developments in the field of superconductivity. Superconductivity is essentially a property of certain metals which means that at very low temperatures they

▶

suddenly lose all electrical resistance. Superconductivity had been first identified in 1911, but from then until the second half of the twentieth century it remained an interesting curiosity with no practical applications. The conference raised the prospect that this might be about to change, with the development of new materials that were sufficiently versatile to be used in practical products. Since electrical resistance dramatically reduces the efficiency of materials for transmitting electricity, any material that has a low resistance (in superconducting materials resistance falls to zero) is likely to have a dramatic impact on the efficiency of transmitting electricity. This was especially important for research laboratories working with high magnetic fields. Hitherto the scope for research had been limited by the need to employ large magnets that required enormous amounts of electrical power and special water-cooling facilities. An indication of the problems posed by the use of conventional magnets can be gauged by the fact that at the time the Clarendon Laboratory at Oxford University had one magnet which alone used 10 per cent of the city of Oxford's electricity. Such was this magnet's requirement for electricity that permission had to be sought from the electricity board every time it was switched on. In essence, developments in superconductivity meant there was the prospect of a completely new generation of magnets (and a whole host of products using magnets) that would no longer be dependent on large quantities of electrical power.

While these developments offered the prospect of a new type of magnet, to develop a commercial superconducting magnet still required many obstacles to be overcome. The new superconducting materials, such as niobium zirconium (NbZr), tended to be inconsistent and unreliable. Not only that, to achieve the very low temperatures required for the material to be come superconducting required a supply of liquid helium, a liquified gas that was costly and hard to obtain.

Undeterred by these problems, Oxford Instruments purchased a pound of niobium zirconium wire from the US and Martin Wood set out to make Europe's first commercial superconducting magnet. Knowledge was extremely sparse. There were no books or manuals on how to manufacture a superconducting magnet. From university sources Martin Wood had access to reports and papers from academic conferences, but even though these told him a lot about the properties of the materials with which he was working, they did not focus on issues directly relevant to manufacturing. He was fortunate in being able to use Oxford University's computer to calculate the quantities of material required and to predict the new magnet's performance characteristics.

With the design completed, the first magnet did not take long to build. Using liquid helium and liquid nitrogen borrowed from the Clarendon Laboratory, the new magnet was carefully tested. Much to everyone's surprise it worked perfectly, producing a magnetic field far more powerful than had previously been possible in a laboratory without major engineering, power and cooling installations. This prototype worked well and was shown and demonstrated at a variety of exhibitions and conferences. Oxford Instruments was soon talking to prospective customers and by the end of the year had had many enquiries and ten orders for the new type of magnet. Though busy manufacturing conventional magnets, Oxford Instruments developed and manufactured its first superconducting magnet, which was delivered to Birkbeck College, at the University of London.

It was at this time that the company moved out of the Woods' garden shed and into a disused stable in North Oxford, rented from a local butcher. To provide additional manufacturing capacity, the company acquired and installed some redundant machine tools bought at auction, and took on its first full-time employee. Within a year the workforce had risen to four. Further equipment such as cryostats for storing liquid helium was manufactured by Clarendon Laboratory technicians working in their spare time.

The need for at least some administration, covering aspects such as payroll and invoicing, personnel and selection, and the purchasing and storing of materials, was now beginning to emerge. Nor was it only a matter of routine administration, since there was also a need to plan production and schedule the delivery of materials so that deliveries were met on time. These were aspects that neither Martin Wood nor his fellow workers with their scientific backgrounds and interest in technical issues knew much about. As a study of the company noted, 'we were rather arrogant, imagining that it would not be difficult for people with a scientific training to grow a company successfully'. Oxford Instruments was fortunate that Martin Wood's wife, Audrey, was able to handle most aspects of the administrative side of the business at this time, using a second-hand caravan that served as an office.

Funding development of the company

Although growing fast the company did not yet need outside finance. Overheads were still low and the operation sufficiently small for costs not to get out of hand. The limited finance that was needed, could be provided by the Woods themselves, supplemented by loans from family members and an overdraft from the bank.

One problem that the company did face at this point in its development was obtaining reliable supplies of raw materials. Liquid helium was a particular problem. Their supplier, British Oxygen Company (BOC), would not deliver supplies of the gas – so that the vacuum flasks containing the gas had to go by train – resulting in delays and losses of gas due to heat and vibration. Not only was this making manufacturing difficult, it was deterring potential users of superconducting magnets because they could not ensure a reliable supply of liquid helium to enable them to operate the magnets. So it was that Oxford Instruments undertook a major new investment, purchasing a helium liquefier from the US. It cost £35 000, a great deal more than the total investment in the company to date. The company was able to finance the purchase by extending its overdraft with the bank, supported by suitable guarantees from members of Martin Wood's family. A separate company was formed, Oxford Cryogenics, and additional premises were acquired to house the new equipment, in the form of a disused laundry rented from Oxfordshire County Council on a temporary basis pending planning decisions in the area. Oxford Cryogenics was soon supplying laboratories across the country and the new service gave a major boost to low-temperature research in the UK.

Meanwhile Oxford Instruments was growing fast. Production of both types of magnet was expanding and the following year the company moved to new larger premises by the river Thames in Oxford. Three years after it had been formerly registered, the company's annual turnover was £41 000 on which the company made a net profit of £2460, representing a return of some 6 per cent. By this point Oxford Instruments had 25 employees almost all of whom were engineers, scientists and technicians.

Within a couple of years however finance was beginning to become a problem. The company relied heavily for profits on systems produced to individual requirements. This stretched the technology to its limits and made profits unpredictable. By this point the existing sources of capital – family loans and guarantees and the bank overdraft – were simply proving inadequate. The company looked at several potential sources of venture capital including:

- Large companies investing in technology
- Merchant banks
- Insurance companies
- Rich private individuals

Although they came very close to agreeing a deal on several occasions throughout this period, the terms were never quite right. Eventually the venture capital firm 3i put in substantial loans and equity finance. This involved 3i purchasing 20 per cent of the equity of Oxford Instruments for £45 000, while at the same time providing a similar sum in debentures (loan finance). With adequate capital Oxford Instruments continued to grow rapidly. Martin Wood finally quit his job at Oxford University. By this time the company had 105 staff on the payroll and the following year turnover reached £350 000. It was a small company no longer.

Figure 10.4 The Range of Superconductivity

Degrees centigrade	Degrees kelvin
C	K
100	373 Boiling water
0	273 Ice
-183	90 Liquid oxygen
-196	77 Liquid nitrogen
* -269	4 Liquid helium *
-273	0 Absolute zero

*Range of superconductivity

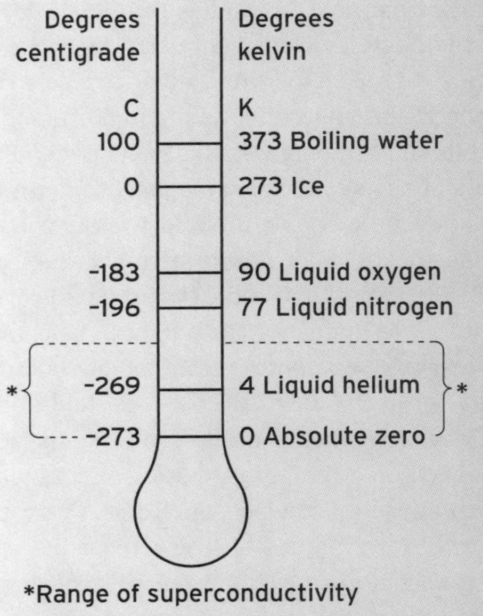

Source: A. Wood (2001) from 'Magnetic Venture: The Story of Oxford Instruments', by permission of Oxford University Press

QUESTIONS

1 What sources of venture capital did Oxford Instruments consider and which source did it ultimately opt for?
2 What is the name usually given to high net-worth individuals who invest in new ventures, especially ones that take the form of technology-based companies?
3 What innovation did Oxford Instruments develop?
4 What kind of entrepreneur would you classify Martin Wood as?
5 What aspects of Martin Wood's previous work experience proved helpful in founding Oxford Instruments and why?
6 What institutional support did Oxford Instruments get in the early years and from whom?
7 What instances of 'financial bootstrapping' can you identify in this case?
8 What sources of venture capital did Oxford Instruments consider and which source did it ultimately opt for?
9 How important do you consider Martin Wood's personal network to have been in the success of this venture?
10 Which organisation was Oxford Instruments 'parent' and to what extent did the company continue to maintain links with it?
11 What other problems besides funding did Oxford Instruments face as it grew and why are these sorts of problems likely to be particularly significant for technical entrepreneurs?

QUESTIONS FOR DISCUSSION

1 Why is cashflow an important issue for technological innovation?

2 What is the cashflow gap that innovators typically face?

3 What is venture capital?

4 What is bootstrapping and why do innovators often resort to it as a way of funding innovation?

5 Why are personal networks often important when it comes to funding innovation?

6 What are business angels?

7 What motivates business angels to risk their capital?

8 What is the rationale behind the government making funds available to support innovation?

9 **What is the SMART scheme?**

10 Why are banks often reluctant to lend to small high-technology businesses?

11 **What is AIM? Why is it likely to be of particular interest to smaller high technology companies?**

12 What is an IPO and why are IPOs attractive to technical entrepreneurs?

13 **What is the difference between a 'passive' and an 'active' informal investor?**

14 What is the difference between formal and informal investors?

ASSIGNMENTS

1 Explain what is meant by the term 'business angel' and explain how such individuals can contribute to the process of innovation.

2 Select a technical entrepreneur who has been active in innovation and outline the problems he/she faced regarding funding and indicate how they overcame these problems.

3 Explain what is meant by the term 'bootstrapping' and then using an account of an innovation show how one innovator (either an individual or an organisation) used it to fund innovation.

4 Analyse the reasons for a cashflow gap when it comes to the provision of 'seedbed capital' for technical entrepreneurs and/or technology-based small firms.

5 'The best place for an innovator to get finance is a bank.' Discuss.

RESOURCES

Books on innovation tend to devote little space to funding. Instead, books on finance in general and capital markets in particular provide an overview of the funding available to business start-ups. However, such books rarely say much about the particular needs of high-technology businesses or innovators. They tend to look at the funding available for a wide range of businesses and fail to deal with the specific needs of innovators. In terms of Figure 10.1 they cover the period after a positive cashflow has been achieved. At best they say something about venture capital but generally ignore pre-venture capital and seed capital. In the absence of texts dealing specifically with the topic, one of the most useful resources is the Bank of England's review of financing technology-based small firms (Bank of England, 2001).

Fortunately there are other resources available. Financial bootstrapping requires a degree of ingenuity. Winborg and Landstrom (2001) provide an overview, while Harrison, Mason and Girling (2004) look at financial bootstrapping in the software industry. Beyond that the best sources are technical entrepreneurs' own accounts of innovation as found in business biographies. Among the most revealing are Jabby Crombac's biography of Colin Chapman, one of the leading innovators in the field of motor racing (Crombac, 2001) and Doug Nye's biography of Charles and John Cooper (Nye, 1991). The early part of Chapman's career shows the innovator's ingenuity at its best.

The field of informal venture capital or business angels is rather better covered by conventional academic sources. A small number of researchers have made a big contribution to this rapidly expanding field. Particularly notable are those of William E. Wetzel in the US and Richard Harrison and Colin Mason in the UK. Among the more important works are Wetzel (1983) and Mason and Harrison (2000b).

For venture capital the best starting point is probably the website of the British Venture Capital Association (http://www.bvca.co.uk).

REFERENCES

Bank of England (2001) *Financing of Technology-based Small Firms*, Bank of England, London.

Caird, S. (1994) 'Sources of technological innovative ideas and their significance for commercial outcomes in small companies', in R. Oakey (ed.) *New Technology-based Firms in the 1990s*, Paul Chapman Publishing, London, pp. 57–67.

Cassidy, J. (2002) *Dot.con: the greatest story ever sold*, Penguin Books, Harmondsworth.

CBI (1999) *Connecting companies: using corporate venturing for growth*, Confederation of British Industry, London.

Coveney, P. and K. Moore (1997) 'A Typology of Angels: A Better Way of Examining the Informal Investment Phenomena', in R. Donckels and A. Miettinen (eds) *Entrepreneurship and SME Research: On its Way to the Next Millennium*, Ashgate, Aldershot.

Crombac, G. (2001) *Colin Chapman: The Man and his Cars*, Haynes Publishing, Sparkford.

Deakins, D. and M. Freel (2002) *Entrepreneurship and Small Firms*, 3rd edn, McGraw-Hill, Maidenhead.

Dodgson, M. (2000) *The Strategic Management of Technical Innovation*, Oxford University Press, Oxford.

Doward, J. (1999) @business: E is for Autonomy, *Observer*, 5 December 1999, Business Section , p. 9.

DTI (1989) *Applications and Guidance Notes for the SMART Competition*, Department of Trade and Industry, London.

Dyson, J. (1997) *Against the Odds*, Orion Business, London.

Harrison, R.T. , C.M. Mason and P. Girling (2004) 'Financial bootstrapping and venture development in the software industry', *Entrepreneurship and Regional Development*, **16,** pp. 307–333.

Howells, J. (2005) *The Management of Innovation and Technology: The Shaping of Technology and the Institutions of the Market Economy*, Sage Publications, London.

Jones-Evans, D. and P. Westhead (1996) 'The high technology small firm sector in the UK', *International Journal of Entrepreneurial Research and Behaviour*, **2** (1), pp. 15-35.

Lawrence, P. (1997) *The business of innovation: How to turn a patentable idea into a profitable product*, Management Books 2000, Chalford.

Linzmayer, O.W. (2004) *Apple Confidential 2.0: The Definitive History of the World's Most Colorful Company*, No Starch Press, San Francisco, CA.

Mason, C.M. and R.T. Harrison (1996) 'Informal venture capital: a study of the investment process, the post-investment experience and investment performance, *Entrepreneurship and Regional Development*', **8**, pp. 105-125.

Mason, C. and R. Harrison (1997) 'Business angels are the answer to an entrepreneur's prayer', in S. Birley and D. Muzyka (eds) *Mastering Entrepreneurship*, FT/Prentice Hall, London, pp. 110-114.

Mason, C. and R. Harrison (2000a) 'The size of the informal venture capital market in the UK', *Small Business Economics*, **15**, pp. 137-148.

Mason, C. and R. Harrison (2000b) 'Informal Venture Capital and the Financing of Emergent Growth Businesses', in D.L. Sexton and H. Langstrom (eds) , *The Blackwell Handbook of Entrepreneurship*, Blackwell, Oxford, pp. 221-239.

Mason, C.M. and R.T. Harrison (2004) 'Improving Access to Early Stage Venture Capital in Regional Economies: A New Approach to Investment Readiness', *Local Economy*, **19**, pp. 159-173.

Moore, B. (1994) 'Financial constraints to the growth and development of small high-technology firms', in A. Hughes and D.J. Storey (eds) *Finance and the Small Firm*, Routledge, London.

Moore, I. (1993) 'Government finance for innovation in small firms: the impact of SMART', *International Journal of Technology Management*, special issue, pp. 104-118.

Moore, I. and E. Garnsey (1993) 'Funding for innovation in small firms: The role of government', *Research Policy*, **22,** pp. 507-519.

Norton, R.D. (2001) *Creating the New Economy: The Entrepreneur and the US Resurgence*, Edward Elgar, Cheltenham.

Nye, D. (1991) *Cooper Cars*, Osprey Publishing, London.

Rorke, M.L., H.L. Livesey and D.S. Lux (1991) *From Invention to Innovation: Commercialisation of New Technology by Independent and Small Businesses*, Mohawk Research Corporation, Rockville, MD; cited in Norton (2001).

Wetzel, W.E. (1983) 'Angels and informal risk capital', *Sloan Management Review*, **24**, pp. 23-34.

Winborg, J. and H. Landstrom (2001) 'Financial bootstrapping in small businesses: examining small business managers' resource acquisition behaviours', *Journal of Business Venturing*, **16**: pp. 235-254.

Wood, A. (2001) *Magnetic Venture: The Story of Oxford Instruments*, Oxford University Press, Oxford.

CHapter 11

ORGANISING for INNOVATION

OBJECTIVES

When you have completed this chapter you will be able to:

- Distinguish different corporate cultures and assess those likely to facilitate innovation
- Analyse the forms of organisation structure that can be used to facilitate innovation
- Evaluate the various organisational roles associated with innovation
- Evaluate the contribution that organisational arrangements can make to the success or failure of an innovation.

INTRODUCTION

For every successful innovation there are probably ten times as many unsuccessful ones. As noted earlier, well-known examples include Corfam, the artificial leather developed by Du Pont in the 1960s, Sinclair's C5 electric tricycle and Rolls-Royce's RB211 jet engine. It is tempting to see innovation failures as the result of poor technology, but while innovations sometimes either do not work or fail to work reliably, often there are other factors in play. Often the fault lies with the market, or rather the failure of the innovation to deliver what the market wants. However, innovation does not occur in a vacuum. Most forms of innovation occur within an organisational context. Hence there are times when the reasons for the failure of an innovation are to do with the organisation responsible for the innovation, in particular the way in which it is managed.

This chapter focuses on the context, especially the organisational context in which innovation occurs. Some organisational contexts appear to be more favourable to innovation than others. A range of different contexts will be reviewed and the characteristics of those that are favourable to innovation will be highlighted.

It is not just the context that organisations provide. They can take more active steps to aid innovation, particularly the implementation of innovation. Such steps can include configuring or structuring an organisation or parts of the organisation in a particular way; also creating positions or roles within an organisation that will assist innovation. Last, organisations can encourage behaviour likely to foster innovation. This chapter reviews some of the organisational arrangements that can facilitate successful innovation.

OVERVIEW

Organisations cannot innovate to order, but there are things that organisations and individuals can do to make successful innovation a more likely outcome. Essentially this means managing innovation effectively and in this context, management covers what Katz (2004) calls 'the human side' of innovation. Space does not permit consideration of all the human aspects of innovation. Instead, this chapter singles out just three aspects:

- Corporate culture
- Architecture/structure
- Roles.

Why these three? First, all are closely connected with organising, that is to say the way in which people work together. The aim is to focus on actions that managers can take to enable people to work together more effectively, particularly when it comes to activities associated with innovation. Second, these three aspects are very different. *Roles* are very specific and closely connected with what people actually do and the part they play in innovation. *Corporate culture* is at the opposite extreme. It is not person-specific, instead, it is about shared understandings which, though they do not relate directly to innovation, can nonetheless be a powerful, if indirect, influence on innovation. *Architecture/structure* lies at the heart of organising, being concerned as it is with the most appropriate ways of grouping people together in order to facilitate innovation.

CORPORATE CULTURE

Corporate culture refers to the internal workings of an organisation. Just as the national culture of a country affects things like attitudes to work, attitudes to and the use of authority, equality and styles of decision-making, so the corporate culture of an organisation influences and affects the process of innovation. Since some organisations appear to be better at innovation than others and since these same organisations often, but not always, appear to have a distinctive corporate culture, there is a case for suggesting that corporate culture can facilitate innovation.

What is corporate culture? It is the internal context of an organisation. At its simplest corporate culture is 'the way we do things around here'. Corporate culture manifests itself in a variety of ways. It comprises the shared values and beliefs of those working in an organisation in the sense of a common (but entirely unwritten) understanding of what is considered important within the organisation. Corporate culture is often manifested in symbols within the organisation. These symbols can take lots of different forms. They might include the titles people are given, or the uniforms (if any) they wear. Physical objects such as prestigious offices and company cars can act as powerful symbols of corporate culture; so equally can the absence of such objects. Often, however, it is the objects closely associated with the everyday life of the organisation that provide the most telling insights into corporate culture. One well-known British manufacturing company at one time had no less than five different grades of dining facility for staff at its main premises, surely a symbol of a corporate culture dominated by formality, hierarchy and structure. Nowhere is an organisation's corporate culture formally laid down or defined. It cannot be prescribed. Corporate culture rests on shared assumptions, assumptions about ways of behaving, assumptions about decision-making and assumptions about what is important.

Handy (1993) provides a useful perspective on corporate culture by classifying corporate cultures into four broad types. The categorisation is not meant to be exhaustive but it does help to illustrate how corporate cultures can differ. According to Handy four recognisable cultures are:

- Power culture
- Role culture
- Task culture
- Personal culture

Organisations with a *power culture* are typically ones led by a strong individual able to stamp his or her ideas firmly on the organisation. Power cultures often give rise to ad hoc decision-making processes, with power and authority centred on a single individual. The individual concerned may well be a charismatic leader. A *role culture* in contrast is highly formulaic with well-defined, rational decision-making processes, highly structured relationships and formal rules and procedures. *Task cultures* tend to be task oriented, with people and systems focused on working together to achieve objectives or solve problems. There may well be considerable informality in terms of how people dress and communicate, and how they relate to each other. With a *personal culture*, individuals and personalities dominate, often to the detriment of decision-making and overall direction. The contrast between these cultures illustrates how different organisations can be when it comes to the values and beliefs members share and the way in which people behave.

Another perspective on corporate culture comes from the work of Miles and Snow (1978). They differentiate three types of culture:

- Prospector
- Defender
- Analyser

Organisations where the *prospector* culture is in evidence will tend to be ones keen to exploit new products and new market opportunities. *Defender* cultures on the other hand stress security and stability. They rely on systems and centralisation to provide a high degree of control, all of which is designed to secure the organisation's present position. *Analysers*, while they are responsive to new ideas, seek to subject them to extensive evaluation, usually with a view to finding out whether the new ideas can be incorporated into the present shape of the business.

Why certain organisations tend to exhibit one culture rather than another is difficult to ascertain. Organisations do not select a particular culture – typically it just emerges over time. Among the main influences on corporate culture are:

- History
- Size
- Technology
- Leadership

History influences culture because the founder may have had a very distinctive way of running the business or there may have been highly significant events early in its life. *Size* on the other hand affects culture because the larger an organisation gets the more it tends to rely on formal systems and controls. There are exceptions as some companies manage to retain some of the characteristics of small firms as they grow. *Technology* impacts on culture depending on the degree of stability it engenders. Hence, where the technology is subject to rapid change it can lead to a more flexible culture, while a stable technology can have the opposite effect. Finally *leadership* influences culture simply because a powerful leader can stamp his or her 'way of doing things' on an organisation.

Since corporate culture influences the internal context of an organisation it can have a significant impact on innovation. Some corporate cultures are generally more conducive to innovation than others. Handy's task culture for instance will probably be more conducive than his role culture. This is not to say that there are certain cultures that promote and lead to innovation. The drive for innovation has to come from individuals or groups of individuals. One cannot innovate to order. Rather it is the case that, if ideas for innovation are present, they are more likely to thrive in some corporate cultures than others. This has been explored by Christensen (1997), who looked at various industries and found that established incumbent firms were often less willing to innovate than new small firms. He attributed this in part to the culture of incumbent firms, particularly their inward-looking focus on existing customers/markets and their reliance on established systems.

In general, organisations with a strong record of innovation will have a corporate culture that is:

- Outward looking and receptive to new ideas, particularly from outside
- Facilitates communication, especially across the organisation
- Is open and receptive to new ideas and approaches
- Challenges established ideas and practices – 'the conventional wisdom'
- Accepts and learns from failure
- Promotes evaluation and reflection

Promoting these qualities will not guarantee innovation. As Nayak and Ketteringham (1993: p. 325) put it, 'breakthroughs can occur in any environment'. However, they do acknowledge that the internal environment (i.e. the corporate culture) within an organisation can make a difference and that some cultures are more conducive to breakthroughs and innovations than others. In organisations with a culture such as that outlined above, those engaged in innovation are likely to find that the barriers they face are likely to be fewer and lower.

MINI-CASE: W.L. GORE AND ASSOCIATES INC

Probably best known for its high-performance waterproof fabric Gore-Tex, W.L. Gore and Associates is one of the world's most innovative companies with a product range that extends from Gore-Tex fabrics to heart patches and synthetic blood vessels, air filters, parts for fuel cells, dental floss and even guitar strings. It is also a company with a very distinctive corporate culture.

W.L. Gore and Associates Inc was formed in 1958 when Bill Gore left the chemical giant Du Pont, when he became frustrated at his employer's unwillingness to take up his ideas about techniques for fabricating materials from the polymer Teflon (Hounshell and Smith, 1988). He and his wife decided to go it alone and with capital raised from their bridge club, they set up a company to manufacture wire insulated with Teflon, which found applications in the electronics and aerospace industries. Gore is still a private company, and the term 'associates' is present in the company title for a sound reason. It refers to the whole workforce, all of whom are known as associates. There is very little hierarchy, with few ranks and titles. There are no job descriptions. Gore is organised into what are effectively autonomous teams of up to 150–200 people. These are small enough for people to know one another and work together with minimal rules, very much in the manner of a task group tackling a crisis. New staff are allocated a sponsor or mentor whose job is to help the newcomer integrate.

The corporate culture is one that facilitates innovation. It is a culture where people feel free to pursue ideas on their own, communicate with one another and collaborate because they want to rather than out of a sense of duty (Deutschman, 2004) Gore encourages associates to engage in what others would probably call 'bootlegging' – that is, spending 10 per cent of their time on speculative new ideas. It is also a company that is very patient with the development of innovations. The company takes the long view where innovations are concerned. As a private company it does not have the pressure of frequently reporting to the financial community and can instead take its time over the process of getting an invention ready for market. As long as there is a possibility that a new idea will lead to an innovation, staff are encouraged to keep a project going.

Source: Deutschman (2004)

THE ARCHITECTURE OF INNOVATION

Can the internal structure or architecture of an organisation assist and facilitate innovation? The answer is probably – yes. Though it is important to appreciate that adopting a particular structure will not automatically bring about innovation, nonetheless an inappropriate structure can easily stifle innovation. The structure or architecture of an organisation clearly has a part to play in innovation.

That there could be a relationship between structure and innovation was recognised way back in the 1960s. Burns and Stalker (1961) in a very influential study of 20 industrial firms in the UK entitled *The Management of Innovation*, reported a relationship between the type of environment a firm operated in and the way it was structured. Burns and Stalker found that firms working in stable environments subject to little innovation and change had conventional hierarchical structures based on functional specialisms. Burns and Stalker termed these 'mechanistic' organisations. On the other hand, organisations operating in unstable environments where change and innovation were common had what Burns and Stalker termed 'organic' structures. Within organic structures, individuals are more flexible and there is less in the way of formal definition of duties and powers. Interaction runs laterally as much as vertically, and communication tends more to lateral communication than vertical command. Burns and Stalker concluded that organic structures were better suited to innovation.

Burns and Stalker's work was supported by further studies, including an influential one by Lawrence and Lorsch (1967) which strongly endorsed Burns and Stalker's findings.

Functional , M-form and Matrix Structures

A common manifestation of the mechanistic organisation is a functional structure. This is the traditional way of structuring an organisation. The essence of this structure is grouping together different functional specialisms. Thus, marketing staff would be grouped together in a marketing department, finance staff in an accounts department and so on. Each department would in turn be led by a functional specialist. While a functional structure can help to provide strong operational control, it normally does not facilitate innovation. With this type of structure lateral communication, essential to knowledge transfer, may be weak. Similarly the emphasis on control may make the organisation unreceptive to new ideas.

To combat possibly excessive emphasis on control and to provide a stronger focus on products and markets, some organisations, especially as they get bigger, opt for a multi-divisional or M-form structure, where the principal grouping switches from functions to product divisions. Functional staff (e.g. marketing and human resources) are attached to each division. The M-form structure also provides for the presence of 'service' departments that provide a central service to all divisions. Sometimes in large organisations among the central services provided in this way is research and development (R & D). With this type of arrangement R & D aims to develop new technologies which then find applications in the product divisions. The strength of the M-form structure is the strong focus on products and markets. However, there is a danger that the innovation will be largely incremental. Radical innovations, perhaps resulting from new technologies, may be ignored because they do not fit into the 'conventional wisdom' that prevails within the organisation. Hence the M-form structure often leads to a mechanistic organisation.

The most obvious example of an organic organisation is the matrix structure. The matrix structure permits dual focus. With this arrangement staff belong to conventionally functional departments such as engineering, but are attached to projects. Sometimes the projects last for several years and membership of a project team is semi-permanent. Clearly this type of structure is really only feasible in industries where there is project work, although the matrix structure might utilise, say, geographical areas instead of projects. The main advantages of this type of structure are: a high degree of flexibility; avoiding the domination of one function over others; improved co-ordination and lateral communication; reduced bureaucracy; and a capacity for handling new developments.

The matrix structure probably comes closest to Burns and Stalker's organic organisation, with good lateral communications, multidisciplinary teams and a strong problem-solving orientation. As such it is the structure most likely to facilitate innovation. Greater flexibility and an absence of strong divisions (sometimes called 'silos') can create an environment in which new ideas are more likely to flourish. However, it must be stressed, no one particular structure is a guarantee of a strong record on innovation.

The three types of organisational structure are in many respects ideal types that serve to illustrate the different ways in which organisations can be structured. These days it is not unusual for organisations to be structured as hybrids combining features of two or more of these types. A large automotive manufacturer for instance is likely to be structured on a multi-divisional (M-form) basis and yet within divisions one might well find a matrix structure.

Network structure

Renewed interest in networks has led some organisations to adopt a network structure. With this arrangement the organisation comprises a small core linked to a number of other external organisations. The core acts as a command centre directing and co-ordinating operations. It determines the product portfolio, it selects markets, it may even be active in designing and developing products, but its activities such as manufacturing, logistics and distribution will be carried out by third parties. Thus a network structure exhibits a high level of vertical disintegration, with many of the normal business activities undertaken by subcontractors. An example of a network organisation is the Italian clothing concern, Benetton SpA, where styling, design, manufacturing, logistics, distribution and sales are all subcontracted (Jarillo, 1993). The footwear firm Nike is very similar. The core of the organisation manages the Nike brand and undertakes design while relying on a network of firms in Asia, India and South America to produce its products (Donaghu and Barff, 1990).

Network organisations are not confined to textiles and footwear. In broadcasting the introduction of Channel 4 in the early 1980s marked the start of increased use of networking (Barnatt and Starkey, 1993). Whereas the BBC was a vertically integrated broadcasting organisation that undertook a wide range of activities including programme making, post-production services and broadcasting, Channel 4 was quite different. From the start Channel 4 comprised a small core that undertook broadcasting but commissioned independent production companies to make its programmes.

MINI-CASE: GORE-TEX AND BERGHAUS

Initially W.L. Gore and Associates produced cables coated with Teflon to provide insulation, but Bill Gore soon began to look around for further applications. He was particularly successful in the medical field developing a range of specialised dressings. These

▶

successes led to the development of Gore-Tex, patented in 1973, a fabric which was both waterproof and with millions of microscopic holes, could breathe. As a fabric Gore-Tex had huge potential in the waterproof clothing market. Waterproof fabrics were not new. Charles Macintosh had patented his rubber-coated waterproof fabric in 1823, Barbour developed paraffin wax-impregnated fabric at the end of the nineteenth century and Burberry patented gabardine in 1879, but all had limitations, particularly when it came to condensation caused by failure of the fabric to breathe.

Gore recognised that for their innovation to succeed they had to collaborate. As a fabric producer, W.L. Gore & Associates was effectively a component manufacturer. As Parsons and Rose (2004) make clear in their study of the outdoor clothing trade, Gore could see that the success of the fabric was dependent on the development of high-performance clothing that would make full use of the new fabric's innovative qualities. Collaboration offered scope not just for the development of such clothing, it also provided an opportunity for the company to learn and thereby enhance its products. Thus was born a strategic alliance with the small specialist outdoor clothing supplier Berghaus, based in Newcastle in north-east England. Under the terms of this alliance Gore not only acted as a fabric supplier but also contributed extensive marketing and advertising support. Large-scale advertising expenditure saw specialist outdoor products promoted in the national press for the first time. This was well beyond the resources of a specialist garment manufacturer. Berghaus' garments displayed the Gore-Tex label and a tag proclaiming, 'Gore-Tex: Guaranteed to Keep You Dry'. In this way Berghaus and Gore-Tex products were promoted simultaneously, raising the profile of both companies. For its part Berghaus was able to provide specialist expertise in the field of garment manufacturing. This proved to be a critical aspect of the innovation process. While the Gore-Tex fabric undoubtedly had its virtues, the first generation of Gore-Tex outdoor garments ran into problems. These centred on contamination from sweat and leaking seams. These problems led to garments being returned and adverse publicity in the outdoor press, which seriously threatened the success of the innovation. Gore solved the problem of contamination by inserting a hydrophilic membrane. However, it was Berghaus that salvaged the reputation of Gore-Tex by solving the problem of leaking seams, first through the use of high-frequency welding and then through the use of taped seams. Thus the development work undertaken by Berghaus proved critical in creating outdoor products made from a breathable synthetic fabric that would perform reliably under demanding conditions.

The strategic alliance helped turn Berghaus into one of the leading suppliers of outdoor clothing and established Gore-Tex as a high performance fabric. In the process it also led to huge growth in the outdoor clothing market, prompting Parsons and Rose (2003: p. 256) to describe the 1980s as 'the Berghaus decade'.

Source: Parsons and Rose (2003: pp 253-256); Parsons and Rose (2004)

Chapter 11: Organising for innovation

Strategic Alliances

While these types of network organisation are still fairly rare except in very specific industrial sectors, they do highlight the increase use being made of an organisational form that is now widely employed by a range of different organisations, and one that has proved to be particularly important in enabling innovation, namely the strategic alliance.

As Figure 11.1 shows, strategic alliances can take a variety of different forms. The essence of a strategic alliance is that it involves some form of collaborative agreement between two or more parties designed to benefit the parties over the long term.

Figure 11.1 A Typology of Strategic Alliances

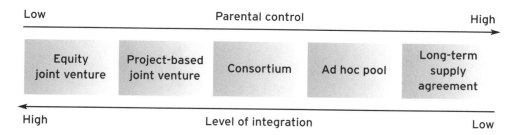

Strategic alliances are not homogeneous. As Figure 11.1 shows, they range between forms with a relatively high degree of integration (i.e. they rely on their own functions) and those with a low level of integration (Lorange and Roos, 1993). In this context integration describes the extent to which activities are left to market transactions with those with a high level of integration leaving comparatively little to the market. They range from equity joint ventures, where a separate organisation is created, owned by the partners, which is likely to have its own identity and be run by its own management with little or no direct control exercised by the partners, to long-term supply agreements which do not involve the creation of a separate entity and are little more than contracts to supply components, though with the supplier normally guaranteed a specific level of business for a given period of time. In between lie a range of collaborative arrangements, including consortia, where the parties agree to pool their resources usually in order to bid for a contract and ad hoc pools where the parties share resources but on an ad hoc basis.

The significance of strategic alliances for innovation is that they provide the necessary 'glue' that allows organisations to work together on an innovation. In this way strategic alliances facilitate the use of a network model of the innovation process (Rothwell, 1994). In the pharmaceutical industry, for instance, large pharmaceutical companies increasingly use strategic alliances to link up with small biotechnology companies as a source for developing new drugs. A similar pattern has emerged in the computer industry. When it wanted to develop a personal computer in the early 1980s, IBM entered an alliance with what was then a very small software house called Microsoft to provide an operating system, MS-DOS.

MINI-CASE: WILLIAMS-ROLLS INC

The jet engine industry has for many years been dominated by the 'Big Three' aero engine manufacturers, comprising Pratt and Whitney and General Electric of the US and Britain's Rolls-Royce. These three firms dominate the commercial jet engine industry producing more than 90 per cent of the industry's output between them. Under these conditions new entry is extremely difficult. Yet during the 1990s a new firm, Williams International, attempted to do just that with an innovative new product.

Williams International was founded in 1954 by Dr Sam Williams, who had worked for Chrysler in the early 1950s on experimental 'gas turbine' powered cars. Frustrated by his employer's lack of interest in commercial developments of the gas turbine, Williams branched out and set up on his own. At the time the conventional wisdom held that small gas turbines were not practical as they were likely to be less efficient than their large-scale counterparts. Williams felt otherwise. He believed that it was possible to design small gas turbines using substantially fewer parts. His logic was that such a design would be cheaper to build and more reliable, which would more than compensate for any lost efficiency. Unfortunately, there were no commercial applications for such an engine. Instead, Williams developed a small gas turbine to power target drones and reconnaissance RPVs (remote piloted vehicles). Williams enjoyed modest success in meeting the needs of this specialised niche market. Then in the 1970s the Department of Defence started development of a new class of missile – the cruise missile. These required a larger engine of 600lbs thrust. However, with this type of application, reliability was crucial, and this was a strong feature of the innovative Williams design.

Engines for cruise missiles enabled Williams to firmly establish a market niche providing small-scale specialist applications for the military. However, Williams was convinced that there was scope for applying the lessons learnt on small gas turbine engines for the military market to powering small corporate aircraft. Up to this point very few corporate aircraft were jet powered. Undaunted, in the early 1980s Williams began developing a bigger engine of 2000lbs thrust, the FJ44. Potentially a jet engine had much to offer over conventional propulsion systems, including higher cruise speed, higher altitude, lack of vibration and reduced noise. Aircraft manufacturers were not slow to recognise the potential and the FJ44 was selected for two new business jet designs, the Cessna CitationJet and the Swearingen SJ30. This success highlighted a problem for Williams. Hitherto his engines had been produced for the military market, where they operated under specialised conditions in which issues like maintenance and product support were of minor concern. For Williams the move to supplying a commercial jet engine, especially one based on an unconventional design employing many fewer components, was a big step.

To overcome its lack of experience and lack of resources in terms of commercial jet engine production, in 1989 Williams International agreed to collaborate with Rolls-Royce in producing the FJ44 engine, through the formation of an equity joint venture: Williams-Rolls Inc. The joint venture was to be owned 85 per cent by Williams International and 15 per cent by Rolls-Royce. Rolls-Royce was able to contribute a worldwide product support network to the joint venture, together with extensive test and manufacturing facilities.

The joint venture brought together complementary expertise and resources to facilitate the process of innovation. The success of the innovation can be gauged by the sales of the FJ44. Entering service in 1992, by 1999 more than 1000 had been sold and the Cessna CitationJet had become the world's best-selling business jet.

Source: Nelms (1991)

Corporate Venturing

Much has been written about the problems that large firms encounter in delivering innovations. These include technological and resource lock-ins and routine and cultural rigidities. While cultural changes and corporate restructuring form generic solutions to these problems, there are also a number of more specific prescriptions which come under the heading of corporate venturing.

We have already encountered the term 'corporate venturing' in the context of funding innovation, where it was used to describe large organisations taking a stake in small ones. In the current context, corporate venturing refers to attempts by large organisations to establish conditions conducive to innovation through a range of initiatives that take the form of internal structural devices that provide a focus for innovation. These usually involve the creation of an internal unit that allows mainstream commercial activity to co-exist side-by-side with more speculative activities associated with innovation.

Tidd and Taurins (1999) suggest that such arrangements are designed to permit a trade-off between learning and leverage. Their typology of internal corporate venturing arrangements is presented in Figure 11.2.

Figure 11.2 A Typology of Corporate Venturing

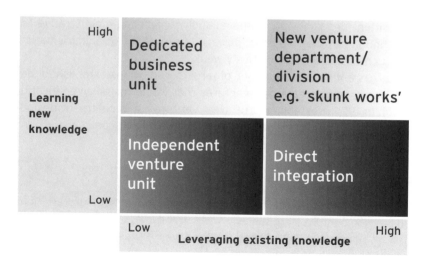

Source: Tidd and Taurins (1999) 'Creativity and Innovation Management', Blackwell Publishing

part III: how do you manage innovation?

Direct Integration

With this arrangement there is no separate entity created to conduct innovation. Rather, individuals drawn from across the organisation are brought together in a team which has responsibility for the development of a new product or service. The individuals may be secondees attached to the team for a given period or they may simply belong to the team and join in its activities alongside their normal work. This type of arrangement has the advantage that it is easy and quick to set up but it is very dependent on the existing corporate culture being 'innovation tolerant'.

Dedicated Business Unit

Although it remains within the existing organisation structure, a dedicated business unit is a separate entity. Such units sit alongside existing business units and may be difficult to distinguish from others. This kind of arrangement is only suitable for innovations with a very short development phase, or ones that are close to being launched, because as a business unit it will be expected to stand on its own feet in terms of profitability. It is also likely to be the case that the innovation will be incremental simply because the organisation will need to be sure that it is likely to break into profit fairly rapidly.

New-Venture Department

In its more extreme form this is the 'skunk works' solution, where a separate, somewhat secret department is created within the organisation charged with coming up with innovations that are probably a significant departure from the existing product/service portfolio. Although it lies within the organisation, differentiation is usually encouraged, which may lead to different sorts of behaviour, such as greater informality and a lack of hierarchy. While it is staffed by people drawn from the organisation, some may be somewhat independently minded with behaviour that might be regarded as problematic in mainstream departments. However, a new-venture department will always remain an organisation within the organisation and as such subject to administrative control from the centre, although if run by a strong, powerful personality, or a charismatic figure, it is possible that such control will be relatively weak.

During the 1970s and 1980s a number of large corporations established 'development departments' designed to create new ventures that would foster diversification. Britain's BP was typical. In 1974 it established BP New Ventures, a department designed to take over responsibility for developing activities other than oil and coal.

MINI-CASE: SKUNK WORKS

The term 'Skunk Works' is the name given to the Lockheed Aircraft Company's Advanced Development Projects Office. It was formed in 1943 to secretly build the company's first jet fighter, the P-80 Shooting Star. Given 180 days to design and build the aircraft, Lockheed's chief designer Kelly Johnson established a small team within Lockheed's main plant at Burbank, California. The small team was located close to a plastics plant whose noxious smell not only kept the curious at bay but helped give the team its nickname (Rich and Janos, 1994). The team met their target building the P-80 in a mere 143 days.

The initial Skunk Works operation set the standard for what followed. Lockheed's senior management agreed that the chief engineer could maintain his small research and development operation as long as it was kept on a shoestring budget and did not distract the chief engineer from his principal duties. The team moved into permanent accommodation in Building 82 within the Burbank plant but their activities were kept highly secret. However, it was not so much that the work was kept secret that distinguished the Skunk Works, it was the way it worked. Staff, many of whom were 'mavericks', were handpicked by Johnson. Dress was informal. Designers, analysts and engineers were all located close to the shop floor. The emphasis was on building prototypes in the shortest possible time. Often this meant 'quick and dirty' solutions to problems such as using stock parts rather than designing new ones.

Among the outputs of the Skunk Works over the years have been the U-2 spy plane of the 1960s, the world's first corporate jet, the Jetstar, and the F-117 Stealth fighter – all new developments, that in their own way, pushed forward the frontiers of aviation.

Source: Rich and Janos (1994)

Independent-Venture Unit

As their name implies independent venture units are separate entities that typically take the form of a company set up for the express purpose of developing innovative new products. The company may be wholly owned by the parent company or it may be a joint venture. It was this type of arrangement that was described in Chapter 10.

Independence gives rise to a number of important features. Most important is that it gives the unit a high degree of autonomy. This stems from the fact that it is a separate legal entity. This should mean that the unit will be free of internal politics and the corporate mindset. Ideally it should have or at least begin to develop its own sub-culture. Autonomy should mean less interference, greater freedom to experiment and try new approaches and less 'baggage' in the sense that there is less need to comply with the requirements of an existing technology or meet the needs of an existing set of consumers; yet, should the unit need further expertise or resources, perhaps for marketing or distribution, these can be tapped into within the parent organisation. As well as greater autonomy, an independent-venture unit should also benefit from greater focus. Without the distraction of corporate policies or internal politics, the team within such a unit should be more cohesive and able to focus specifically on innovation. They ought not to be distracted by the need to update, improve and revise existing products. In short, they should be able to focus on radical innovation free of the need to engage in incremental innovations. Tidd *et al.* (2001) also note that within a separate, independent-venture unit managers may be more highly motivated because they feel in control of their own destiny.

If the unit is a joint venture then there will also be advantages in terms of additional resources and the sharing of risk.

ORGANISATIONAL ROLES

So far we have looked at 'organisational' arrangements designed to facilitate innovation . In fact, managers do not necessarily have to rely on such 'macro level' arrangements. There are some 'micro level', people-related arrangements that can facilitate innovation. These tend to be associated with assigning individuals to particular roles.

There are a number of roles within organisations which, while not specific to innovation, have nonetheless been found to contribute to bringing about successful innovation. Some of the roles are formal in that they often take the form of designated posts, but most are not. Rather they are informal roles carrying no title and not being formally designated. They are no less important for that. The roles, both formal and informal, include:

- Project leader
- Product champion
- Gatekeeper
- Godfather

Project Leader

The role of project leaders is a formal one. He or she is likely to be a figurehead, the person probably most closely associated with the project. Their job is to take responsibility for the project and manage it. Obviously project leaders need to have a strong technical knowledge but they also need a breadth of knowledge and experience to enable them to co-ordinate and draw together the various functions required to bring an innovation to market successfully.

Alongside the leadership element of this role, the project leader also needs to be a planner, able to methodically chart what needs to be done, by whom and when. Together with the scheduling capability required to do this he or she also needs to be able to exercise control by monitoring performance and taking action as necessary.

Thus, a project leader is likely to possess a mix of talents, combining the communicating and motivating skills required to ensure a cohesive and effective multidisciplinary team alongside the analytical skills required to ensure effective organisation and management.

Product Champion

The idea of a product champion was first put forward by Schon (1963) in the early 1960s. He noted how new developments especially innovations within large corporations frequently run into trouble. Why? Schon (1963: p. 83) argued that there could be a variety of reasons for this. Often the novelty of an innovation challenges 'accepted ways of doing things and long-established skills'. Sometimes senior managers will feel threatened by their lack of knowledge of the technology, or possibly, if changes in technology lead to changes in social organisation, staff will feel threatened by possible potential structural changes. Whatever the reasons, innovations can easily come up against powerful vested interests, who see innovation as a threat to the current status quo. Often out of fear such interests will seek to block or at least hinder the innovation. Nor according to Schon are vested interests the end of the story, as often within large corporations the systems and procedures designed to screen new ideas can also provide a series of formidable obstacles for innovations. Requiring levels of detail that it is often very difficult to provide in the early stages of a new development screening procedures can easily act as a deter-rent, discouraging innovation.

In order to assist new developments in battling their way through the corporate minefield, Schon proposed the idea of someone who would act as a champion doing all in their power to promote the innovation in order to ensure its success. Essentially the role of product champion implies someone who will act as an advocate for the innovation prepared to sup-port and defend it even in the most difficult circumstances. In Schon's (1963: p. 84) words,

> ...the champion must be a man willing to put himself on the line for an idea of doubtful success. He is willing to fail. But he is capable of using any and every means of informal sales and pressure in order to succeed.

To carry out the role, product champions clearly need political support within the organisation. More importantly the product champion has to identify with the innovation. He or she has to regard it as their 'child' to be defended and protected at all times. They also need to be individuals who are familiar with the organisation, who know their way around and in particular know where the power lies. Finally, the product champion has to be an excellent communicator, with the ability to win over others to the cause.

Gatekeeper

There is an increasing awareness that knowledge is critical to innovation. Studies have shown that in particular it is an organisation's ability to transfer knowledge that leads to innovation. Cohen and Levinthal's (1990) theory of absorptive capacity, for instance, high-lights the importance of external sources of knowledge to the process of innovation. In this context the phrase 'it is not what you know but who' has a ring of truth about it. Certainly individuals can play a key part in the networking that forms part of the knowl-edge-transfer process. In the process they are acting as gatekeepers. Individuals taking on this gatekeeper role effectively hold the key to accessing knowledge. Quite how they operate is likely to vary but it might include:

- Acting as a repository of knowledge
- Knowing who possesses knowledge
- Exercising skill in making connections
- Acting as a 'go-between' for parts of the organisation or between organisations

At the simplest level gatekeepers may be repositories of knowledge. This is rarely a matter of formal codified knowledge; it is much more likely to be tacit knowledge, the sort that is informal, unstructured and difficult to capture in a structured way. Often with gatekeepers their actual knowledge is limited. Instead, their value lies in what they know about others, especially the knowledge that these others hold. It is not the gatekeeper's knowledge that matters but rather their ability to access others, specifically those who do possess the nec-essary knowledge. This is likely to be less a matter of knowledge and more of skill, especially skills in making connections to others. Such skills are likely to be social ones that enable individuals to make very effective use of the informal structure of an organisation. Finally, it is worth observing that gatekeepers often act as a bridge between different parts of an organisation. This may have nothing to do with knowledge or skill. Instead, it may be more a matter of culture or background or perhaps social ties. Whatever the reason, gate-keepers of this type can be very valuable as they can act as a conduit to facilitate knowledge transfer. Allen (1977) in a study of the Apollo space programme noted the importance of communication and information flows to the innovation process and he par-ticularly highlighted not just formal communications but informal ones. The latter were rarely linked to formal positions within an organisation; instead they were a function of individuals who were well placed within the informal structure of the organisation. Allen termed them 'gatekeeper' to pinpoint their role in accessing knowledge.

MINI-CASE: LEONARD LORD AND THE MINI

'God damn those bloody awful bubble cars. We must drive them off the streets by design-ing a proper miniature car.' According to Alec Issigonis, the designer of the Mini, the innovative small car that transformed many aspects of motoring and added a new word to the dictionary, it was an intense dislike of so-called bubble cars on the part of Leonard Lord the boss of the British Motor Corporation that led to the creation of the Mini. The oil crisis of the late 1950s resulted in motorists being rationed to between 30 and 55 litres of petrol per month and created a massive increase in the popularity of 'bubble cars', small, three-wheel cars capable of carrying two people and powered by fuel-efficient motorcycle engines. In March 1957 Leonard Lord took the bold step of shelving all the new car proj-ects his company then had under development in order to commission a new small car. Issigonis formed a small team and set about designing a new small car. Leonard Lord's brief was that it should be able to carry four people on four wheels using a standard four-cylinder engine, while offering better fuel economy than existing British small cars. Thus began development of what ultimately became the Mini, a car that remained in production largely unchanged until 2000, when more than 5.5 million had been built, and whose inno-vations were to have a profound impact on car design around the world.

From the start Leonard Lord's backing of the project was to prove vital. It enabled the development team to get the car into production in a mere two years, less than half the time normally taken. With Lord's blessing the development team was issued with a special priority number for prototype parts. Consequently, at the British Motor Corporation's Longbridge fac-tory in Birmingham 'everything stopped when they saw that number on a drawing'. The machine shop would even make parts for the prototype overnight using freehand sketches, providing they were given the required dimensions. By October 1957 the team had two hand-built prototypes, known as 'orange boxes', running. These prototypes proved unexpectedly fast, so the decision was made to reduce the size of the engine and widen the car by 2 inches. A mere 16 months after the project had started Leonard Lord was able drive one of the proto-types round the Longbridge site. After five minutes at the wheel he got out and, turning to Issigonis, said: 'Alec this is it. I want it in production within twelve months.' And it was. In April 1959 two production Minis were hand built on the Longbridge production line. By June pro-duction had started and the car was launched on the 26 August 1959.

Source: Nahum (2004)

Godfather

The godfather role is probably the least formal of these four roles. After all, even in the most formal and status conscious of organisations, one is hardly likely to see a sign on the door to somebody's office bearing the word 'Godfather', still less a similarly designated car parking space. However, in large organisations in particular, this is likely to be a crucial role in terms of the success of an innovation. Many innovations, including some very well-known ones like Wilbert L. Gore's Gore-Tex, James Dyson's bagless vacuum cleaner and Ron Hickman's Workmate®, bear witness to the reluctance of large organisations to take up innovations.

The godfather role is one taken by senior managers, preferably working at board level. The role is essentially one of providing 'behind-the-scenes' support. To be effective, a godfather has to be able to exercise power and influence within the organisation. Sometimes described as the capacity to 'pull strings' (Tidd *et al.*, 2001), only those who can do this can function as a godfather. Support can be exercised in a variety of different ways, but will almost always be exercised within the organisation. Support may mean looking out for the innovation and affording it protection, particularly from reactionary forces within the organisation. Such forces might include those who are risk averse, those possessed of a 'not-invented-here' perspective, those who find it difficult to see future potential, or those just who see their powerbase threatened. Again support may mean protecting the people working on the innovation and making sure they do not get moved elsewhere. Since most large organisations endeavour to weed out potentially unsuccessful projects, innovations have typically to navigate a range of hurdles/procedures designed to evaluate and assess their potential, before the organisation finally commits to full production. Clearing these hurdles can be extremely difficult especially for innovations where the market is new or ill-defined or in situations where there are a number of projects all competing for funds. The godfather may be able to help the innovation 'navigate the rapids' of project evaluation, especially if they have inside knowledge of where the worst and most dangerous rocks lie. As well as acting in a defensive capacity, the godfather can take a more proactive stance. Typically this might mean removing potential obstacles, be they people or potential hurdles. It might mean providing access to resources. These could be financial, but are probably more likely to be people or equipment or facilities. Finally, a godfather may simply exercise moral support for the innovation team. Difficult to quantify, it may nonetheless be vitally important in maintaining the motivation of the innovation team. In any event it is highly likely that the godfather will not have direct contact with the innovation or the project team, precisely because the role is a 'divine' one.

What qualities are needed to be a godfather? Clearly qualities such as discretion, and a capacity for networking help, but what really matters is the ability to wield power and influence within the organisation.

MAKING IT HAPPEN

This chapter is essentially about organisational 'prescriptions' designed to facilitate and foster successful innovation. These prescriptions relate to what was described earlier as 'the human side ' of innovation. Only three facets of the human dimension - culture, architecture/structure and roles - have been explored here. In reality there are other important aspects of organisation including teams/teamwork; communication; motivation; and leadership. (For these aspects the reader is referred to the collection of readings by Katz (2004) which covers the human side of innovation in detail.) However, the three aspects that have been covered perhaps also highlight the importance of the human side of innovation.

The emphasis has tended to be on highlighting benefits to be derived from various arrangements, whether in the form of a particular organisational architecture/structure or a particular role. In reality none of these prescriptions will necessarily lead to successful innovation. However, while their presence will not guarantee success, instances of innovations that have failed suggest that their absence may contribute to failure.

One example will serve to illustrate this. Virgin Trains recently introduced new high-speed 'Pendelino' trains which tilt to enable them to take curves at high speed. Although the Pendelino trains were designed and built in Italy, the technology was pioneered in the UK. British Rail, the former state-owned railway organisation, was the first to use tilt technology with its Advanced Passenger Train (APT).

Designed to run at speeds of up to 260 km/hour the APT incorporated a number of technological innovations: a tilting mechanism; hydro-kinetic brakes; lightweight aluminium construction; and articulated bogies. The prototype APT made a number of high-speed runs at speeds in excess of 250km/hour. However, there were problems with both the tilt mechanism and the brakes. Undeterred, British Rail pressed on with passenger trials operating a limited service between London and Glasgow. This proved a public-relations disaster as trains broke down, failed to arrive on time or were forced to operate at reduced speed. Unwilling to see a prestigious project labelled as unreliable, British Rail withdrew the APT from passenger service and eventually cancelled the project.

A subsequent investigation by a firm of management consultants found that, while the APT did encounter technical failures, it was in the organisation and management of this innovative train that 'the true seeds of failure lay' (Roy *et al.*, 1999: p. 35). These included a lack of support for, and isolation of, the APT team within the functionally organised CM&EE department of British Rail, and a 'workshop culture' (Roy *et al.*, 1999: p. 35) where the craft tradition prevalent in BREL at Derby where the trains were built had shop-floor staff, unfamiliar with the production standards required, changing design details without informing the design team. The report highlighted the need to appoint a 'heavyweight' project manager reporting directly to senior management.

This shows that, while appropriate organisation and management of innovation will not assure success, the APT case provides powerful evidence that ignoring the human side of innovation can all too easily lead to failure.

CASE STUDY: CLEARBLUE PREGNANCY TESTING KIT

A recent study (Jones and Kraft, 2004: p. 101) notes that, 'large established corporations face particular challenges' when it comes to innovation. Although detergents, soap and margarine remained at the heart of Unilever's operations, in the postwar years the company began a process of diversification, aware that its principal products were all in

chapter 11: organising for innovation

mature markets. The company expanded its presence in the food sector, regarded as its 'third leg', through developing the frozen-food market in Britain and Europe in the 1960s and 1970s (Cox *et al.*, 2003). Unilever's aim was to grow via a combination of innovation and acquisitions. While the development of frozen food provided an example of the former, the company's expansion of its ice cream interests came about through the latter.

In pursuit of innovation Unilever expanded its research function from the 1950s onwards. This reflected, 'a climate in which there were high expectations that research would lead to innovation and so provide a source of growth and new business' (Jones and Kraft, 2004: p. 106). Although Unilever's research effort contributed to developments in frozen foods, detergents and toothpaste, innovation tended to be incremental and lie within the company's core markets. According to Jones and Kraft (2004: p. 107): 'Radical innovations remained unlikely not least because operating companies were usually not interested in developing and marketing concepts far beyond their existing product lines.' Unilever had a strong record for research, but it was in fact a marketing-led company, where marketing and research were divorced from one another. The situation was not helped by the fact that Unilever's corporate culture emphasised consensus rather than risk taking. During the 1960s and 1970s the company recorded a number of innovation failures including long-life yoghurt and disposable feminine hygiene products. These failures only served to highlight the difficult of innovating outside the company's existing product portfolio.

Against this background, Unilever's success during the 1980s in developing a new medical diagnostic business stands in sharp contrast. The Clearblue pregnancy testing kit was a 'flagship' product that enabled Unilever to enter an entirely new field, the 'over-the-counter' (OTC) healthcare market using the fruits of its own research.

That it was able to do this owes much to the organisational arrangements surrounding the innovation. The origins of Clearblue went back to the immunological research undertaken at Unilever's Colworth laboratory in the company's animal feeds division in the 1970s. Although this work led to a number of successful antibody products for young livestock, the company failed to build on this work. However, at about this time the company's United Africa Company (UAC) division, which operated pharmacies in West Africa, began to diversify into medical products and the Colworth laboratory was active in helping its new medical division launch a number of diagnostic kits for use in hospitals and doctors' surgeries to identify bacterial infections. In time this work led to a number of important patents being filed in the monoclonal antibodies (MCA) field. Despite the company's leading research in the immunology field at this time, changes in company strategy in the early 1980s, in particular the rapid deterioration of UAC's trading position caused by political events in Africa, and the company's decision to withdraw from the animal feeds business outside the UK, meant that further innovations in this field looked unlikely. Fortunately, however, the company's Chemical Co-ordinator, T. Thomas, who was on Unilever's main board, recognised the potential in the company's science base. Following a visit to the Massachusetts Institute of Technology (MIT) he became convinced that Unilever should enter the medical diagnostics field. So it was that in 1983 Thomas, with the strong support of Unilever's research director, Sir Geoffrey Allen, was

▶

able to convince the company to establish a new Medical Products Group (MPG). The following year the MPG relaunched the medical diagnostics business as a separate company, Unipath. Described by Jones and Kraft (2004: p. 113) as, 'a "walled off" venture within Unilever', Unipath not only enjoyed support at the highest levels within the company, it also enjoyed a high degree of autonomy. Similarly, the MPG had a culture that emphasised both the commercial and the scientific. Research went forward in a number of directions, much of it undertaken in collaboration with universities and medical institutes. Wary of entering the pharmaceutical field, Unilever's research director strongly supported a focus not on products for hospitals and clinics but on the growing over-the-counter (OTC) diagnostics market, where MPG's biotechnology expertise could be applied to diagnostic products for pregnancy, fertility status, cholesterol and infectious disease status.

Strategy centred on achieving a significant innovation in a short space of time in order to ensure that MPG was not perceived internally as an expensive 'research boutique' (Jones and Kraft, 2004: p. 113). An OTC pregnancy test kit, derived from Unilever's research into monoclonal antibodies and ready-to-use immunological reagents, and developed by the Unipath subsidiary, quickly emerged as a potential 'flagship product'. Although there were diagnostic pregnancy test kits on the market, existing products were cumbersome, slow and often unreliable.

The Clearblue pregnancy test kit developed by Unipath and launched in 1985, less than two years after MPG had been established, represented a significant advance on anything then on the market. Clearblue was more sensitive, faster and more reliable than any competing product.

At Unipath particular attention was directed towards consumer feedback which identified problems in using existing products. As a result, the findings from consumer feedback were incorporated into the design process to yield not only a very effective but also a thoroughly user-friendly product. Unipath's structure as a relatively small but independent unit made it possible to maintain close links between research and marketing. This was helped by the fact that Unipath had its own marketing team, which included many science graduates who were at home in both scientific and marketing environments. Similarly, the marketing team was led by an individual who subscribed to the view that, 'science is no good without matching it to its users' (Jones and Kraft, 2004: p. 115), a philosophy which underpinned Unipath's marketing strategy. Hence the science-led, but innovation-driven culture facilitated the flow of information across the research/marketing interface. The strength of internal links also permitted a more rapid process of innovation. This contrasted with the more cautious, consensus-building culture that was typical elsewhere in Unilever.

From its unique position, Unipath was able to draw on a range of valuable capabilities within the Unilever organisation. These included expertise in packaging that drew on Unilever's vast experience in food packaging, as well as expertise in chemistry for coatings derived from work at the soap and detergents division at Port Sunlight. In addition, Unipath also benefited from having a culture that was both team oriented and outward looking, maintaining strong links with the external scientific community. In this environment and given the relatively youthful nature of the enterprise, innovation was seen as a priority. As Jones and Kraft (2004: p. 116) note, 'the company only had a future, and nothing to defend'.

Clearblue was an immediate success. Within three months of its launch in 1985 it had become the market leader in Britain with a one-third share of the market. By 1988 Clearblue was generating an annual profit of £8 million. Further innovations followed. In the same year Clearblue One Stop was launched featuring a simplified and easier-to-use process. Other OTC diagnostic testing kits were developed including Clearplan launched in 1989, which tested for the onset of ovulation, and Clearview which tested for Chlamydia, and Persona a fertility-monitoring kit both launched in the mid-1990s.

Source: Jones and Kraft (2004)

QUESTIONS

1 Why was Clearblue a radical innovation as far as Unilever was concerned?
2 What reasons can you put forward for the failure of a number of Unilever's earlier innovations?
3 Which roles do you think contributed to this successful innovation and why?
4 What evidence is there of individuals able to play the godfather role and how did this help the Clearblue project?
5 What evidence is there of what Burns and Stalker (1961) term 'mechanistic' and 'organic' organisations?
6 How would you describe the culture of Unipath and how did it assist the process of innovation?
7 What form of corporate venturing did Unilever use to to develop innovations in the medical self-diagnosis field and what did it contribute to innovation?
8 What was Unilever's overall structure and where did research fit into it?
9 Why do large corporations appear to find it difficult to undertake innovation especially innovation that leans more towards the radical rather than the incremental variety?
10 In your view to what extent did organisational structures help and/or hinder innovation in this case?

QUESTIONS FOR DISCUSSION

1 What do you understand by the godfather role and why is it important for innovation?

2 Why have scholars such as Schumpeter and Chandler been convinced that only large organisations could undertake innovation successfully?

3 What evidence is there to support the perspective of those such as Schumpeter and Chandler and in what ways may it be flawed?

part iii: how do you manage innovation?

4 What is meant by corporate culture and how does it differ from, say, national culture?

5 **What sort of corporate culture is likely to be conducive to innovation?**

6 What is meant by corporate venturing and what are some of the forms that it can take?

7 **Why have large organisations increasingly turned to corporate venturing as a means of fostering innovation?**

8 What are the potential drawbacks of corporate venturing?

9 **What is a strategic alliance and why have organisations increasingly turned to them as a means of facilitating innovation?**

10 What are the potential drawbacks of the traditional functional structure of an organisation in terms of facilitating innovation?

ASSIGNMENTS

1 Explain what is meant by the term 'gatekeeper' and show why this role is important for effective innovation.

2 Select an organisation, analyse its corporate culture and show how it contributes to successful innovation.

3 Show how an organisation's architecture or structure can help or hinder innovation.

4 Explain why a relatively large US corporation like W.L. Gore and Associates would want to team up with a specialist supplier of outdoor wear like Berghaus.

RESOURCES

Books

Insights into organisational aspects of innovation come from a variety of sources. This variety is reflected in the collection of readings by Katz (2004). This collection covers several different facets of organising innovation including: Culture/Climate; Roles; Teams/Groups; Communication; Leadership: Creativity; and Architecture/Structures. Among the more than 50 readings in the collection are a number that are directly relevant to the aspects covered in this chapter. Hence corporate culture at 3M is analysed by Nayak and Ketteringham and at Du Pont by Funderburg. In terms of roles, the gatekeeper role is analysed by Katz and Tushman, while Roberts and Fusfield provide an overview of the range of roles that may be associated with innovation. Corporate venturing is covered by papers from Peters and Pinchot dealing with new-venture departments and entrepreneurship.

Further details of the development of the APT and innovations in tilting train technology can be found in studies by Potter (1987; 1989). These studies provide a truly fascinating account of how organisational aspects of innovation can lead to failure. The study particularly highlights the destructive potential of an inappropriate corporate culture and what can happen when management fails to define roles that will facilitate innovation. Nor is this all, for these studies also highlight problems with architectural/structural arrangements and the importance of teams. Even though the studies describe events that took place nearly two decades ago, they still provide valuable insights into 'people management' aspects of innovation.

Other studies that provide insights into corporate culture include, James Dyson 's (1997) autobiography which provides a perspective on the corporate culture of an organisation, Dyson Domestic Appliances Ltd, with a strong innovation record and Audrey Wood's (2001) business history of Oxford Instruments plc which does the same for another organisation with a very strong record of successful innovation.

When it comes to corporate venturing Rich and Janos (1994) provide an insider's account of the operation of the original and much imitated 'Skunk Works'. Rich may not be Kelly Johnson, but his 36-year career meant he knew it better than most.

REFERENCES

Allen, T. (1977) *Managing the Flow of Technology*, MIT Press, Cambridge, Mass.

Barnatt, C. and K. Starkey (1994) 'The Emergence of Flexible Networks in the UK Television Industry', *British Journal of Management*, **5** (4), pp. 251-260.

Burns, T. and G. Stalker (1961) *The Management of Innovation*, Tavistock, London.

Christensen, C.M. (1997) *The Innovator's Dilemma: When Technologies Cause Great Firms to Fail*, Harvard Business School Press, Boston, MA.

Cohen, W.M. and D.A. Levinthal (1990) 'Absorptive Capacity: A New on Learning and Innovation', *Administrative Science Quarterly*, **35**, pp. 128-152.

Cox, H., S. Mowatt and M. Prevezer (2003) 'New Product Development and Product Supply within a Network Setting: The Chilled Ready-Meal Industry in the UK', *Industry and Innovation*, **10** (2), pp. 197-217.

Deutschman, A. (2004) 'The Fabric of Creativity', *Fast Company*, December 2004.

Donaghu, M. and R. Barff (1990) 'Nike just did it – International Subcontracting and Flexibility in Athletic Footwear Production', *Regional Studies*, **24** (6), pp. 537-552.

Dyson, J. (1997) *Against the Odds*, Orion Business, London.

Handy, C. (1993) *Understanding Organisations*, 4th edn, Penguin Books, Harmondsworth.

Hounshell, D.A. and J.K. Smith (1988) *Science and Corporate Strategy: Du Pont R & D, 1902-1980*, Cambridge University Press, Cambridge.

Jarillo, J.L. (1993) *Strategic Networks: Creating the Borderless Organisation*, Butterworth-Heinemann, Oxford.

Jones, G. and A. Kraft (2004) 'Corporate Venturing: The Origins of Unilever's Pregnancy Test', *Business History*, **46** (1), pp. 100-122.

Katz, R. (2004) *The Human Side of Technological Innovation*, 2nd edn, Oxford University Press, Oxford.

Lawrence, P.R. and J.W. Lorsch (1967) *Organisation and Environment: Managing Differentiation and Integration*, Harvard University Press, Boston, MA.

Lorange, P. and J. Roos (1993) *Strategic Alliances: Formation, Implementation and Evolution*, Blackwell, Oxford.

Miles, R.E. and C.C. Snow (1978) *Organizational Strategy, Structure and Process*, McGraw-Hill, New York.

Nahum, A. (2004) *Issigonis and the Mini*, Icon Books, Cambridge.

Nayak, P.R. and M. Ketteringham (1993) *Breakthroughs!*, Mercury Business, London.

Nelms, D. (1991) 'The little engine with a big future', *The Rolls-Royce Magazine*, **50**, September 1991, pp. 7-10.

Parsons, M.C. and M.B. Rose (2003) *Invisible on Everest: Innovation and the Gear Makers*, Northern Liberties Press, Philadelphia, PA.

Parsons, M.C. and M.B. Rose (2004) 'Communities of Knowledge: Entrepreneurship, Innovation and Networks in the British Outdoor Trade, 1960-1990', *Business History*, **46** (4), pp. 609-639.

Potter, S. (1987) *On The Right Lines? The Limits of Technological Innovation*, Frances Pinter, London, p. 48.

Potter, S. (1989) 'High-speed Rail Technology in the UK, France and Japan: Managing Innovation – the Neglected Factor', *Technology Analysis and Strategic Management*, **1** (1) pp. 99-121.

Rich, B.R. and L. Janos (1994) *Skunk Works: A Personal Memoir of My Years at Lockheed*, Warner Books, London.

Rothwell, R. (1994) 'Towards the Fifth Generation Innovation Process', *International Marketing Review*, **11** (1) pp. 7-31.

Roy, R., S. Potter and D. Wield (1999) *Innovation, Design, Environment And Strategy*, T302 Technology, Block 4 Case Studies, Open University, Milton Keynes.

Schon, D.A. (1963) 'Champions for Radical New Inventions', *Harvard Business Review*, **41**, March-April, pp. 77-86.

Tidd, J. , J. Bessant and K. Pavitt (2001) *Managing Innovation: Integrating Technological, Market and Organizational Change*, 2nd edn, J. Wiley and Sons, Chichester.

Tidd, J. and S. Taurins (1999) 'Learn or leverage? Strategic diversification and organizational learning through corporate ventures', *Creativity and Innovation Management*, **8** (2), pp. 122-129.

Wood, A. (2001) *Magnetic Venture: The Story of Oxford Instruments*, Oxford University Press, Oxford.

HOW DO you foster innovation?

CHAPTER 12

INNOVATION POLICY

OBJECTIVES

When you have completed this chapter you will be able to:

- Analyse the factors that lead governments to foster innovation
- Analyse the range of policy instruments employed in the UK
- Distinguish the range of public and private agencies involved in innovation
- Evaluate the effectiveness of a range of policy instruments.

INTRODUCTION

Most governments are favourably disposed towards innovation. It is viewed positively as something to be encouraged and fostered. This is because innovation is linked to economic growth through the creation of new firms and new jobs. Also, innovation is closely associated with technology which has similar positive overtones and is linked to high skill/high quality/high income jobs. For politicians and policy-makers innovation forms part of a virtuous circle of more and better products and services leading to more and better jobs. Consequently, innovation is something that governments and other public agencies try to foster and encourage and they do this through the use of policies employing a variety of policy instruments. This chapter explores the policies and policy instruments employed by governments and endeavours to provide some evaluation of their effectiveness. It should be noted that this chapter does not cover broader government policies that can and do have an impact on the 'environment' within which innovation takes place. These aspects are covered in Chapter 14 which deals with National Innovation Systems, leaving the emphasis in this chapter firmly on specific policy measures targeted at innovation.

RATIONALE FOR GOVERNMENT INTERVENTION

The case for government intervention to aid and assist innovation rests upon a variety of factors:

- Public nature of knowledge
- Uncertainty
- Complementary assets
- Network externalities
- Market inflexibility

Scientific knowledge especially has a number of public properties, that make it difficult for firms or individuals to extract rent from it (Afuah, 2003). This is very much a problem of appropriability in the sense that it can be difficult for those with knowledge and expertise to appropriate benefit from it. The problems that can arise when the innovator tries to appropriate his or her knowledge include:

- Imitation or copying
- Non-rivalrous aspects
- Spillovers and leakages
- Knowledge in the public domain

Imitation may simply mean that a third party has 'reverse engineered' an innovation and acquired the knowledge that way. A non-rivalrous aspect of knowledge occurs where A sells the knowledge to B, but A still has it and can use it. Spillovers or leakages occur where knowledge cannot be retained within an organisation perhaps through staff leaving. Finally, much knowledge, particularly explicit, codified knowledge, is in the public domain anyway and can be accessed by all. Factors such as these reduce the incentive for commercial organisations to invest in knowledge creation particularly in the form of R & D. Under these circumstances the level of innovation may be lower than it would otherwise be and there is a rationale for some form of government intervention.

Similarly, there is often a high level of uncertainty surrounding innovation. The uncertainty can be of the technical or market varieties. Technical uncertainty arises because research and development may not lead to innovation. There may be uncertainty whether an invention can be commercialised, or whether a device that works well in the laboratory will operate effectively in the hands of the consumer, or whether there is a potential safety problem as yet not identified. The market presents a similar range of uncertainties. Will consumers want the innovation? Will consumers be willing to pay for it? Will the market have changed by the time the device enters service? Hence, for a range of reasons associated with technical and market uncertainties, firms may choose not to proceed with an innovation. If this means a loss to consumers, there is again a rationale for government intervention.

Complementary assets are an important but often neglected aspect of innovation. They consist of things like market knowledge, access to distribution channels and facilities for product support. While no one would normally suggest that government should provide assets such as these, there are others which are more public by nature. Examples include transport facilities, power, sites and infrastructure facilities. The absence of such complementary assets may defer innovation, leading again to a case for government intervention either to provide them or assist in providing them.

Network externalities arise where the more people use a technology, the greater they value it. Afuah (2003) gives the example of an old technology which is valued because large numbers use it. There may then be a reluctance to switch to a new technology. Under these circumstances there can be a case for government intervention perhaps as a lead user.

Finally, market inflexibilities can mean that the institutions available to support and service a technology are outmoded when a new technology comes along. Market conditions should induce institutional changes, but markets can be slow and unresponsive to change and under these conditions governments may wish to induce institutional changes.

Hence, there are a variety of reasons for governments engaging in interventions designed to facilitate innovation. However, the forms of intervention can vary greatly as the following section indicates.

POLICY INITIATIVES

Support for innovation can have a variety of different objectives and take a variety of different forms. The policy objectives pursued by the UK government in recent years include:

- Technology forecasting
- Knowledge transfer
- Location
- Research and development
- Exploitation/licensing
- Lead user

The objective of *technology forecasting* is to increase awareness of current trends in technology development and to highlight the future implications. The 'Foresight' programme introduced in 1994 is an example of a policy initiative designed to achieve this objective. In terms of *knowledge transfer* the objectives are self-evident, being the transfer of knowledge between organisations and the 'Knowledge Transfer Partnership' programme run by the DTI and the UK Research Councils exemplifies policy initiatives in this field. The *location* objective is to affect the location of innovative firms with a view to them being clustered together. Science parks are an example of the sort of policy initiative designed to achieve this. *Research and development* involves encouraging and stimulating research and development by firms. The SMART award scheme and R & D tax credits are examples of policy initiatives in this field. The penultimate objective is *exploitation/licensing* of technology and, while it is not a policy initiative as such today, the activities of BTG plc exemplify work to try and achieve this objective. Finally there is the *lead user* objective. Research by Von Hippel (1988) highlighted the crucial role that users can play in the process of innovation. However, it is the case that potential users do not always know that they need a product. This may particularly be the case if the technology is entirely new or is a radical innovation. In situations such as these the uses of an innovation or potential users may be unclear. Under such circumstances governments can have an important part to play, taking the role of lead user. In this role they can undertake activities that will help to establish an innovation. These activities can include:

- Allowing time for new/potential users to develop
- Acting as a 'demonstrator' showing possible uses
- Interacting with the innovator to facilitate further development and improvement
- Stimulating demand

Afuah (2003) refers to these activites as 'shepherding'. He cites the case of the transistor where the US Department of Defense played a crucial role in establishing its viability. At the time the technological paradigm of electronics was dominated by the thermionic valve. However, the Department of Defense recognised that the transistor offered significant benefits in terms of: lightness; less heat; greater reliability; lower power consumption and small size. By awarding contracts for the application of transistor technology to military projects, the Department of Defense was able to procure new uses for the transistor, thereby enabling this innovation to become established.

Foresight

The Foresight Programme is a UK government initiative introduced following the White Paper 'Realising our Potential' (DTI, 1993). It aims to identify potential opportunities for the economy and society that may be present in new science and technologies and to consider how future science and technology might address future challenges for society with a view to initiating actions to realise these opportunities. The programme brings together key people, knowledge and ideas to look beyond the normal planning horizons to identify potential opportunities.

The first round of Foresight was launched in 1994 and brought together experts from industry, government and academia into a series of sector-based panels:

1 Agriculture, Horticulture and Forestry
2 Chemicals
3 Construction
4 Defence and Aerospace
5 Energy
6 Financial Services
7 Food and Drink
8 Health and Life Services
9 IT, Electronics and Communications
10 Leisure and Learning
11 Manufacturing, Production & Business Processes
12 Marine
13 Materials
14 Natural Resources and Environment
15 Retail and Distribution
16 Transport

During the main analysis phase, these panels considered emerging market and technological opportunities over a 20-year timescale, emerging priorities for research, and actions needed to exploit them. Following widespread consultation the panels published their findings in 1995. The reports identified likely social, economic and market trends over the next 10-20 years and the developments in science, engineering, technology and infrastructure required to best address future needs. The reports included 300+ recommendations for action.

The Materials Panel played a key role in gaining acceptance of nanotechnology as one area, if not the key area, of technology for the future. In terms of action it encouraged support by research councils such as the Engineering and Physical Sciences Research Council (EPSRC) for the establishment of two Interdisciplinary Research Centres in nanotechnology.

Similarly, the Foresight Toolkit allowed the benefits of Foresight to be delivered to SMEs. The Toolkit is a training tool that can be used by training facilitators to encourage companies to plan towards a future vision and anticipate challenges and opportunities.

The outputs of the Foresight programme are meant to inform and influence those who make decisions about research funding including business, government departments and charities.

The criteria for selecting topics as Foresight projects include:

- Significant current developments in science or technology, with potential to bring radical change, crossing the boundaries of established disciples
- Important challenges for society or the economy, to which science and technology have the potential to make a substantial contribution
- Scope to put together a group of people, with an interest in exploring the science and technology and ways of making it useful

Knowledge Transfer Partnerships

The Knowledge Transfer Partnership (KTP) programme (formally known as the Teaching Company Scheme (TCS)) is a government department/research council scheme that aims to transfer technology (in the broadest sense) between universities/colleges and small companies. This is achieved by the KTP programme providing funding for specific projects of up to three years 'duration' as part of which an 'Associate' who is normally a recent graduate is recruited to undertake the development work managing the project, applying their knowledge and ensuring that the expertise of the university/college is transferred to the business. The Associate works full time in the company and is paid a salary with most of the cost being met by the KTP funding, which also provides for academics to work on the project. The academics in the university/college remain closely involved throughout the project, working with senior managers in the business in order to contribute their knowledge and experience. One of the benefits of the programme is that, as well as improving links between universities and industry, it gives the SME access to the resources, especially in terms of knowledge, of a university. In some 70 per cent of cases the Associate goes on to work permanently for the company.

The projects undertaken by KTP Associates include:

- Improving existing products
- Developing new products
- Streamlining manufacturing processes
- Improving logistics processes
- Developing a marketing strategy

Figure 12.1 Knowledge Transfer Partnerships

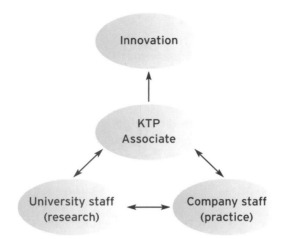

As Figure 12.1 shows the aim of KTP programmes is for research staff from the university/college to work with the KTP Associate and company staff on a project that will lead to an innovation which could be a new product, new service or new process.

MINI-CASE STUDY: AN INSTRUMENT MANUFACTURER'S KTP

The company made hand-held instruments for monitoring potentially harmful chemicals that could pose a threat to health safety. A chemistry graduate was employed as a KTP Associate to improve their existing products and create a completely new one. The new product was an ethanol analyser that measured alcohol levels in beer, wine and cider. The new instrument was used by breweries, trading standards departments and Customs and Excise. These were entirely new markets for the company. During the course of the 2-year project employment at the company rose from three to eight and the company now exports to more than 30 countries.

Science Parks

The concept of a science park is based on the idea of high-technology firms undertaking similar work being located close to each other and close to a knowledge base, usually in the form of a university. This model originated in the US. Frederick Terman, the dean of electrical engineering at Stanford University at Palo Alto in California, established one of the first science parks on a 600-acre site adjoining the university in the early 1950s. Terman was keen to develop what he termed a 'community of interest between the University and local industry' (Saxenian, 1983: p. 9). The park was landscaped and leases granted only to high-technology firms. The first company to move to the park was Varian. Hewlett-Packard moved to the park in 1954. By now Terman's notion of a 'community of technical scholars' had become a reality and there was the nucleus of what was to become Silicon Valley. By 1955 there were seven firms; by 1960, 32; by 1970, 70; and by the 1980s, 90; with total employment amounting to 25 000 (Castells and Hall, 1994).

Encouraged by the success of Stanford University science park, similar facilities began to spring up around the world in the 1970s and 1980s. One of the first science parks in the UK was established at Cambridge. Established in 1970 following the Mott Report of 1969 (Castells and Hall, 1994), Cambridge Science Park was in fact developed by Trinity College, Cambridge. It received its first occupants in 1976.

Since then it has grown dramatically and has been joined by two more science parks set up in Cambridge. By 1999 the number of high-tech establishments had grown to 959 employing more than 31 000 people (Athreye, 2000). Many other universities have now set up their own science parks including Cranfield, Nottingham and Warwick.

While Terman's notion of a 'community of technical scholars' lies at the heart of the science park concept, other benefits that flow to high-technology firms located there include:

- Access to a university knowledge base – i.e. researchers
- Scope for university spin-offs and joint ventures
- Access to specialist institutions – e.g. venture capitalists
- Role models
- Access to a pool of specialist labour

All these factors provide support for firms that are actively engaged in innovation.

SMART awards

The Small Firms Merit Award for Research and Technology (SMART) scheme, which is analysed in depth in Chapter 10, was initiated by the UK government's Department of Trade and Industry (DTI) in 1986 (Dodgson and Bessant, 1996), to provide financial support to small firms for research and product development.

As Dodgson and Bessant (1996) note, the rationale of the SMART scheme is to bridge the funding gap faced by small firms seeking to undertake innovation, particularly a shortage of seedcorn capital and investment funding for firms in the early stages of growth (Moore and Garnsey, 1993). This gap arises for two reasons. First, the presence of an information gap covering technological and commercial aspects, which makes it difficult for venture capitalists to evaluate the technical strengths and development potential of a project. This is exacerbated by an asymmetry of interest between innovators who do not know want to lose control and venture capitalists who typically demand a substantial stake.

Evaluations of the SMART scheme have proved positive and highlighted a number of important benefits including impetus to small-firm formation, stimulating firm growth and making the innovation more attractive for potential investors (Moore, 1993; Moore and Garnsey, 1993).

The SMART scheme has benefited from a number of high-profile innovation successes, including Bookham Technologies which developed a range of components utilising optical communications (ASOV) technology.

part iv: how do you foster innovation?

MINI-CASE: LASERS ARE US

Husband and wife team Simon and Debbie Lau set up Laser Application Unlimited in 1999. Their partnership aimed to develop a new laser marking technique that would be both faster and applicable to a wider range of plastics than previously. As an experienced laser physicist, it was Simon who handled technical development, while Debbie began test marketing the concept among industrial clients. 'We were impatient with the performance and high cost of existing laser systems and reckoned we could develop a better system at half price' said Simon. Early development was carried out in a spare bedroom of the couple's home in Porthcawl, South Wales and it was at this stage that the partnership sought and won a £45 000 SMART award to cover technical development and market research. Lasers Are Us Ltd was formed in May 2001 and began operations from a factory unit in Bridgend in November 2001.

Source: DTI (2005)

MINI-CASE: ISKRA WIND TURBINES

Redlands Primary School in Worksop, Nottinghamshire is unusual. It is one of the first schools in the country to install its own wind turbine in order to generate power. The 12-metre high turbine uses a 5-foot diameter rotor that generates 5kw, enough electricity to light the school. The wind turbine is produced by Iskra Wind Turbine Manufacturers Ltd based in Nottingham. In October 1999 Iskra won a DTI SMART award for innovative technology which was used to fund development of the turbine, which is a small-scale version of the wind turbines installed in so-called 'wind farms' now springing up in windy spots around the UK such as the Lake District. This phase of development work culminated in the installation and testing of a prototype turbine in the Derbyshire Dales. Following encouraging results of these trials, Iskra gained a further SMART award to develop a production version. Product development aimed at reducing:

- Cost
- Complexitiy
- Weight

while at the same time improving:

- Reliability
- Corrosion protection

The result was the AT5-1 wind turbine which is now in production and in 2003 achieved a 'Highly Commended' in the commercial enterprises category of the Eurosolar UK awards for inspiring renewable energy projects.

Source: *Worksop Guardian* (20 December 2004)

R & D Tax Credits

The R & D Tax Credit is another recent policy initiative from the UK government that aims to help and assist SMEs working on innovations. The scheme applies to SMEs that are registered limited companies spending more than £25 000 per year on research and development. Three different types of tax relief are available:

- Basic R & D tax relief: treats expenditure as equal to 150 per cent of actual
- Pre-trading R & D tax relief: expenditure creates a tax loss
- R & D tax credit: a tax refund of 16 per cent of unused loss

R & D is defined as 'creative' work undertaken on a systematic basis in order to increase the stock of knowledge and the use of this knowledge to devise new applications.

The scheme is aimed specifically at SMEs endeavouring to 'break new ground' by developing new products. Unlike the SMART scheme which provides a grant, it provides tax relief instead. It is a relatively flexible scheme that aims to benefit not only companies that have already started trading but also those which are at a point where they have not yet started. Since it is a relatively new scheme, it is too early for there to have been any formal evaluation of its effectiveness.

MINI-CASE: BTG AND ISIS TO DEVELOP DIAGNOSTIC TESTS AND TREATMENTS FOR WHEAT INTOLERANCE

Coeliac disease (CD) is a lifelong illness caused principally by intolerance to protein in wheat and other grains. Also known as gluten intolerance, it is a genetic disorder that has symptoms that include stomach pain, diarrhoea, vomiting and failure to thrive, and correlate with inflammation and subsequently destruction of the surface of the intestine.

The current blood tests for coeliac disease are frequently non-specific, leaving an intestinal biopsy as the only specific diagnostic test. The only treatment is the complete withdrawal of the toxic protein from the diet.

The Oxford researchers, led by gastroenterologist and principal investigator Dr Robert Anderson, have carried out extensive studies into the 'toxic' protein fractions in cereal crops and have identified the particular part of the wheat protein that, following modifications by gut enzymes, causes the immune reaction associated with coeliac disease.

This opens the way for a specific diagnostic test for the disease as well as new prevention and treatment strategies, and even the possibility of producing wheat that does not contain the rogue sequence.

BTG and Isis, the technology transfer arm of Oxford University, recently announced an agreement that granted BTG exclusive rights to Oxford's proprietary technology, which provides for new prevention, treatment and diagnosis strategies for coeliac disease. The new technology has the potential to be the world's only therapeutic treatment for the disease which may affect up to 1 per cent of the population of the UK.

Under the terms of the Isis agreement BTG will have exclusive access to the university's technology for use in the diagnosis, prevention and treatment of coeliac disease. The technology is based on identification of the particular epitopes that cause

▶

priming of the immune system in coeliac disease. BTG will underwrite the development and commercialisation of the technology and will share any revenue from commercialisation of the technology with Isis and the university.

Source: BTG Press Release (16 October 2002)

AGENCIES PROMOTING INNOVATION

Unlike the situation 20 years ago there are now a number of agencies in both the public and private sectors that are able to provide a variety of forms of assistance to organisations and individuals engaged in innovation. It would not be possible to detail them all but two very different agencies – Business Link and BTG plc – indicate the range of assistance available.

Business Link

Business Links are publicly funded agencies set up as a 'one-stop-shop' to provide advice and training primarily for small businesses and start-ups. Business Links operate in close collaboration with the regional development agencies in their area. While their principal function is small-business support, this does not extend to innovation. Most Business Links have a Technology and Innovation adviser (Lawrence, 1997). They can advise on any grants (e.g. SMART) that may be available, they can also advise on the availability of local sources of professional services (e.g. patent agents) as well as the services provided by national agencies (e.g. Patent Office).

BTG plc

BTG was established in 1981 through a merger of two government bodies – the National Research and Development Corporation (NRDC) and the National Enterprise Board (NEB) (Dodgson and Bessant, 1996: p. 147). Its role is to help exploit technology drawn from both public and private sector organisations. BTG finds, develops and commercialises emerging technologies in the life and physical sciences. These innovations are protected by a strong portfolio of intellectual property (IP) that BTG develops and enhances. BTG then captures the value of these technologies through licensing and venturing activities. BTG was privatised in 1992. From its inception back in 1949, BTG has commercialised several major innovations including magnetic resonance imaging (MRI), recombinant factor IX blood-clotting protein, Campath® (calemtuzumab) and Multi Level Cell (MLC) memory.

BTG's function is essentially to evaluate inventions, ensure appropriate IPR protection is in place through patents and then achieve effective exploitation normally through some form of licensing agreement. Dodgson and Bessant (1996) summarise its particular strengths as a willingness to:

- Take a long-term view – in some cases a 10–15 year time horizon
- Provide a depth of knowledge in the protection of IPR on a worldwide basis
- Provide technical, legal, commercial and patent resources

In recent years it has focused particularly on 'latent innovation' – that is, helping larger firms make effective use of underused or unused technologies.

CASE STUDY: MALAYSIA'S MULTIMEDIA SUPER CORRIDOR

In the space of a generation Malaysia has been transformed from a developing country exporting primary products into an industrially oriented economy. Throughout the 1970s government industrial policies sought to diversify the economy in order to diminish its reliance on primary commodities (e.g. rubber, palm nut oil and tin) and encourage manufacturing. In the 1980s the government became ever more interventionist, launching an industrial strategy that set priorities for specific industry sectors. Manufactured goods which made up only 19 per cent of exports in 1980 had risen to 77.4 per cent of a much larger total by 1995 (Jomo and Felker, 1999).

However, by the 1990s Malaysia had reached a critical point in its development. The growth in manufacturing had occurred largely in the field of labour-intensive manufacturing. Malaysia presented a curious picture of burgeoning high-technology exports (mainly electronics) with little local innovation activity. The country's industrial structure was somewhat shallow with poor inter-industry linkages and an underdeveloped capital-goods sector. Much of this was a function of foreign direct investment (FDI) attracted to the country on the basis of low wages and resource availability. There was an increasing awareness on the part of policy-makers that for economic development the country had to move towards an industry structure based on higher value-added, technology-intensive production. Vision 2020, the Malaysian Prime Minister's ambitious plan for the country published in 1991, highlighted the importance of technology development. It stressed the need for Malaysia to enhance the scope of its industrial activities beyond simple assembly and production through the development of indigenous technology.

The Industrial Master Plan (IMP) 1986–1995 had already identified a weak indigenous technology base as a threat to future growth. It noted that what technological competence existed was largely foreign owned, with the country heavily dependent on external sources of technology. The IMP recommended aggressive strategic investment in key industry sectors to build up local capabilities.

Then in 1996 the Malaysian Prime Minister Dr Mahathir announced the country's most ambitious industrial technology policy initiative - the establishment of the Multimedia Super Corridor (MSC).

The MSC is a 50 x 50-km zone stretching from the capital Kuala Lumpur to the newly built Kuala Lumpur International Airport at Sepang. It comprises a number of 'clusters', including Putrajaya - a newly built federal seat of government, Cyberjaya a new high-tech city housing multimedia industries, research centres and the Multimedia University, and Technology Park Malaysia - a zone providing engineering and IT facilities for entrepreneurs and industrial organisations.

The government's role in the project has been absolutely central. In the first place it has ensured a first-class environment. For instance, Putrajaya the new federal seat of government has been constructed as a garden city and the hub of the government's E-government project. Cyberjaya offers not only high-class multimedia working facilities including a very high level of connectivity but also residential and civic facilities. At the Petronas Twin Towers complex which serves as the northern gateway to the MSC there

▶

are extensive commercial, recreational, entertainment and retail facilities in a park-like setting. At the southern end of the MSC is the Kuala Lumpur International Airport built at a cost of more than US$3.5 billion, which acts as regional logistics hub and has the capacity to cater for 25–50 million passengers per year.

As well as providing an appropriate infrastructure for the MSC, the Malaysian government has also provided a range of financial incentives designed to attract investment from foreign companies that meet appropriate high technology criteria. These financial incentives include:

- A 5-year exemption from Malaysian Income tax renewable to 10 years or a 100 per cent investment tax allowance
- Duty-free importation of multimedia equipment
- Eligibility for R & D grants for Malaysian-owned MSC companies
- Special guidelines to regulate foreign currency transactions and loans

As well as financial incentives the Malaysian government has also revised its regulatory requirement for MSC-status companies to facilitate innovation. The measures taken include:

- Unrestricted employment of foreign knowledge workers
- Freedom of ownership within the MSC (i.e avoiding the requirement for local Malay participation)
- Extensive intellectual property protection

In addition the Malaysian government has promoted a number of 'flagship application' projects designed to promote new uses for new technologies. There are seven flagship applications in total:

- Electronic government
- Smart schools
- Smart card (My Kad)
- Telemedicine
- R & D clusters
- Borderless marketing centres
- Manufacturing co-ordination

The flagship applications are designed to transform core elements of Malaysia's technology infrastructure and social systems in areas where there is normally a high level of public-sector involvement. Driving the development of the seven flagship applications are government agencies that report directly to the MSC Implementation Council chaired by the Prime Minister. These agencies work in close collaboration with leading international and local companies on concept planning and project implementation. Pilot schemes are identified jointly by project team members through a mechanism known as the Concept Request for Proposal (CRFP). This provides guidelines for bidding for consortia interested in undertaking multimedia projects within the MSC.

The Smart card is typical of the flagship application projects. The aim is to provide a multi-purpose card that will provide the holder with a range of applications. The primary purpose of the card is to act as an identity card for all adults over the age of 12 and this is why the lead agency is the National Registration Department. Other public agencies involved include: the Road Transport Department, the Immigration Department, the Health Ministry and the Royal Malaysian Police. The other dimension to this flagship application is that the card should also act as a payment card to provide access to a wide range of financial services, e.g. as a credit card.

Source: Ramasamy *et al.* (2004)

QUESTIONS

1 Why is the Malaysian government interested in innovation?
2 Why had Malaysia's industrial development up until the 1990s done little to embrace the country's innovation capability?
3 What is foreign direct investment?
4 What is the purpose of the Multimedia Super Corridor?
5 Which well-known location famous for its capacity for innovation is the Multimedia Super Corridor trying to emulate?
6 Towards which of the innovation policy objectives of the Malaysian government is the Multimedia Super Corridor targeted?
7 To what extent is the Multimedia Super Corridor an example of the Malaysian government pursuing the lead user objective?
8 What is the Smart card contributing to innovation in Malaysia?

QUESTIONS FOR DISCUSSION

1 Why should governments provide assistance to innovators?

2 Why is market failure sometimes associated with innovation?

3 What are the UK government's objectives in promoting innovation?

4 What does Afuah (2003: p. 312) mean when he says that governments have a role to play 'shepherding' innovations by being a lead user?

5 When has the UK government acted as a lead user?

6 How do knowledge-transfer partnerships transfer knowledge?

7 How can science parks help to facilitate innovation?

8 Which university invented the science park and why?

▶

9 What is the function of the SMART award scheme?

10 How do SMART awards help innovators?

11 How do R & D tax credits help innovators?

12 Why are regional development agencies increasingly involved in innovation?

ASSIGNMENTS

1 Choose one government initiative designed to foster innovation. Describe the initiative, outline its purpose and provide an evaluation of its effectiveness.

2 Why do governments try to stimulate innovation?

3 With the aid of appropriate examples, explain what is meant by the term 'lead user'.

4 Explain why in the early life of an innovation there is typically a lot of technological and market uncertainty.

5 Prepare a presentation explaining the nature of SMART awards and showing how they can help innovators.

6 Prepare a briefing document explaining the role that BTG plc undertakes in connection with innovation.

7 Explain the nature and purpose of the DTI's Knowledge Transfer Partnership (KTP) scheme.

8 Evaluate current UK national policies for innovation, illustrating your answer with reference to recent policy initiatives.

RESOURCES

Books

The literature on innovation policy can be divided into two categories. There are texts that deal with the formulation of innovation policy, while in a rather different vein there are texts concerned with the implementation of policy that provide practical help and guidance for innovations.

Examples of the former include Dodgson and Bessant (1996) which provides both an overview of innovation policy in the UK and an evaluation of the effectiveness of some policy instruments. For those interested in innovation policy outside the UK, Branscomb and Keller (1998) cover innovation policy in the US.

When it comes to policy implementation, especially practical guidance on the availability of policy measures, Lawrence (1997) is invaluable. This text not only provides details of the policy measures relevant to innovation, it also provides a briefing on the wide range of different agencies available, ranging from invention brokers and design consultants to technology brokers and venture capitalists. It even goes as far as giving contact details.

REFERENCES

Afuah, A. (2003) *Innovation Management: Strategies, Implementation and Profits*, Oxford University Press, New York.

Athreye, S. (2000) 'Agglomeration and Growth: A Study of the Cambridge High Tech Cluster', in T. Bresnahan and A. Gambardella (eds) *Building High Tech Clusters: Silicon Valley and Beyond*, Cambridge University Press, Cambridge.

Branscomb, L. and J. Keller (eds) (1998) *Investing in Innovation*, MIT Press, Cambridge, MA.

Castells, M. and P. Hall (1994) *Technopoles of the World: The making of 21st Century Industrial Complexes*, Routledge, London.

Department of Trade and Industry (DTI) (1993) *Realising Our Potential: A Strategy for Science Engineering and Technology*, DTI, London.

Department of Trade and Industry (DTI) (2005) 'R & D case studies', available online at: www.dti.gov.uk/r-d/studies/lasers.htm.

Dodgson, M. and J. Bessant (1996) *Effective Innovation Policy: A new approach*, International Thomson Business Press, London.

Jomo, K.S. and G. Felker (eds) (1999) *Technology, Competitiveness and the State: Malaysia's industrial technology policies*, Routledge, London.

Lawrence, P. (1997) *The Business of Innovation*, Management Books 2000, Chalford.

Moore, I. (1993) 'Government Finance for Innovation in Small Firms: The Impact of SMART', *International Journal of Technology Management*, special issue pp. 104-118.

Moore, I. and E. Garnsey (1993) 'Funding innovation in small firms: The role of government', *Research Policy*, **22,** pp. 507–519.

Ramasamy, B., A. Chakrabarty and M. Cheah (2004) 'Malaysia's leap into the future: an evaluation of the multimedia super corridor', *Technovation*, **24,** pp. 871-883.

Saxenian, A. (1983) 'The Genesis of Silicon Valley', *Built Environment*, **9** (1) pp. 7-17.

Tidd, J. and M. Brocklehurst (1999) 'Routes to Technological Learning and Development: An Assessment of Malaysia's Innovation Policy and Performance', *Technological Forecasting and Social Change*, **62,** pp. 239-257.

Von Hippel, E. (1988) *The Sources of Innovation*, Oxford University Press, New York.

CHAPTER 13

INNOVATION CLUSTERS

OBJECTIVES

When you have completed this chapter you will be able to:

- Distinguish different types of cluster
- Describe the characteristics and attributes of clusters
- Analyse the factors that cause clusters to generate high rates of innovation
- Evaluate the range of policies available to promote the growth and development of clusters.

INTRODUCTION

Silicon Valley, or to give it its real name, Santa Clara County, on the southern flank of San Francisco Bay in California, is synonymous with innovation and high-technology. It comprises the densest concentration of high technology firms in the world. Once a sparsely populated agricultural area, today more than 8000 firms, most of them employing less than 50, provide employment for more than quarter of a million people. Since the 1970s Silicon Valley has exhibited an extraordinarily high rate of new-firm formation. Many of these firms are 'spin-off' companies where new firms are formed by employees leaving their existing employer in order to go it alone, by founding their own company. This has helped Silicon Valley become associated with an entrepreneurial culture where, to quote one leading study (Castells and Hall, 1994: p. 12), 'new ideas born in a garage can make teenagers into millionaires, while changing the way we think, we live, and we work'.

Focusing originally on electronics, Silicon Valley evolved to become a centre for information industries. Some of the most important innovations of the second half of the twentieth century, including the integrated circuit and the personal computer, originated in Silicon Valley. The region also specialises in breeding highly successful and innovative high-technology companies including Hewlett-Packard, Apple, Intel, Oracle and eBay (Lee *et al.*, 2000).

The extraordinary success of Silicon Valley has attracted the interest of politicians, policy-makers and industrialists around the world, all anxious to reproduce the 'Silicon Valley effect' by establishing concentrations or clusters of high-technology firms through direct government action. In France the government has created a high-technology cluster at Sophia Antipolis near Nice where large numbers of multinational companies have established manufacturing plants and research laboratories (Longhi, 1999). In Taiwan the government sponsored the Hsinchu Science Park which is home to large numbers of indigenous IT companies (Castells and Hall, 1994; Saxenian, 2004). In India the government has been instrumental, through a range of fiscal incentives in establishing the southern city of Bangalore as an internationally recognised software cluster of more than 200 firms comprising both foreign MNEs and locally owned software companies (Balasubramanyan and Balasubramanyan, 2000). Malaysia in the late 1990s saw the launch one of the most ambitious attempts to replicate Silicon Valley in the form of the Multimedia Super Corridor, a 50km by 50km high-technology zone, stretching southwards from the federal capital Kuala Lumpur, that houses a range of information-related digital businesses (Bunnell, 2002). In each case policy-makers hoped not only to create large numbers of high-technology jobs, but also to stimulate the national rate of innovation in pursuit of a more dynamic and expansive economy.

Clusters like Silicon Valley have been variously described as 'high- technology clusters', 'innovative milieux' and 'innovation clusters'. This reflects the fact that the growth and development of clusters is associated with innovation and the development of a strong knowledge base (Armstrong and Taylor, 2000). Using the generic term 'innovation cluster' to describe clusters of high-technology firms, this chapter looks at the link between innovation and such clusters. The characteristics of these clusters are analysed and the factors reputed to contribute to innovation identified. Not only does this enable us to understand more about the geography of innovation and why it is that some locations appear to be extraordinarily successful as centres of innovation, it also serves to deepen our knowledge of innovation and the factors that can both enhance and inhibit the process of successful innovation.

THE NATURE OF CLUSTERS

Whether they are termed 'innovation clusters', 'high-technology clusters' or 'innovative milieux', there are a number of attributes or characteristics associated with this kind of cluster. These attributes include:

- Geographical concentration
- High degree of specialisation
- Large number of mainly small and medium-sized firms
- Ease of entry and exit
- High rate of innovation

Geographical concentration is the defining attribute behind the concept of a cluster. Porter (1998: p. 78) for instance describes clusters as, 'geographic concentrations of interconnected companies and institutions in a particular field'.

In this sense 'geography matters' because a cluster consists of a number of firms belonging to the same or related industries, grouped in relatively close proximity within a particular location. The proximity of firms, and the fact that they work on related activities, differentiates the cluster from other forms of industrial location.

At the same time one normally finds that the presence of a number of firms, most of which are probably small, gives rise to a high degree of specialisation. In this context 'specialisation' describes the way in which firms are narrowly focused in terms of the range of outputs they provide. This is in contrast to a situation where firms offer a broad range of products and might be described as 'general purpose' concerns. The significance of specialisation is not merely that it is an attribute that serves to define the nature of the cluster. It is linked to the nature of a cluster because it is the presence of a significant number of firms that enables specialisation to occur. The presence of a number of firms justifies specialisation.

The requirement that there should be a number of firms is self-evident. Too few firms and there simply is no grouping. However, the characteristics of the firms that make up a cluster are also important. Though there can be variations, in general, clusters comprise small firms. There may be large firms present within a cluster, but for a location to be recognised as a cluster one would normally expect many if not most of the firms to be small.

Another attribute of clusters is that there should be ease of entry and exit. In other words it should be possible for start-up and spin-off companies to spring up and join a cluster with relative ease. In economic terms, barriers to entry should be absent. At the same time it should also be possible for firms to leave an industry and therefore the location.

Finally, clusters are usually associated with a high rate of innovation. New ideas, new concepts, new designs and new processes can be commercialised quickly and easily, by setting up a new company or spinning one off from an existing one. This does not necessarily make innovation easy, but it does facilitate the process of innovation, as there are likely to be fewer hurdles to be overcome, such as criteria for funding, for continuation or for the use of resources, as one might find in a large firm.

THE CLUSTER CONCEPT

The notion of a cluster was first put forward by Alfred Marshall (1890). He used the term 'industrial district' to describe agglomerations of small specialised firms found in particular localities. He cited as examples the cotton industry in Lancashire and the cutlery trade in Sheffield. He explained the success of these industrial agglomerations in terms of external economies of scale, where the close proximity of large numbers of small firms generated a market for increasingly specialised services.

According to Marshall (1890), agglomeration economies centre around three sources of collective efficiency, namely:

■ A local pool of specialised labour
■ Firms specialising in the intermediate stages of production
■ Knowledge spillovers

The availability of a pool of specialised labour occurs within industrial districts because the existence of a large number of similar firms encourages the concentration of supplies of skilled labour. For firms this can mean lower costs because they can poach skilled labour from other firms rather than going to the expense of investing in training and skills development. The scope for firms specialising in the intermediate stages of production occurs because agglomeration can result in a significant demand from local firms. Knowledge spillovers occur informally. Proximity makes it relatively easy for firms to observe better business practices and copy working methods from their neighbours. Similarly, new ideas and knowledge of new technologies can be shared through informal contact between the employees of different firms.

While the logic of industrial districts rests primarily on economies derived from the proximity of large numbers of similar firms leading to lower costs, Marshall himself recognised that agglomeration could be a spur to innovation, where the level of expertise available leads via human ingenuity to some firms exploring different ways of doing things.

For much of the twentieth century, however, large firms and internal economies resulting from size were dominant. A revival of interest in economic localisation, particularly on the part of economic geographers, occurred in the 1980s, spurred among other things by studies of small specialised firms in Northern Italy. Work by Piore and Sabel (1984) and others identified 'flexible specialisation' as an alternative to the prevailing logic of mass production. Since flexible specialisation was a form of production pursued by small firms, particularly concentrations of small firms, this led to a revival of interest in industrial districts.

However, although the 'new industrial districts' as they were termed by economic geographers, subsumed Marshall's ideas from a century earlier they also added new ones. These focused particularly on the importance of 'institutional factors' associated with concentrations of small firms. These institutional factors included the presence of:

- supportive socio-cultural attributes associated with working practices
- a network of public and private institutions supporting firms in the locality
- an intense set of backward, forward and horizontal linkages between firms based on non-market as well as market exchanges

The socio-cultural attributes comprise a range of local conventions, rules, routines, and norms that influence behaviour particularly in a work context. These attributes are often informal and non-systematic being based on tradition and practice. They often involve a certain 'way of doing things' within a locality. As such they serve to define a 'local world of production' and will often involve tacit knowledge. The institutions comprise a variety of organisations including training agencies, financial institutions, development agencies, marketing board and the like. A feature of these institutions is that they are for the most part collaborative, not-for-profit bodies. The presence of such institutions is termed 'institutional thickness' and its significance is that it supports and assists the firms present in the locality. They not only provide support to potential entrepreneurs contemplating the establishing of new firms, they also serve to embed existing firms in the locality. The significance of the linkages, which economic geographers term 'untraded interdependencies', is that they assist knowledge transfer and dissemination.

Hence, this new strand of thinking has helped revive the concept of the industrial district. This owes much to the nature of the agglomerations to which the analysis has been applied. A feature of many modern agglomerations is that, like Silicon Valley, they comprise high-technology firms that are active in innovation. At the same time new ideas about innovation, particularly those that stress the importance of networking (e.g. Rothwell's (1992) fifth-generation innovation process) and knowledge assimilation (e.g. Cohen and Levinthal's (1990) absorptive capacity) have emphasised the importance of transfer and learning. As a result neo-Marshallian industrial districts and the institutional factors associated with them have been seen as vehicles for stimulating and enhancing innovation. Neo-Marshallian industrial districts are often associated with innovative milieux and innovation clusters which in turn have been pursued by policy-makers as a way of reproducing the Silicon Valley effect.

However, the biggest factor behind the return of industrial districts to the forefront of the policy agenda and in particular interest in and promotion of clusters has come from quite another source, namely the work of two economists, Michael Porter (1990) and Paul Krugman (1991). They did much to re-ignite interest in industrial agglomerations, now re-branded as clusters. Porter's work on clusters, coming a decade after his earlier work had drawn attention to the importance of competitiveness, proved particularly influential. Policy-makers the world over seized on Porter's notion of business clusters as a tool for promoting both national and regional competitiveness as well as growth and innovation.

TYPES OF CLUSTER

MINI-CASE: GALWAY'S ICT CLUSTER

Over the last 30 years Ireland's economy has developed rapidly, principally by attracting large amounts of FDI. In Galway the process began in the 1970s when the American computer company Digital established a manufacturing operation. At this point Galway was effectively a greenfield site with virtually no previous experience of the computer industry. Digital's plant in Galway expanded rapidly and through FDI was joined by other branch plants of American multinationals including Compaq and Nortel in the computing field and Medtronics AVE and Boston Scientific in the medical instruments field. While local linkages were at first extremely limited, over time these plants have become more firmly embedded in the local economy of Galway. Digital became not only a major provider of advanced training and development for the largely locally recruited workforce, it also became a source of broader research linkages and collaboration including a major research project on computer integrated manufacturing with Renault, Compaq and NUI Galway (Green et al., 2001). Similarly the shortage of experienced professional ICT staff in the area forced inward investors to co-operate with each other for skills, infrastructure and market opportunities, rather than just compete.

The local linkages were such that, when Digital closed its Galway plant in 1993, the repercussions reverberated throughout the local economy. However, efforts by the local business chamber and national and regional development agencies ensured that the pool of skills and professionalism within Digital's workforce was not lost. The provision of business support, training and 'incubator' facilities working together with informal networks among ex-Digital staff contributed to further development of Galway's software cluster. While some ex-Digital staff moved to other ICT-based branch plants in the area, others chose to go it alone. As a result, helped by some early successes such as Toucan Technology, the number of small start-up companies within the local economy of Galway has increased substantially. While the cluster continues to be dominated by branch plants, it has become more diverse over the course of the last decade as an increasing number of small plants working in the computing sector have been established.

Source: Green *et al.* (2001)

Renewed interest in clusters and industrial agglomeration has led to the identification of a number of variations on the conventional Marshallian industrial district. Work by Markusen (1996) differentiates four distinct types of cluster (see Figure 13.1):

1 Neo-Marshallian industrial district (ID) cluster
2 Hub-and-spoke cluster
3 Satellite platform cluster
4 State-anchored cluster

Neo-Marshallian ID Cluster

The first of these four types of cluster is the conventional industrial district as identified by Marshall and extended in recent years to include Italian-style industrial districts. This type of cluster possesses a structure characterised by the presence of large numbers of small, locally owned firms. Firms being locally owned, investment and production decisions are typically made within the region. There will be a substantial amount of local trading, with inputs bought locally, resulting in strong inter-firm linkages within the cluster. Outputs on the other hand may well be exported from the region, reflecting the specialist nature of production in the region. The labour force too will be recruited locally with both immigration and outmigration largely absent. According to Markusen (1996: p. 301), the presence of a high proportion of workers engaged in design- and development-type activities ensures that such clusters are 'seedbeds of innovation'. Given its record for innovation and the presence of a large number of small firms with strong inter-firm linkages, specialist suppliers and local sources of capital, Silicon Valley in California represents an example of a technologically oriented, neo-Marshallian industrial district. In Britain, Motor Sport Valley in Oxfordshire provides another example of this type of cluster. It comprises a world-leading agglomeration of small and medium-sized firms clustered in a 50-mile radius around Oxford (Henry and Pinch, 2000a). Within this area are more than 600 firms engaged in the design, development and manufacture of racing cars and their components (Aston and Williams, 1996). What makes the cluster a neo-Marshallian industrial district reminiscent of Silicon Valley (Beck-Burridge and Walton, 2000) is not merely the presence of many leading racing car constructors, but also of large numbers of specialist component suppliers, a pool of skilled staff and a 'knowledge community' (Henry and Pinch, 2000b: p. 127) comprising inter-firm links that facilitate knowledge generation and diffusion leading to a high rate of innovation.

Figure 13.1 Types of Cluster

Marshallian ID cluster

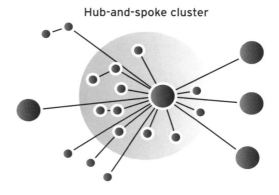

Suppliers   Customers

Hub-and-spoke cluster

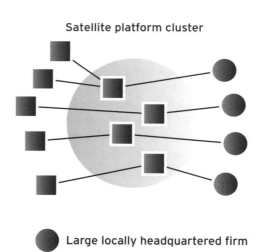

Satellite platform cluster

● Large locally headquartered firm

● Small local firm

■ Branch office, plant

Source: Markusen (1996), Economic Geography, reprinted with kind permission of Clark University

Hub-and-Spoke Cluster

While small firms predominate in neo-Marshallian industrial districts, hub-and-spoke clusters are characterised by having a number of large firms or facilities that act as a hub around which are to be found many small firms that may well be tied to the hub firm by virtue either of origin or of ongoing exchange relationship. Hub firms are typically large, with an international outlook that may involve exchange relationships with branch plants, suppliers, customers and competitors outside the region as well as locally. They may be vertically integrated, incorporating several distinct stages of the production process. According to Markusen (1996) they are also likely to be oligopolistic and dominate a single industry. Linkages between the hub firms and the small firms that make up the spokes typically take the form of supply contracts. Unlike relations within a neo-Marshallian cluster, relations within a hub-and-spoke cluster are unlikely to be collaborative and co-operative in nature, but instead will tend to be confined to being contractual. The disparity in size between the hub firms and the others means that the hub firms tend to play a dominant role within the local economy, giving rise to a unique local culture. Gray, *et al.* (1996) suggest that the success of hub-and-spoke clusters is dependent on the ability of a hub that emerges in one industrial era to tolerate and encourage the development of hub organisations in another era.

Gray *et al.* (1996) cite the Seattle region of the US as exemplifying the hub-and-spoke configuration. In this instance, Boeing, the world's largest aerospace company, acts as a hub. As a hub firm, Boeing's focus is firmly international as it has dominated the market for commercial airliners for many years and this international orientation has resulted in arm's length, purely contractual relationships with local suppliers (Gray *et al.*, 1996). In fact only a small proportion of Boeing's suppliers are locally based. Despite this, Boeing has been a powerful agent in shaping the regional economic infrastructure and the labour market of Seattle. Just how influential can be gauged by reaction to lay-offs at Boeing. In the early 1970s during one of the periodic downturns in the civil airliner market, Boeing cut its Seattle workforce by two-thirds, from 150 000 to 50 000. In April 1971 two Seattle residents put up a sign on a large roadside advertising hoarding that read: 'Will the last person leaving SEATTLE – Turn out the lights' (Rodgers, 1996: p. 302).

On a more positive note, Gray *et al.* (1996: p. 658) comment that,

> ❝ Boeing has helped to make the region an attractive place to live and work, which translates palpably into strong affinities for the region on the part of both long term residents and newcomers ❞.

Significantly, Boeing's role in defining the region as a high-technology centre has been a tolerant one that has permitted the rise of other hubs, most notably in the field of computing, with Seattle becoming home to more than 500 software companies, including Microsoft.

Satellite Platform Cluster

The third variant of the cluster form is the satellite platform cluster characterised by a concentration of branch plants of large externally owned and headquartered organisations. The firms to be found within satellite platform clusters can range from routine assembly functions to relatively sophisticated research. However, one distinguishing characteristic is that they are likely to form what Markusen (1996) describes as 'stand alone' facilities detached in spatial terms, with few linkages to other firms in the cluster. As a result commitments to local suppliers are likely to be for the most part 'conspicuously absent'. Linkages where they exist will be to the parent corporation or other branch plants of the same concern located in other regions. Because the plants located within satellite platform clusters tend to be owned by absent multinational corporations, their destiny, especially with regard to things like key investment decisions, will largely be determined outside the region. The main sources of finance, technical expertise and business services will tend to be external to the region. In contrast to the strong external focus associated with the provision of capital, local government and local development agencies typically play an important role in the provision of infrastructure, tax incentives and generic support services. Despite the absence of many of the cluster features that foster innovation, satellite platforms can nonetheless prove to be important centres of high-technology activity.

With an economic structure dominated by branch plants of major, high-technology companies (Longhi and Keeble, 2000), Sophia Antipolis in France provides a good example of a satellite platform cluster (Longhi, 1999). Among the companies represented are IBM, Digital, Thomson, Rockwell and Dow Corning, attracted to Sophia Antipolis as a *parc de prestige*. Attempts to encourage the growth of SMEs through incubator schemes (Castells and Hall, 1994) have generally met with only limited success, resulting in a cluster polarised into large and small organisations. However, recent research by Longhi (1999) indicates a move away from exogenous growth in favour of endogenous growth as the SME sector at last began to flourish in the late 1990s and form linkages with some of the large foreign firms.

Another example of the satellite platform type of cluster is the software cluster in Bangalore in southern India. This is another government-sponsored cluster (Balasubramanyam and Balasubramanyam, 2000) comprising more than 60 foreign MNEs including Motorola, Texas Instruments, Hewlett-Packard, Oracle and Siemens, as well as many indigenous software firms. However, in this case linkages between large foreign firms and small local ones are a feature of the cluster. According to Balasubramanyam and Balasubramanyam (2000) this is thanks to the nature of the software which consists primarily of bespoke systems tailored to specific client needs that demand a high degree of collaboration.

State-Anchored Cluster

The fourth type of cluster in Markusen's (1996) typology is the state-anchored cluster. This occurs where some form of public sector or not-for-profit organisation acts as an 'anchor tenant' within a region. Examples of the sorts of organisation that might undertake this anchoring function include:

- A military base
- A large defence plant
- A public laboratory
- Government offices

As examples Markusen (1996) cites Denver in Colorado which houses the second largest concentration of government offices in the US and Ann Arbor in Michigan, where a university acts as the anchor. The structure of the state-anchored cluster is likely to resemble a hub-and-spoke cluster in the sense that one organisation, or a small number of organisations, will tend to dominate the economic landscape, so much so that the local business structure is likely to reflect strongly the anchor organisation(s). Because the anchor organisation is not a conventional commercial organisation (i.e. it is in the public or not-for-profit sectors) it is likely to be relatively immune from the threat of a sudden exodus caused by corporate failure or a downturn in the business cycle. Indigenous firms are likely to play a smaller role than in neo-Marshallian industrial districts or hub-and-spoke clusters, although over an extended period of time supplier sectors will often grow up to meet the needs of the anchor organisation. In the case of universities and military and scientific establishments, labour markets will generally reflect national, or in the case of universities, even international market conditions.

Examples of state-anchored clusters to be found in the UK include Cambridge and the so called 'M4 Corridor', an area of high-technology firms that stretches from the western side of London to Bristol following the westbound motorway and covering counties such as Berkshire, Hampshire and Wiltshire. The M4 Corridor's claim to be this type of cluster configuration rests on the presence of several large Government Research Establishments (GREs) in the counties to the west of London (Castells and Hall, 1994; Hall, et al., 1987). Mainly established in the years before the Second World War, they expanded rapidly during the Cold War years when their links to high-technology industries were extended. The Royal Aircraft Establishment at Farnborough in Hampshire, the Atomic Weapons Research Establishment at Aldermaston in Berkshire and the National Physical Laboratory at Bushy Park in Middlesex are all publicly funded laboratories located in this area. Similarly, the M4 Corridor is also home to a number of major military establishments including the British Army's home base of Aldershot and GCHQ at Cheltenham. The rapid development of the military-industrial complex in the postwar years led the GREs to 'form the nodes of an intricate web defence procurement in which close and intimate contacts between research establishments and high technology clusters became the everyday rule' (Hall et al., 1987: p. 121).

Cambridge similarly represents a state-anchored cluster but by virtue of its university. Cambridge University dates back to medieval times and has had a profound influence on the development of the local economy. Since the 1960s the university's global reputation for research and scientific activity has spawned a substantial cluster of technology-based SMEs in the Cambridge region. Sometimes described as the 'Cambridge phenomenon', the number of local high-technology firms has mushroomed from 30 in 1960 to more than 700 by the late 1990s (Keeble et al., 1999). This growth has occurred as start-up firms have been spun off from the university or been attracted in because of it. The technologies represented in the cluster are diverse but include computing, electronics and biotechnology, the latter linked to the university's pioneering work in the field of genetics. A significant proportion of the high-technology firms in the cluster maintain close links with the university. In the manner of state-anchored clusters, the university's generally liberal attitude towards 'research collaboration, sharing and the development of new knowledge' has, according to Keeble *et al.* (1999), spilled over and helped to shape attitudes in the local business community. Such attitudes have been particularly conducive to innovation.

THE INNOVATIVE MILIEUX

In 1841, Richard Birkin, a Nottingham lace machinery manufacturer, giving evidence before a Parliamentary committee on the export of machinery, noted (Church, 1966: p. 78),

> ❝ …it is not the machines themselves in which value exists, but the great practice we in Nottingham have had and the ideas we have from being congregated together, that enables us to apply our various improvements to the machines. ❞

Nottingham, through the cluster of lace manufacturers in the lace market area of the city at this time, formed a major centre of lace production and was the source of a string of innovations that created the machine-made lace industry. Hence Richard Birkin was suggesting that the rate of innovation was a direct function of the lace manufacturing cluster in and around Nottingham's lace market.

What is the connection between clusters and innovation? How do clusters facilitate innovation?

There are a number of features of clusters that are particularly conducive to innovation. These features include:

- Networking
- Specialisation
- Ease of entry and exit
- Resource mobility

Networking

Research in the 1990s (Rothwell, 1992) highlighted the importance of networking. Innovation is portrayed not as something that is completely internalised within the organisation, but as something which is carried on as a multi-actor process that requires a high level of inter-firm integration. This model notes how firms do not operate in isolation but tend to draw on other firms for ideas, knowledge and services. In so far as they comprise small tight-knit groups of people working in the same field but within a number of different firms located in close proximity, clusters can facilitate networking. Proximity gives rise to a 'community of practice' where different occupations rub shoulders with their counterparts in rival firms enabling knowledge to circulate on the back of shared practice, thus allowing each firm to draw upon a wider knowledge base than it would otherwise have access to. Research by Henry and Pinch (2000a) into Motor Sport Valley in Oxfordshire has shown just how important such a 'knowledge community' can be in enabling a cluster to maintain its position as an internationally recognised centre of innovation. Nor are the benefits of networking confined to accessing knowledge. As a community of practice, clusters ensure that firms know a lot about other firms, especially what their competitors are doing. It becomes relatively common knowledge which firms are good at particular tasks. In this way a form of collective benchmarking, where firms are constantly looking over their shoulders at what their competitors are doing, drives them to innovate.

Specialisation

A second feature of clusters that helps to stimulate and encourage innovation is the scope they provide for specialisation. The presence of a large number of firms in the same industry or at least in similar sectors, located in close proximity to each other, enables firms to specialise in those activities they are good at, and to provide specialist products or services. As a result firms located in a cluster can draw on a range of specialist suppliers. These can extend to subcontractors and fabricators who can perform production tasks vital to innovation such as those associated with the development of prototypes. In addition, specialisation is likely to lead to a range of 'subsidiary trades' being present. These are likely to include firms providing specialist services that support innovation and new-venture development such as venture capitalists and patent agents. Hence, through specialisation clusters can provide the sort of infrastructure that will support and enable innovation.

Ease of Entry and Exit

Most high-technology clusters feature the presence of significant numbers of small and relatively young firms. This is a function of the conditions within the cluster which allow for ease of entry into the cluster. Thus it is relatively easy for enterprising individuals to break away from their existing employer and 'go it alone'. While it used to be argued that large firms were more likely to innovate (Galbraith, 1956), on the grounds that they had the resources to sustain the level of R & D required for new discoveries and thence innovation, an increasing body of evidence points to small firms being a more suitable environment for innovation. Why? Essentially they lack the baggage associated with old technology. Second, small firms can provide rapid decision-making in a way that large firms, with several tiers of management, find hard to match. Last, knowledge can circulate more rapidly within small firms. Thus, where clusters have a tradition of new start-ups combined with the facilities to make the process of start-up relatively easy, they will provide a climate in which innovation can flourish.

Resource Mobility

Resource mobility, especially where the resource is people, is a feature of many high-technology clusters. Several studies of such clusters have noted that they are frequently characterised by a high level of labour mobility. This mobility encourages innovation in a number of ways. First, if people move then so too will ideas. Second, the high level of mobility is likely to to mean that there is an active and responsive market for human skills. Under these conditions rewards will be an accurate reflection of skills and individuals, especially those with a high level of skill, can gain rewards that accurately reflect their abilities. Finally, the mobility of labour means that firms are likely to be very well aware of what others firms are doing, thus providing a powerful spur to innovation.

CASE STUDY: THE BRITISH HI-FI INDUSTRY CLUSTER

A Hi-fi cluster?

The market for audio equipment in Britain as in other parts of the world is dominated by producers from the Far East, particularly well-known Japanese brands such as Sony, Pioneer, Sanyo and Panasonic (Milne, 1989). Despite this, British manufacturers continue not only to survive but even to prosper. A key feature of their relative success is that they occupy a niche within the global audio equipment market, dedicated to producing hi-fi equipment that provides the best possible sound quality. In this market segment the emphasis is on quality rather than cost, and in the quest for the best possible sound quality British manufacturers actively embraced innovation in order to maintain their position in the marketplace.

The British hi-fi manufacturers are distinctive in a number of ways. First, there are a remarkably large number of them. A recent study by May *et al*. (2001) identified no less than 65 British hi-fi manufacturers. Not only are there a lot of them, but most of them are relatively small. Typically they employing 25 people or less, which makes them a great deal smaller than their Japanese competitors. Third, they are for the most part concentrated in the South East of England, being clustered in an arc that stretches from Cambridge through London and on into Kent, Sussex and Hampshire. Unlike their Japanese counterparts who mass-produce a range of consumer electronics products, the British hi-fi manufacturers specialise in hi-fi products. These extend to the design and manufacture of:

- Compact audio systems
- Amplifiers
- Specialist audio products
- CD players
- Loudspeaker systems

Recently some manufacturers have moved into the nascent market for home cinema amplifiers and loudspeaker systems. The extent to which British manufacturers engage in the quality end of the audio equipment market can be gauged by the way their products do not tend to make use of modern display aids. Most renounce such design gimmicks as a distraction, preferring instead to rely upon the sound quality their products offer.

Though small, the British hi-fi manufacturers are very active in terms of innovation. While some of the major technological breakthroughs that have transformed the audio equipment field in recent years, such as video and CD formats, have been introduced by global consumer electronics companies like Philips and Sony, May *et al*. (2001) in their study of the industry showed that the small British hi-fi companies have nonetheless been responsible for many significant product innovations including: stereo sound, small bookshelf speakers and flat-panel loudspeakers. The highly innovative nature of the small British hi-fi companies is reflected in the large number of international awards they have gained in recent years. May *et al*. (2001) point out that between 1992/1993 and

1996/1997 the European Imaging and Sound Association placed them second to the Japanese in terms of the number of awards granted. In the awards organised by the US Consumer Electronics Show the small British companies were close behind the US and Japanese companies that dominated the event. British hi-fi companies in 1994/1995 and 1995/1996 received as many awards as all their European counterparts combined. In addition, many British firms are regularly used as consultants by large Japanese concerns such as Sony, Canon and Pioneer. The British firm NXT has been at the forefront of developing the flat-panel loudspeaker technology being used by Japanese companies, and Meridian developed the 'home cinema' speaker system used by Canon. The record of small British companies in terms of innovation led May *et al.* (2001: p. 367) to observe that,

> **"**Other nations also have many significant specialist hi-fi firms (such as Audio Access, Cello, Krell, Mark Levinson, Proceed and Revel to name but a few from the US) but British manufacturers certainly seem to excel in this sector.**"**

How did the cluster arise?

In some respects it is surprising to find a hi-fi industry in Britain, since most of the early innovations were made elsewhere. The phonograph for instance was invented by an American, Thomas Edison, in 1877. Later improvements such as the disc phonograph, the forerunner of the vinyl record player, also came from America. However, the country's interest in audio technology gained a boost when Marconi, the pioneer of radio, moved to Britain in the early part of the twentieth century. He established the Wireless and Telegraphy Company at Chelmsford near London. Despite a late start Britain has in recent years been responsible for several major innovations in the field of audio technology. A.D. Blumlein pioneered the early development of two channels on a single disc to produce stereo recording. Similarly D.T.N Williamson and Harold Leak were responsible for the development of amplifiers. Developments such as these owed much to government support, often indirect rather than direct from the interwar years onwards.

The formation of the BBC in the 1920s not only fostered interest in radio, but also provided trained engineers. Similarly, another government initiative, the setting up of the National Grid in 1926, helped to open the market for wireless sets, by extending the provision of electricity to all parts of the UK and ensuring a common standard for electrical supply. Both of these publicly funded schemes contributed to the development of a thriving amateur radio community in Britain in the interwar years and after. According to May *et al.* (2001), the amateur input to the industry, in which a strong *esprit de corps* results from the hobbyist and enthusiast origins of the industry, has close parallels with motor sport (Pinch *et al.*, 1999), another industry in which innovation clusters are much in evidence.

While a small number of the British hi-fi manufacturers, such as Wharfdale, were founded in the 1930s, it was in the 1960s that there arose the first of a series of waves of new company formations associated with the creation of a cluster of small specialist hi-fi manufacturers in the South East of England. The Second World War had seen a big investment by the government in research into radio and associated electronic technology. Much of this work was connected with the development of radar designed to check the threat

▶

posed by German U-boats and aerial bombing by the Luftwaffe. This produced a generation of engineers trained in the armed forces, who went on to develop civilian audio products in later years. Typical were John Bowers and Roy Wilkins who, having been trained in radio technology in the armed forces, went on to found B & W Loudspeakers in the 1960s, a company that has developed a range of innovations in the field of loudspeaker design.

This trend towards new company formation was boosted in the 1970s as major British mass market consumer electronics firms such as Thorn, EMI, GEC and Rank, all of which had extensive research laboratories in the London area, began to rationalise their activities in the face of intense competition from Japan. As these large concerns rationalised, so skilled engineers left and set up their own businesses focusing on a specialist field of audio equipment. The process of ever-more specialised companies being spun off in this way continued throughout the 1980s and 1990s as employees of existing companies spotted market niches for new products and services and decided to set up on their own. May *et al.* (2001) give the example of Myriad, a manufacturer of amplifiers and tuners, spun off from NAD Electronics (which was itself established in the 1970s) a manufacturer of similar products, but aiming for a more expensive market segment. As Curran and Blackburn (1994) have noted, those involved in spinning off new firms can be very dependent in the early stages on linkages, in the form of contacts and market knowledge, to their previous employment.

When it comes to market and industry knowledge, those setting up new firms are also able to benefit from the numerous hi-fi shows that are held in the UK each year. These provide an opportunity for those who have set up on their own to meet old friends and maintain contacts as well as making new connections. As with all trade fairs, gossip and rumour serve to disseminate information about market trends and technological innovations. However, it should be noted that for the most part technological innovation by hi-fi firms is an 'internalised and self-sufficient enterprise' (May *et al.*, 2001: p. 372) using the resources of their own R & D departments. This is because hi-fi is a field where the pursuit of technical excellence provides scope for taking different approaches.

Source: *May et al.* (2001)

QUESTIONS

1 What might be the basis for suggesting that there is a hi-fi industry cluster in Britain?
2 What kind of a cluster is it?
3 What evidence is there to suggest that innovation is an important feature of this cluster?
4 What evidence is there to suggest that the success of the hi-fi cluster is based on what Marshall (1890) terms 'agglomeration economies'?
5 Recent work on clusters has focused particularly on innovation clusters or innovative milieux. What is meant by the term 'innovative milieux'?
6 What other examples of innovative milieux are there in the UK?
7 Recent research into innovation clusters has suggested that knowledge, particularly institutional factors associated with knowledge-transfer mechanisms, is a crucial feature of successful innovation clusters. What evidence is there of such institutional factors at work in this case?

QUESTIONS FOR DISCUSSION

1 Explain what is meant by the term 'agglomeration economies'.

2 What is 'flexible specialisation' and how has it contributed to a revival of interest in clusters/industrial districts?

3 Explain the value of typologies, such as Markusen's typology of clusters.

4 What are knowledge spillovers, how do they arise and how do they contribute to collective learning within a local economy?

5 What is the link between clusters and competitiveness?

6 Why should would-be innovators be aware of clusters?

7 Using an example of your choice explain what is meant by the term 'untraded interdependencies'.

8 Indicate what you consider to be high-technology industries. Provide a rationale for your choice.

9 In an era of globalisation, how can a local economy be competitive?

ASSIGNMENTS

1 Using any innovation cluster of your choice, prepare a presentation that provides an overview of the cluster in terms of its location, scale, and activities and also highlights the benefits for firms of locating within the cluster.

2 Compare and contrast an innovation cluster established as a result of deliberate government policy with one that has emerged without any public policy intervention.

3 Explain what is meant by the term 'institutional thickness' and show how it can contribute to stimulating and encouraging innovation.

4 Prepare a presentation for a group of policy-makers (e.g. a regional development agency) showing how institutional factors can contribute to economic dynamism within a local economy.

5 Using a biography of any innovator, show how he or she has been assisted by being located within an innovative cluster.

6 Using Markusen's typology of clusters, analyse an innovation cluster of your choice.

PART IV: HOW DO YOU foster INNOVATION?

RESOURCES

There is a rich literature on clusters, particularly high-technology clusters which exhibit a high rate of innovation. Books by Castells and Hall (1994) and Bresnahan and Gambardella (2004) provide surveys of some of the world's best-known, high-technology clusters. The merit of these works is that they not only offer insights into individual clusters but also provide an important comparative element which permits the reader to evaluate some of the wilder claims that appear in the media.

Individual clusters have been the subject of extensive research. Extraordinarily this is a relatively recent phenomenon. Little appeared on Silicon Valley until the late 1980s while the 1990s and the following decade saw a flurry of research activity. Saxenian (1983; 2004) shows how the nature of Silicon Valley has in fact changed. Valuable insider accounts of the development of Silicon Valley and insights into the nature of its entrepreneurial culture come from Lee et al. (2000) and Moore and Davis (2004). Two of the UK's leading high-technology clusters, Cambridge and Motor Sport Valley in Oxfordshire, have been extensively researched. Studies of Cambridge include: Athreye (2004); Keeble et al. (1999); and Segal et al. (2000). Studies of Motor Sport Valley are more diverse and include both academic and non-academic sources. Academic studies include Henry et al. (1996); Henry and Pinch (2000a); and Henry and Pinch (2000b), while the non-academic studies include a variety of industry surveys (Aston and Williams, 1996; Beck-Burridge and Walton, 2000; Collings, 2001) and wide range of individual biographies of innovators and entrepreneurs in Formula One motor racing.

REFERENCES

Armstrong, H. and J. Taylor (2000) *Regional Economics and Policy*, 3rd edn, Blackwell, Oxford.

Aston, B. and M. Williams (1996) *Playing to Win: The success of UK motorsport engineering*, Institute for Public Policy Research, London.

Athreye, S. (2004) 'Agglomeration and Growth: A Study of the Cambridge High Tech Cluster', in T. Bresnahan, and A. Gambardella (eds) *Building High Tech Clusters: Silicon Valley and Beyond*, Cambridge University Press, Cambridge.

Balasubramanyam, V.N. and A. Balasubramanyam (2000) 'The Software Cluster in Bangalore', in J.H. Dunning (ed.) *Regions, Globalization and the Knowledge Economy*, Oxford University Press, Oxford.

Beck-Burridge, M. and J. Walton (2000) *Britain's Winning Formula: Achieving World Leadership in Motorsports*, Macmillan, Basingstoke.

Bresnahan, T. and A. Gambardella (eds) (2004) *Building High-Tech Clusters: Silicon Valley and Beyond*, Cambridge University Press, Cambridge.

Bunnell, T. (2002) 'Multimedia Utopia? A Geographical Critique of High-Tech Development in Malaysia's Multimedia Super Corridor', *Antipode*, **34** (2), pp. 265-295.

Castells, M. and P. Hall (1994) *Technopoles of the World: The Making of 21st Century Industrial Complexes*, Routledge, London.

Church, R.A. (1966) *Economic and Social Change in a Midland Town: Victorian Nottingham 1815-1900*, Frank Cass, London.

Cohen, W.M. and D.A. Leventhal (1990) 'Absorptive Capacity: A New Perspective on Learning and Innovation', *Administrative Science Quarterly*, **35** pp. 128-152.

Collings, T. (2001) *The Piranha Club: Power and influence in Formula One*, Virgin Books, London.

Curran, J. and R. Blackburn (1994) *Small Firms and Local Economic Networks*, Paul Chapman Publishing, London.

Galbraith, J.K. (1956) *American Capitalism*, Houghton Mifflin, Boston.

Gray, M., E. Golob and A. Markusen (1996) 'Big Firms, Long Arms, Wide Shoulders: The 'Hub-and-Spoke' Industrial District in the Seattle Region', *Regional Studies*, **30** (7), pp. 651-666.

Green, R., J. Cunningham, I. Duggan, M. Giblin, M. Morony and L. Smyth (2001) 'Boundaryless Clusters: Information and Communications Technology in Ireland', *The Future of Innovation Conference*, ECIS, Eindehoven, Netherlands, 20-23 September 2001.

Hall, P., M. Breheny, R. McQuaid and D. Hart (1987) *Western Sunrise: The Genesis and Growth of Britain's Major High Technology Corridor*, Allen and Unwin, London.

Henry, N. and S. Pinch (2000a) 'Spatialising knowledge: placing the knowledge community of Motor Sport Valley', *Geoforum*, **31,** pp. 191-208.

Henry, N. and S. Pinch (2000b) '(The) industrial agglomeration (of Motor Sport Valley): A knowledge, space, economy approach', in J. Bryson, P. Daniels, N. Henry and J. Pollard (eds) *Knowledge, Space, Economy*, Routledge, London.

Henry, N., S. Pinch and S. Russell (1996) 'In pole position? Untraded interdependencies, new industrial spaces and the British Motor Sport Industry', *Area*, **28** (1), pp. 25-36.

Keeble, D., C. Lawson, B. Moore and F. Wilkinson (1999) 'Collective Learning Processes, Networking and "Institutional Thickness" in the Cambridge Region', *Regional Studies*, **33** (4), pp. 319-332.

Krugman, P. (1991) *Geography and Trade*, MIT Press, Cambridge, Mass.

Lee, C.M., W.F. Miller, M.G. Hancock and H.S. Rowan (2000) *The Silicon Valley Edge*, Stanford University Press, Stanford, CA.

Longhi, C. (1999) 'Networks, Collective Learning and Technology Development in Innovative High Technology Regions: The Case of Sophia Antipolis', *Regional Studies*, **33** (4), pp. 333-342.

Longhi, C. and D. Keeble (2000) 'High Technology Clusters and Evolutionary Trends in the 1990s', in D. Keeble and F. Wilkinson (eds) *High Technology Clusters, Networking and Collective Learning in Europe*, Ashgate Publishing, Aldershot.

Markusen, A. (1996) 'Sticky places in slippery space: a typology of industrial districts', *Economic Geography*, **72**, pp. 293-313.

Marshall, A. (1890) *Principles of Economics*, Macmillan, London.

Martin, R. and P. Sunley (2003) 'Deconstructing clusters: chaotic concept or policy panacea?', *Journal of Economic Geography*, **3**, pp. 5-35.

May, W., C. Mason and S. Pinch (2001) 'Explaining industrial agglomeration: the case of the British high-fidelity industry', *Geoforum*, **32**, pp. 363-376.

Milne, S. (1989) 'Small firms, industrial reorganisation and space: the case of the high fidelity audio sector', *Environment and Planning A*, **21**, pp. 833-852.

Moore, G. and K. Davies (2004) 'Learning the Silicon Valley Way', in T. Bresnahan and A. Gambardella (eds) *Building High Tech Clusters: Silicon Valley and Beyond*, Cambridge University Press, Cambridge.

Pinch, S. and N. Henry (1999) 'Paul Krugman's Geographical Economics, Industrial Clustering and the British Motor Sport Industry', *Regional Studies*, **33** (9), pp. 815-827.

Piore, M. and C. Sabel (1984) *The Second Industrial Divide: Possibilities for Prosperity*, Basic Books, New York.

Porter, M. E. (1990) *The Competitive Advantage of Nations*, Macmillan, London.

Porter, M. E. (1998) 'Clusters and the New Economics of Competition', *Harvard Business Review*, November–December, pp. 77–90.

Rodgers, E. (1996) *Flying High: The Story of Boeing and the Rise of the Jetliner Industry*, The Atlantic Monthly Press, New York.

Rothwell, R. (1992) 'Successful industrial innovation: Critical success factors for the 1990s', *R & D Management*, **22** (3) pp. 221–239.

Saxenian, A.L. (1983) 'The Genesis of Silicon Valley', *Built Environment*, **9** (1) pp. 7–17.

Saxenian, A. L. (2004) 'Taiwan's Hsinchu Region: Imitator and Partner of Silicon Valley', in T. Bresnahan and A. Gambardella (eds) *Building High-Tech Clusters: Silicon Valley and Beyond*, Cambridge University Press, Cambridge.

Segal, Quince and Wicksteed (2000) *The Cambridge Phenomenon Revisited*, Segal, Quince and Wicksteed Ltd, Market Street, Cambridge.

CHAPTER 14

NATIONAL INNOVATION systems

OBJECTIVES

When you have completed this module you will be able to:

- Distinguish those nations with a stronger record of innovation
- Distinguish the institutions and policies that influence the rate of innovation
- Analyse the nature of the institutions that contribute to innovation
- Evaluate the policies that contribute to innovation.

INTRODUCTION

Some countries appear to be better at innovation than others. In the late eighteenth and early nineteenth centuries Britain was widely recognised as the leading innovator amongst nations. However, the country's success in innovation was, as Freeman (1987: p. 313) points out,

> ❝ not just a succession of remarkable inventions in the textile and iron industries. Rather it can be attributed to a unique combination of interacting social and economic and technical changes within the national economic space ❞

Britain at the time was notable because of the strong links which existed between scientists and entrepreneurs;

organisational structures (e.g. partnerships) that enabled inventors to raise capital; efficient capital markets at national and local levels that brought capital to where it could be used for innovation; policies of deregulation that helped to reduce restrictions on trade; and reduction in the power of medieval guilds that restricted the movement of labour. These factors all helped to contribute to innovation. They supported and facilitated innovation and they were at the time peculiarly British. Today we would identify these factors as forming part of a national system of innovation. This chapter aims to explain what national systems of innovation are and to show how such systems can encourage or constrain innovation within a particular country.

THE PUBLIC NATURE OF INNOVATION

In the popular imagination, innovation tends to be portrayed as a private activity conducted by individuals. Accounts of well-known innovations, particular in familiar fields such as household products, tend to endorse this picture. James Dyson and the bagless vacuum cleaner, Ron Hickman and the Workmate® workbench are prime examples of the innovator-entrepreneur – individuals working seemingly in isolation to bring about successful innovation. Even when one is dealing with corporate innovation, individuals tend to surface in accounts of innovation. Accounts of the development of 'Post-it®' notes by the 3M company give a prominent role to Art Fry, a product designer with 3M, and his realisation that an adhesive developed by a colleague some years earlier might have commercial potential precisely because of its poor sticking qualities.

Nonetheless, whether one is dealing with corporate innovation or individual innovation, there is an important public dimension to innovation. The public dimension arises because many of the inputs and outputs of innovation are in fact public, that is to say they accrue to the public at large. The outputs become public when a successful innovation is diffused through the economy. Then it begins to generate new jobs and new firms, higher incomes and greater prosperity as well as increased tax revenues and know-how. The inputs include such things as education and training, knowledge and intellectual property rights. These assets tend to have 'public' properties (Afuah, 2003). Much knowledge for instance is publicly available. Even if it has been developed by individuals or organisations, it enters the public domain and becomes publicly available. In a knowledge-based economy specialised knowledge tends to be held by an ever-increasing number of institutions, including many that are public by nature. This trend means that individuals and firms have increasingly to be able to access a range of sources of knowledge and apply them to their own needs if they wish to be successful.

A recent report by the OECD (1997) has extended the public dimension still further. It notes that the overall performance of an economy depends not so much on how specific institutions (firms, research institutes, universities) perform, but on how they interact with each other as elements of a 'collective' system of knowledge creation and use, and their interplay with social institutions such as values, norms and legal frameworks.

The notion of a collective system recognises that, although innovation is often introduced by private individuals and private firms, it is ultimately the product of a system of institutions that are closely linked together by economic and social relationships (Simonetti, 2001). The term 'system of innovation' describes the configuration of institutions and the resulting flows of knowledge. Such systems can operate at the level of the nation state, the region or the industry sector:

- *National innovation systems*
  i.e. a country-specific system such as the national innovation system of the UK or Japan

- *Regional innovation systems*
  i.e. an innovation based on a specific location or place such as Silicon Valley in California or Motor Sport Valley in Oxfordshire in the UK

- *Sector innovation system*
  i.e. an industry-specific innovation system such as those of the pharmaceutical or aerospace industries

NATIONAL INNOVATION SYSTEM

Dahleman (1994: p. 554) defines a national innovation system as,

> 66 the network of agents and set of policies and institutions that affect the introduction of technology that is new to the economy ... [including] policies toward foreign direct investment, arm's length technology transfer, intellectual property rights, and importation of capital goods. The innovation system also comprises the network of public and private institutions and agents supporting or undertaking scientific and technological activities, including research and development, technology diffusion, and creation of technical human capital. 99

The idea of a national innovation system is that it describes the arrangements pertinent to innovation that prevail within a particular nation state. 'National' implies that we are dealing here with aspects influenced by the nation state. Thus, we are dealing primarily with institutions established by or specific to a particular nation state. In an age of globalisation, the nation state still exerts a high degree of influence. There are big differences in the innovative performance of nations and this is increasingly recognised as having links to the characteristics of particular national innovation systems. Innovation here describes the processes by which firms market and set into practice product designs and manufacturing processes that are new to them (Barker and Goto, 1998). In terms of knowledge, innovation in this context includes both the creation of new knowledge and existing knowledge. In either case it refers to knowledge from inside the firm and outside and, where the latter is concerned, primarily knowledge that resides within the country where the firm is based (though it could also be from outside).

The term 'system' refers to the set of institutions whose actions and interactions have a bearing on the innovative performance of national firms. As the quotation from Dahleman (1994) indicates, it includes aspects of the financial system, the educational system, the attitudes and behaviours of firms and the role of government organisations.

Simonetti (2001) stresses that the notion of a system of innovation recognises that tacit knowledge has a vitally important role to play in the innovation process. If all knowledge was explicit and codified, firms could just purchase it like any other factor of production. However, the presence of tacit knowledge means that firms have to interact with other firms and a variety of different organisations in order to acquire knowledge and use it effectively.

GOVERNANCE

The pressure of interactions and linkages in a system raises the issue of governance – in other words how the system itself is governed (Simonetti, 2001). National innovation systems tend to exhibit three types of governance mechanism.

- Corporate governance
- Political governance
- Network governance

Corporate governance describes the mechanism surrounding the exercise and control of corporate ownership. Within capitalist economies it operates through the market mechanism. Two main approaches to corporate governance are generally recognised. Albert (1992) terms the first 'Anglo-Saxon' and the second 'Nippon Rhineland'. Under the Anglo-Saxon model which is found in the US and the UK there is a clear separation of ownership of corporate undertakings from their control. Ownership lies in the hands of stakeholders, many of whom are institutional innovators such as pension funds while control is left to a cadre of professional managers. The two are mediated through the stock market. Shareholders know little about the internal operation of the firms in which they invest. An efficient stock market means they can switch investments quickly and easily and investors' motivation to hold a particular stock is high due to increasing dividend payments, generally based on short-term profits. Professional managers are trained to deliver such profits through MBA courses in the US and accountancy training in the UK. This model of corporate governance tends to favour the active management of a portfolio of businesses with the harvesting of assets within a fairly short timescale. Some argue that it is not conducive to innovation since the short-term perspective means that it may not provide the 'patient capital' required for innovation.

Under the Nippon-Rhineland model of governance, which as its name implies is found in countries such as Japan and Germany, there is no sharp division between ownership and control. The presence of multiple stakeholders means that banks, suppliers and even employees have a stake in the management of the firm. In Germany, for instance, banks have been important as sources of investment capital and as a consequence have often been closely involved in the day-to-day governance of the firm (Moran, 2001). As well as having a stake in the firm these stakeholder groups have usually had detailed knowledge of, and taken a detailed interest in, internal aspects of the firm including its technological competence. This has important consequences as far as innovation is concerned. The stakeholders are in a much better position to evaluate the long-term commercial potential of R & D activities. This is reflected in a tendency for firms in countries strongly influenced by the Nippon-Rhineland model to invest more strongly in R & D than those in the UK and the US (Tidd and Brocklehurst, 1999).

However as Tidd and Brocklehurst (1999) point out, during the 1990s there have been growing doubts about the supposed superiority of the Nippon-Rhineland model in terms of R & D. In particular the US appears to have reasserted its lead in information technology and biotechnology.

Political governance refers to the role of government in fostering innovation. This role centres around two functions: the policy function and regulating function.

The regulatory function covers intellectual property rights and environmental protection. Intellectual property rights (IPR) are critical for innovation. IPR regimes are generally designed to promote innovation. Without a strong IPR regime, it is difficult for innovators to appropriate the benefits of their endeavours. Instead imitation, copying and even counterfeiting tend to proliferate. Clearly, if the rewards for genuine innovation appear to be low, this will discourage innovation and a country's overall performance will be poor.

While the remit of government in providing environmental protection falls outside of the remit of national innovation systems, it can influence innovation in that a requirement to meet environmental standards can act as a useful spur to innovation.

MINI-CASE: THE NATIONAL LITERACY STRATEGY

In 1998 the UK government introduced the National Literacy Strategy into primary schools in England. Using ideas and teaching methods developed in New Zealand and Australia, the strategy was the first prescribed scheme for the teaching of literacy to be used in England. The strategy introduced learning objectives for three aspects of literacy:

- Text level (reading and writing)
- Sentence level (grammar)
- Word level (spelling and vocabulary)

These formed a part of a comprehensive scheme of objectives for each term for each primary school year group, from Reception (aged 4–5 years) to Year 6 (aged 10–11 years). Included in the strategy was the introduction of a 'literacy hour', a sequence of one-hour-per-day lessons involving whole class teaching, independent work and a plenary session.

The strategy was implemented in primary schools in England the following year. As an education initiative it was the first time a nationwide comprehensive scheme for literacy had been adopted in England, where the education system had developed at local level and left the delivery of the curriculum to individual schools.

The strategy was supported by resources and support developed at a national level. These included high-quality teaching and learning resources that extended to:

- Intervention programmes for children who are just below the age expected levels
- Lesson plans
- Materials to support areas of teaching, e.g. grammar for writing, spelling banks and progression in phonetics

The support included locally based consultants able to provide guidance and support as well as running regular training sessions.

A year later a National Numeracy Strategy followed.

Network governance recognises that increasingly innovation takes place through firms working together with other organisations. These organisations are likely to include many of the institutions that play an important role within national systems of innovation such as universities, educational establishments, research bodies and financial institutions. Within such networks relationships will not necessarily be governed by the market mechanism. While some services within the network may be bought and sold, some, in particular those associated with advice and guidance, may be provided on a non-market basis through common interest and mutual trust reinforced by some form of common cultural background.

Simonetti (2001) gives the example of links and relationships between firms and universities. These links can take a variety of different forms. Firms and universities may work together on government-sponsored research programmes. Scientists working in firms may well retain informal contact with the universities where they were trained. Scientists working in firms will attend conferences where they meet and have contact with their counterparts in universities. Links such as these all provide opportunities for knowledge transfer, especially the transfer of tacit knowledge which can play a vital role in the process of innovation. Nor need the contact be confined to the scientific, for activities such as design, marketing and production may all benefit from links with universities.

Arrangements such as these will tend to be self-organising and self-governing. In some instances where shared interests emerge between the parties in a network then mutual interest bodies such as trade associations and professional bodies may have a part to play in governing relationships. Such bodies may work to pool information, identify common interests and agree strategies designed to promote collaboration. Sometimes the collaboration can extend to the setting of standards. These can be important actions as a new technological paradigm emerges. The early phases of such a paradigm will typically see a number of competing product architectures or configurations. Competition of this sort can lead to compatibility problems for consumers and users that ultimately slow the diffusion of an innovation. By agreeing and setting standards, trade associations can help to further innovation.

INSTITUTIONS

At the heart of any system of innovation lies what Freeman (1987) describes as: 'The network of institutions in the public and private sectors whose activities and interactions initiate, import, modify and diffuse technologies'.

Inevitably the range of institutions that this encompasses is extremely broad as it includes all those that in some way influence the creation and use of technology (Dodgson and Bessant 1996) (see Figure 14.1).

A broad categorisation of those institutions would include:

- Industrial institutions e.g. firms and industry sectors
- Financial institutions e.g. banks and venture capitalists
- Science and technology institutions e.g. universities and public research laboratories
- Educational institutions e.g. schools, colleges and training providers

Figure 14.1 National Systems of Innovation

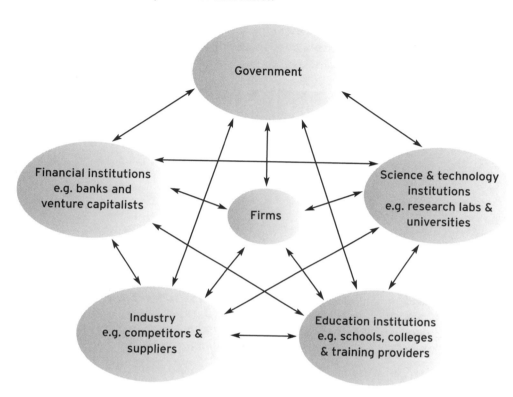

Industrial institutions

Industrial institutions encompass both the broad industry structure that prevails within an economy and the range of industrial undertakings with which a firm competes and from which it obtains components and materials. Industry structure covers the mix of industries present within an economy, especially the proportion of high-technology industries and the size distribution of firms. Both of these dimensions can influence the level of innovation.

In terms of competition a key factor is likely to be the overall level of competitiveness within an economy. While Schumpeter argued that monopoly conditions might actually be beneficial for innovation, in that monopoly profits could provide scope for funding high levels of R & D – generally competitive conditions are associated with economic dynamism, which in turn facilitates the diffusion of innovation.

Similarly, supply conditions can be important in facilitating innovation. The availability of high-quality components at appropriate prices can be a powerful boost to innovation. The development of the personal computer industry in the US in the late 1970s would undoubtedly have been much slower had it not been for the ready availability of micro-processors, disk drives and other computer accessories.

The significance of industrial institutions comes into sharp relief when one compares national innovation systems. Table 14.1 compares the national systems of innovation of Japan and the USSR in the 1970s. Although the USSR spent more, in fact much more, on R & D than Japan, it had a relatively poor record of innovation. A key factor was the USSR's poorly developed industrial system.

Japan	USSR
High R & D/GNP Ratio (2.5%)	Very high R & D/GNP Ratio (4%)
Low proportion of Military/Space R & D (<2% of R & D)	Very high proportion of Military/Space R & D (>70% R & D)
High proportion of R & D at firm level (>60%)	Low proportion of R & D at firm level (<10%)
R & D and imported technology integrated into production	R & D, imported technology and production poorly integrated with weak linkages
Strong user-producer and subcontractor linkages	Weak or non-existent linkages between marketing, production and procurement
Strong incentives to innovate at the firm level	Largely top-down innovation following government policy objectives
Intensive experience of competition in international markets	Relatively weak exposure to international competition

Table 14.1 National Systems of Innovation: Japan and USSR (1970s)
Source: Freeman and Soete (1997) 'The Economics of Industrial Innovation', Continuum International Publishing Group

MINI-CASE: CERN AND THE WORLD WIDE WEB

CERN is the European particle physics laboratory. It is a public research laboratory devoted to basic research in physics. It is located close to Mont Blanc just outside Geneva in Switzerland. CERN employs scientists from all over Europe. In 1980 these scientists included a young software engineer who had recently graduated from Oxford University where he studied physics. His name was Tim Berners-Lee.

Berners-Lee soon noticed that his fellow scientists tended to bring their own computers and their own software with them when they joined CERN. Given that they came from all over Europe, this often made it difficult to track what they were doing. Berners-Lee began to think about creating a space in which every computer at CERN would be available to him and to others in the laboratory. When he returned to CERN in 1984, this time as a full-time researcher, he began to pursue this idea further.

By 1990 he had developed the necessary software. It would sit on top of the Internet using its computer networking facilities. It comprised three elements:

- A computer language for formatting hypertext files: Hypertext Markup Language (HTML)
- A protocol for switching between files: Hypertext Transfer Protocol (HTTP)
- A unique address code: Universal Resource Locator (URL)

By the end of 1990 the software was running successfully on the computers of CERN. It now needed a name. Unhappy with his first attempt 'The Information Mine' which produced the acronym TIM, Berners-Lee eventually opted for World Wide Web. Because he was an employee of CERN, the laboratory owned the intellectual property rights to the World Wide Web. The bosses at CERN were unsure what to do with these intellectual property rights. Eventually they came round to Berners-Lee's view that as a public research facility they should release the rights and allow the public access to the World Wide Web without charge.

Source: Cassidy (2002)

MINI-CASE: OXFORD INSTRUMENTS

It was while working as a Senior Research Officer at the Clarendon Laboratory at Oxford University that Martin Wood, together with his wife Audrey, founded Oxford Instruments. The company began as very much a part-time venture in the early years, with Martin Wood advising on the equipping of laboratories and most of the work coming through his contacts at the Clarendon Laboratory. In time he and his wife came to realise that many of these laboratories such as the Royal Radar Establishment (RRE) in Malvern and the Harwell Laboratory of the UK Atomic Energy Authority (UKAEA) had a requirement for powerful magnets that they were finding hard to fulfil. As a result Oxford Instruments moved into the manufacture of industrial magnets. Using their garden shed as a workshop, they bought an old lathe at auction and arranged to borrow the Clarendon Laboratory's special machine for winding magnetic coils. Material supplies were less of a problem because Martin Wood was dealing with the same suppliers that he dealt with regularly at the university. The new company employed a retired Clarendon Laboratory technician on a part-time basis with the administration being undertaken by Martin's wife Audrey.

Source: Wood (2001)

Science and Technology Institutions

We have already seen that innovation is in many respects an example of 'public goods' where the benefits accrue not just to private individuals but to society at large. This causes problems in funding R & D, because individuals or firms may be reluctant to fund it if they feel they are unlikely to appropriate the benefits. One way round this difficulty is for the government to step in and fund R & D.

Public funding of R & D in this way can provide additional benefits beyond finding a way around the problem of appropriability. Afuah (2003) identifies four such benefits:

- Training the workforce in skills needed for innovation
- Stimulating private firms to invest in related innovations
- Gaining economies of scale in R & D
- Reducing the cost of firms entering new markets

The training function occurs where large publicly funded research projects require large numbers of scientists/technicians to be trained in particular skills. Afuah (2003) gives the example of the US Defense Department's Advanced Research Projects Agency (DARPA) sponsoring research into computer networks, with the result that when the Internet began to take off it was relatively easy for firms to find employees with the necessary computing skills. Silicon Valley provides a good example of stimulating private firms to invest in related innovations. Much of the early development of Silicon Valley was based on government-funded research into electronics in the aerospace field in the 1950s and 1960s which produced a string of related innovations in the computing field in the 1970s (Castells and Hall, 1994). Space programmes provide an example of economies of scale in R & D. The European Space Agency (ESA), for example, funds research that would be prohibitively expensive if undertaken by individual European nations or single firms. Finally, governments can facilitate market entry as in the case of the US's entry into the commercial jet airliner market where the R & D costs associated with developing the Boeing 707 airliner were largely paid for by the KC-135 tanker aircraft developed for the US Air Force.

Government sponsorship of R & D can take a variety of contexts. Afuah (2003: p. 311) identifies four:

- Public research laboratories
- Universities
- Firms
- Consortia

Public research bodies include organisations like the National Physical Laboratory at Bushey Park in Hertfordshire, the Atomic Energy Research Establishment at Harwell in Oxfordshire and the Royal Aircraft Establishment at Farnborough in Hampshire. They are funded by the UK government and undertake both basic and applied research. A good example of research carried out in a government research laboratory is the work undertaken in the 1960s by the Royal Aircraft Establishment at Farnborough into carbon fibre. This work not only resulted in an in-depth understanding of the process for producing carbon fibre, but led to a patent being granted in 1968 and the negotiation of licences for commercial manufacture to three firms: Courtaulds, Morgan Crucible and Rolls-Royce.

Government-sponsored research is also carried out by universities. In the UK this can take a variety of forms of which the commonest are research projects carried out directly for government departments and those undertaken by the government-funded research councils such as the Engineering and Physical Science Research Council (EPSRC) and the Medical Research Council (MRC). As with government research laboratories this can involve both basic and applied research. An outstanding example of the latter would be the discovery of the double helix structure of DNA by Watson and Crick at Cambridge University in conjunction with work by Wilkes and Franklin at King's College, London (Maddox, 2002).

The government also sponsors R & D that is carried out by firms. This sort of work tends to be applied rather than basic research with a view to the development of specific products. The 'Launch Aid' provided to the UK aerospace industry to facilitate the development of new aircraft and engines is typically of this type of R & D. For instance, when Rolls-Royce developed the all new 25 000-lb thrust V2500 engine for the Airbus A320, it received £70 million in launch aid from the UK government (Kavianto, 1997).

The fourth category of government-sponsored research is the funding of consortia, which typically comprise government research laboratories, universities and industrial firms or some combination of the three.

Finally, Afuah (2003) reminds us that the range of innovations that have benefited from government funding of R & D includes a wide range of high-technology products including: the Internet, UNIX, computers, semiconductors, RISC technology and the jet engine.

Financial Institutions

As we saw in Chapter 10, innovation requires finance. Within a market economy, the capital market will be the source of funding for innovation. The function of the capital market is to link savers and investors. The latter in this instance will be individuals and organisations undertaking innovation. The capital market is made up of a variety of financial institutions which act as intermediaries. These financial institutions fall within the following categories:

- Commercial banks
- Investment banks
- Venture capitalists

Banks as we have already seen play a role in innovation. However, the support they provide tends to be in the form of banking services rather than capital. This reflects the generic nature of commercial banks in the UK which serve the full range of industry sectors rather than providing specialist finance for innovators. The generic nature of commercial banks means they do not normally fund innovation directly. Most banks simply do not have the capability to evaluate the risks – both market and technical – involved.

Investment banks are specialist financial institutions. They specialise in corporate finance, particularly equity finance through the stock market (e.g. initial public offerings (IPOs) and mergers and acquisitions. This sort of finance can play an important part in innovation. However, it has to be said that historically this sort of finance has tended to be used later in the innovation cycle, when an innovation has already achieved at least some market success. This is because, in order to access equity finance, it is normally necessary to have a track record in terms of sales and profitability.

In terms of institutions, this leaves venture capitalist organisations. These are specialist financial institutions which take an equity stake in firms in order to provide funding for major developments such as innovation. In the UK the leading venture capital organisation is 3i. Since it was founded in 1945 it has funded a large number of firms, though usually only when they have reached a certain size. Venture capitalist organisations like

3i typically invest substantial sums of £100 000 or more. Hence it is mainly innovation in larger firms that is funded by venture capitalists.

Educational Institutions

Human capital is vital to innovation. It is a source of both knowledge and skills. The institutions that provide knowledge and skills are:

- Schools
- Colleges
- Universities

The school system provides the generic skills required for innovation. These include skills in literacy and numeracy as well as a range of generic skills associated with learning, problem-solving and creativity.

Colleges generally provide technical and vocational training to develop a range of work-related skills. Historically such training in the UK has been highly fragmented. There have been attempts to re-organise and re-structure the system, such as the creation of a Technician Education Council (TEC) in the 1970s and the introduction of National Vocational Qualifications (NVQs) in the 1990s. Although influenced by the highly structured and very effective German model of technical education, none of these reforms had a significant impact on the relatively low skills levels that prevailed in the UK. As Howells notes (2005: p. 241): 'In this regard, the British "reform" had continuity with the old-style, voluntary institutions of informal British apprenticeship.'

The third level within this institutional framework comprises the universities, who provide degree-level and postgraduate courses. In terms of their contribution to innovation, there are two crucial aspects of university education. First, the proportion of the population that attends such institutions and, second, the proportion of those who attend that take science, engineering and technology courses. In the UK the university system has historically been highly elitist, with only about 10 per cent of the population attending university. This has resulted in an outstanding record of achievement in terms of scientific breakthroughs; but it has equally been blamed by some for Britain's apparently poor record in terms of successful innovations.

CASE STUDY: CARBON FIBRE TECHNOLOGY

Carbon fibre, or composite as it is sometimes termed, is a high-performance plastic that combines great strength with lightness. Today it is found in a range of products that extends from the wings of commercial airlines and the fan blades of jet engines to Formula One racing cars and sporting equipment such as tennis racquets and golf clubs. The theoretical properties of carbon fibre had been known for a long time, but it was only in the 1960s that it began to attract interest.

Although a number of researchers contributed to the development of carbon fibre, the work carried out by a government research laboratory in the UK, the Royal Aircraft Establishment (RAE) at Farnborough in Hampshire, is generally recognised as having made the principal contribution.

A team at the RAE began their work in the early 1960s. The search for ways of making a strong, stiff material like carbon fibre reflected the view that such a material would have useful aerospace applications where strength and lightness were at a premium.

The RAE team were able to get markedly better results than a rival Japanese team led by Akio Shindo working at the Japanese Government Industrial Research Institute in Osaka. The main factor was that the research carried out at RAE resulted in a better understanding of key aspects of the process for making carbon fibres.

The RAE team discovered that varying the final heat treatment temperature used in the carbonisation process resulted in variations in the strength of the final product. In particular, using lower temperatures resulted in a different type of carbon fibre with even greater strength. It was this type of carbon fibre which ultimately became the dominant material used in most carbon fibre applications such as sporting goods and aircraft structures. Further work by the RAE team also highlighted the way impurities could compromise the strength of carbon fibre and the discovery that carrying out the oxidisation and carbonisation processes in 'very clean conditions' could yield big improvements in strength and reliability.

Having successfully patented their work, the RAE then negotiated licences with organisations interested in manufacturing or using carbon fibre. Licences were negotiated with three firms – Courtaulds, Morgan Crucible and Rolls-Royce – through the UK government's National Research and Development Corporation (now known as BTG plc), a public body charged with exploiting the intellectual property rights contained in government-funded research. Unlike the other two concerns Rolls-Royce was only interested in producing carbon fibre for use in components of its engines.

The RAE breakthrough came at a time when Rolls-Royce was trying to break into the American market with a large high-thrust engine, the RB211. Rolls-Royce faced intense competition from its two American rivals, Pratt and Whitney and General Electric. As an outsider, Rolls-Royce was banking on the use of an advanced design that was both lighter and more powerful than those of its rivals. A fan using blades made of carbon fibre offered exactly the right qualities – it was substantially lighter and yet stronger than the equivalent fan with titanium blades. A lighter fan meant a more fuel-efficient engine – something that had a great appeal to the world's airlines.

Unfortunately, the prototype engines' carbon fibre fan blades proved insufficiently robust to withstand bird strikes and the company was forced to revert to a back-up design using titanium blades.

Courtaulds, one of the other licensees, proved more successful. It negotiated an agreement with Hercules Aerospace of the US, a company that had extensive experience of plastics applications in aerospace. Under the licensing agreement with the NRDC, Hercules staff were able to visit both the Courtaulds plant at Coventry and the RAE team at Farnborough. Hercules was successful in getting US aerospace manufacturers to adopt carbon fibre. Early applications included the AV8B jump jet and the F-18 Hornet fighter in the military field and the Boeing 767 twinjet airliner in the commercial field. Hercules also led the way in developing motorsport applications. When Britain's McLaren Formula One team wanted to construct a racing car employing a chassis made entirely of

▶

carbon fibre, it was Hercules, based in Salt Lake City, Utah, working with McLaren's designer John Barnard that built it. The design rapidly proved successful and other teams soon followed, leading to the development of an important new market. With its early success on programmes such as the AV8B, Hercules became the dominant US carbon fibre manufacturer supplying about 75 per cent of the Department of Defense's requirements during the 1970s and 1980s. Meanwhile, Courtaulds made substantial in-roads into the sporting goods market where manufacturers in South East Asia, particularly Japan and Taiwan, proved willing to adopt the new material for high-performance equipment. Between 1978 and 1983 Courtaulds' carbon fibre division grew by 50 per cent every year. By then Courtaulds was by far the largest carbon fibre producer in Europe, accounting for about 90 per cent of European carbon fibre production.

Source Spinardi (2002)

QUESTIONS

1 Why was the Royal Aircraft Establishment (RAE) Farnborough an appropriate body to conduct research into carbon fibre?
2 What sort of research did the RAE undertake and why?
3 Why, having obtained patents for carbon fibre, did the RAE choose to license its technology to manufacturers such as Courtaulds and Rolls-Royce?
4 What does this case tell us about the role of public research laboratories such as the RAE?
5 Why did Courtaulds not develop aerospace applications itself?
6 What were the attractions of carbon fibre for aerospace firms like Rolls-Royce?
7 What does the case tell us about the value of linkages?
8 What sort of knowledge would have been crucial in adopting carbon fibre for aerospace applications?
9 What were the attractions of carbon fibre for motor sport firms such as McLaren?
10 What does the case of carbon fibre tell us about the nature of the process of innovation?

QUESTIONS FOR DISCUSSION

1 If innovation is conducted by individuals or organisations, how can there be a public domain to innovation?

2 Who are the agents involved in the introduction of new technology?

3 What are the institutions associated with intellectual property rights?

4 What is the difference between the Anglo-Saxon and Nippon-Rhineland systems of corporate governance?

5 **What are the institutions typically associated with science and technology?**

6 How can corporate governance affect innovation?

7 **What is a regional system of innovation?**

8 How do financial institutions contribute to a national system of innovation?

9 **What part do schools play in a national system of innovation?**

10 Using examples show how changes in higher education have affected the UK's national innovation system.

11 **What part do financial institutions play within a national system of innovation?**

12 What is the national science base and how does it contribute to innovation?

ASSIGNMENTS

1 Using an example of a public research institution of your choice, show how such institutions contribute to the UK's national system of innovation.

2 Critically review the UK's education system in terms of its contribution to the national system of innovation.

3 Take any country of your choice and identify the main features of its national system of innovation.

4 From the perspective of an innovator, how useful is the concept of a regional system of innovation?

5 What has tacit knowledge to do with a system of innovation?

RESOURCES

Books

The notion of a national system of innovation is a fairly academic one and as a result the resources are almost entirely confined to literature. Two key texts that provide an overview are Nelson (1993) *National Innovation Systems* and OECD's (1997) text of the same title. While it does not focus specifically on innovation, Porter's (1990) *The Competitive Advantage of Nations* is a major contribution to the field. For regional innovation systems, Cooke *et al.*'s (2004) *Regional Innovation Systems* is the definitive work.

REFERENCES

Afuah, A. (2003) *Innovation Management: Strategies, Implementation, Profits*, 2nd edn, Oxford University Press, New York.

Albert, M. (1992) *Capitalism against Capitalism*, Whurr, London.

Barker, B. and A. Goto (1998) 'Technological systems, innovation and transfers', in G. Thompson (ed.) *Economic Dynamism in the Asia-Pacific*, Routledge, London.

Cassidy, J. (2002) *Dot.con: the greatest story ever sold*, Penguin Books, Harmondsworth.

Castells, M. and P. Hall (1994) *Technopoles of the World: The Making of the 21st Century Industrial Complexes*, Routledge, London.

Cooke, P., M. Heidenreich and H. Braczyk (2004) *Regional Innovation Systems*, 2nd edn, Routledge, London.

Dahleman, C.J. (1994) 'Technology Strategy in East Asian Developing Economies', *Journal of Asian Economics*, **5** (4) pp. 541-572.

Dodgson, M. and J. Bessant (1996) *Effective Innovation Policy: A New Approach*, International Thomson Business Press, London.

Freeman, C. (1987) *Technology Policy and Economic Performance*, Frances Pinter, London.

Freeman, C. and L. Soete (1997) *The Economics of Industrial Innovation*, Continuum, London.

Howells, J. (2005) *The Management of Innovation and Technology*, Sage Publications, London.

Kavianto, K. (1997) *UK Launch Aid Experience*, Research Paper No. 260, Warwick Business School, Warwick.

Maddox, B. (2002) *Rosalind Franklin: The Dark Lady of DNA*, HarperCollins, London.

Moran, M. (2001) 'Governing European Corporate Life' in G. Thompson (ed.) *Governing the European Economy*, Sage Publications, London.

Nelson, R.R. (1993) *National Innovation Systems: A Comprehensive Analysis*, Oxford University Press, NY.

OECD (1997) *National Innovation Systems*, OECD, Paris.

Porter, M. (1990) *The Competitive Advantage of Nations*, Macmillan, Basingstoke.

Simonetti, R (2001) 'Governing European Technology and Innovation', in G. Thompson (ed.) *Governing the European Economy*, Sage Publications, London.

Spinardi, G. (2002) 'Industrial Exploitation of Carbon Fibre in the UK, USA and Japan', *Technology Analysis and Strategic Management* **14** (4), pp. 381-398.

Tidd, J. and M. Brocklehurst (1999) 'Routes to Technological Learning and Development, An Assessment of Malaysia's Innovation Policy and Performance', *Technological Forecasting and Social Change*, **62**, pp. 239-257.

Wood, A. (2001) *Magnetic Venture: The Story of Oxford Instruments*, Oxford University Press, Oxford.

INDEX

INDEX

311